T0212765

More information about this series at http://www.springer.com/series/7911

Witold Abramowicz (Ed.)

Business Information Systems

18th International Conference, BIS 2015
Poznań, Poland, June 24–26, 2015
Proceedings

 Springer

Editor
Witold Abramowicz
Department of Information Systems
Poznań University of Economics
Poznań
Poland

ISSN 1865-1348   ISSN 1865-1356 (electronic)
Lecture Notes in Business Information Processing
ISBN 978-3-319-19026-6   ISBN 978-3-319-19027-3 (eBook)
DOI 10.1007/978-3-319-19027-3

Library of Congress Control Number: 2015939674

Springer Cham Heidelberg New York Dordrecht London
© Springer International Publishing Switzerland 2015

Printed on acid-free paper

Springer International Publishing AG Switzerland is part of Springer Science+Business Media
(www.springer.com)

Preface

The 18th International Conference on Business Information Systems was held in Poznań, Poland. Since the first BIS edition in 1997, every other conference is held at Poznań University of Economics. The BIS conference is a well-renowned event of the scientific community, where researchers and business practitioners conduct scientific discussions on up-to-date issues that include the development, implementation, and application of business information systems.

BIS conference follows trends in the academia and business research. The theme of the BIS 2015 conference was, "Making Big Data Smarter." For the past few years, we have been observing an increasing interest in Big Data, accompanied by development of new, more efficient data analysis methods. Big Data is now a fairly mature concept, recognized and widely used by both research and industry. They work on developing more adequate and efficient tools for data processing and analyzing, making Big Data Smart Data. Possibilities that emerge from Smart Data are unlimited, but one will not be able to take advantage of it without appropriate technologies, applications, methods, and business processes. Thus, Smart Data links many fields of computer science and information systems together.

The two first sessions of BIS 2015 were dedicated to Big and Smart Data research. However, during the conference other research directions were also discussed, including Business Process Management and mining, semantic technologies, collaboration, content retrieval and filtering, enterprise architecture and BITA as well as specific BIS applications. The volume ends with two invited papers. The first is dedicated to the analysis of Open Data and Open API adoption based on software developer contests. The second paper shortly describes one application developed during such a contest.

The Program Committee consisted of 130 members that carefully evaluated all the submitted papers. Based on their extensive reviews, a set of 26 papers were selected, grouped into 8 sessions. The volume closes with 9th session, "Open Data for BIS."

I would like to thank the Track Chairs and reviewers for their time and effort. I wish to thank all keynote speakers who delivered enlightening and interesting speeches. Finally, I would like to thank all authors who submitted their papers and who build an active scientific community around BIS conference.

June 2015   Witold Abramowicz

Conference Organization

BIS 2015 was organized by Poznan University of Economics, Department of Information Systems.

Organizing Committee

Elżbieta Lewańska (chair)	Poznan University of Economics, Poland
Barbara Gołębiewska	Poznan University of Economics, Poland
Włodzimierz Lewoniewski	Poznan University of Economics, Poland
Bartosz Perkowski	Poznan University of Economics, Poland
Wioletta Sokołowska	Poznan University of Economics, Poland
Milena Stróżyna	Poznan University of Economics, Poland

Program Committee

Witold Abramowicz	Poznan University of Economics, Poland
Stephan Aier	University of St. Gallen, Switzerland
Antonia Albani	University of St. Gallen, Switzerland
Rainer Alt	University of Leipzig, Germany
Dimitris Apostolou	University of Piraeus, Greece
Timothy Arndt	Cleveland State University, USA
Maurizio Atzori	University of Cagliari, Italy
David Aveiro	University of Madeira, Portugal
Eduard Babkin	National Research University - Higher School of Economics, Russia
Morad Benyoucef	University of Ottawa, Canada
Markus Bick	ESCP Berlin, Germany
Maria Bielikova	Slovak University of Technology, Slovak Republic
Tiziana Catarci	Sapienza University of Rome, Italy
Michelangelo Ceci	University of Bari, Italy
Wojciech Cellary	Poznan University of Economics, Poland
Francois Charoy	Université de Lorraine, France
Dickson K. W. Chiu	Dickson Computer Systems, Hong Kong
Tony Clark	Middlesex University, UK
Enrique De La Hoz	University of Alcalá, Spain
Andrea De Lucia	University of Salerno, Italy
Stefan Decker	Digital Enterprise Research Institute, Ireland
Zhihong Deng	Peking University, China
Tommaso Di Noia	Technical University of Bari, Italy
Ciprian Dobre	University Politehnica of Bucharest, Romania
Josep Domingo-Ferrer	Universitat Rovira i Virgili, Catalonia, Spain
Suzanne Embury	University of Manchester, UK
Vadim Ermolayev	Zaporozhye National University, Ukraine

Werner Esswein	TU Dresden, Germany
Dieter Fensel	University of Innsbruck, Austria
Agata Filipowska	Poznan University of Economics, Poland
Adrian Florea	Lucian Blaga University of Sibiu, Romania
Vladimir Fomichov	Higher School of Economics, Russia
Ulrich Frank	University of Duisburg-Essen, Germany
Flavius Frasincar	Erasmus School of Economics, The Netherlands
Johann-Christoph Freytag	Humboldt University, Germany
Katsuhide Fujita	Tokyo University of Agriculture and Technology, Japan
Naoki Fukuta	Shizuoka University, Japan
Henner Gimpel	University of Augsburg, Germany
Rüdiger Grimm	University of Koblenz and Landau, Germany
Norbert Gronau	Universität Potsdam, Germany
Volker Gruhn	Universität Duisburg-Essen, Germany
Francesco Guerra	University of Modena and Reggio Emilia, Italy
Hele-Mai Haav	Tallinn University of Technology, Estonia
Martin Hepp	Bundeswehr University Munich, Germany
Frank Hogrebe	Hessische Hochschule für Polizei und Verwaltung, Germany
Stijn Hoppenbrouwers	HAN University of Applied Sciences and Radboud University Nijmegen, The Netherlands
Constantin Houy	German Research Center for Artificial Intelligence, Germany
Maria-Eugenia Iacob	University of Twente, The Netherlands
Björn Johansson	Lund University School of Economics and Management, Sweden
Monika Kaczmarek	University of Duisburg-Essen, Germany
Paweł J. Kalczyński	California State University Fullerton, USA
Kalinka Kaloyanova	University of Sofia, Bulgaria
Hariklea Kazeli	CYTA, Cyprus
Marite Kirikova	Riga Technical University, Latvia
Gary Klein	University of Colorado at Colorado Springs, USA
Ralf Klischewski	German University in Cairo, Egypt
Ralf Knackstedt	University of Hildesheim, Germany
Jacek Kopecky	University of Portsmouth, UK
Marek Kowalkiewicz	SAP, USA
Helmut Krcmar	Technical University of Munich, Germany
Dalia Kriksciuniene	Vilnius University, Lithuania
John Krogstie	Norwegian University of Science and Technology, Norway
Nor Laila Md Noor	Universiti Teknologi MARA, Malaysia
Winfried Lamersdorf	University of Hamburg, Germany
Christine Legner	L'Université de Lausanne, Switzerland
Maurizio Lenzerini	Sapienza Università di Roma, Italy
Peter Loos	Saarland University, Germany

André Ludwig	University of Leipzig, Germany
Qiang Ma	Kyoto University, Japan
Wolfgang Maaß	Saarland University, Germany
Maria Mach-Król	University of Economics in Katowice, Poland
Leszek Maciaszek	Wrocław University of Economics, Poland
Alexander Mädche	University of Mannheim, Germany
Florian Matthes	Technical University of Munich, Germany
Raimundas Matulevičius	University of Tartu, Estonia
Heinrich C. Mayr	University of Klagenfurt, Austria
Massimo Mecella	Sapienza University of Rome, Italy
Jan Mendling	Wirtschaftsuniversität Wien, Austria
Günter Müller	University of Freiburg, Germany
Markus Nüttgens	University of Hamburg, Germany
Andreas Oberweis	University of Karlsruhe, Germany
Marcin Paprzycki	Polish Academy of Sciences, Poland
Eric Paquet	National Research Council, Canada
Dana Petcu	West University of Timisoara, Romania
Geert Poels	Ghent University, Belgium
Jaroslav Pokorný	Charles University, Czech Republic
Birgit Pröll	Johannes Kepler Universität Linz, Austria
Elke Pulvermueller	University Osnabrueck, Germany
Fenghui Ren	University of Wollongong, Australia
Stefanie Rinderle-Ma	University of Vienna, Austria
Antonio Rito Silva	Instituto Superior Técnico, Portugal
Dumitru Roman	SINTEF/University of Oslo, Norway
Michael Rosemann	Queensland University of Technology, Australia
Stefan Sackmann	Martin Luther University, Germany
Virgilijus Sakalauskas	Vilnius University, Lithuania
Sherif Sakr	University of New South Wales, Australia
Demetrios Sampson	University of Piraeus, Greece
Kurt Sandkuhl	University of Rostock, Germany
Jürgen Sauer	University of Oldenburg, Germany
Stefan Schulte	Vienna University of Technology, Austria
Matthias Schumann	Georg-August-Universität Göttingen, Germany
Ulf Seigerroth	Jönköping University, Sweden
Gheorghe Cosmin Silaghi	Babes-Bolyai University of Cluj-Napoca, Romania
Elmar J. Sinz	University of Bamberg, Germany
Janice C. Sipior	Villanova University, USA
Alexander Smirnov	St. Petersburg Institute for Informatics and Automation of the Russian Academy of Sciences, Russia
Stefan Smolnik	Fern Universitaet in Hagen, Germany
Henk Sol	University of Groningen, The Netherlands
Andreas Speck	Kiel University, Germany
Athena Stassopoulou	University of Nicosia, Cyprus
Stefan Stieglitz	University of Münster, Germany
Janis Stirna	Stockholm University, Sweden

Darijus Strasunskas	Norwegian University of Science and Technology, Norway
Jerzy Surma	Warsaw School of Economics, Poland
Bernhard Thalheim	Universitat Kiel, Germany
Barbara Thönssen	University of Applied Sciences Northwestern Switzerland, Switzerland
Ramayah Thurasamy	Universiti Sains Malaysia, Malaysia
Robert Tolksdorf	Free University Berlin, Germany
Genny Tortora	University of Salerno, Italy
Bruno Vallespir	University of Bordeaux, France
Herve Verjus	University of Savoie, France
Herna Viktor	University of Ottawa, Canada
Mathias Weske	Hasso Plattner Institute for IT-Systems Engineering, Germany
Krzysztof Węcel	Poznan University of Economics, Poland
Anna Wingkvist	Linnaeus University, Sweden
Robert Winter	University of St. Gallen, Switzerland
Guido Wirtz	University of Bamberg, Germany
Qi Yu	Rochester Institute of Technology, USA
Slawomir Zadrozny	Polish Academy of Sciences, Poland
John Zeleznikow	Victoria University, Australia
Jozef Zurada	University of Louisville, USA

Additional Reviewers

Akkaya, Cigdem
Baader, Galina
Batoulis, Kimon
Baur, Aaron
Beese, Jannis
Blesik, Till
Bock, Alexander
Böhmer, Kristof
Bühler, Julian
Dadashnia, Sharam
Dellepiane, Umberto
Dessi, Andrea
Deufemia, Vincenzo
Di Nucci, Dario
Dittes, Sven
Ebner, Katharina
Fasano, Fausto
Geiger, Matthias
Graupner, Enrico

Gulden, Jens
Haake, Phillip
Hamann, Kristof
Harrer, Simon
Heppner, Konstantin
Imran Daud, Malik
Indiono, Conrad
Jander, Kai
Kaes, Georg
Kretzer, Martin
Kunze, Matthias
Kärle, Elias
Lasierra Beamonte, Nelia
Orsini, Gabriel
Overbeek, Sietse
Palomba, Fabio
Peris, Martina
Radloff, Michael
Ribes González, Jordi

Roessler, Richard
Schneider, Alexander W.
Serafino, Francesco
Seyffarth, Tobias
Sorgenfrei, Christian

Thaler, Tom
Tucci, Maurizio
Waldmann, Alexander
Waltl, Bernhard
Wendler, Hannes

Contents

XIV   Contents

Business Process Management and Mining

Collaboration

Enterprise Architecture and Business-IT-Alignment

Specific BIS Applications

Open Data for BIS

Big and Smart Data

A Parallel Approach for Decision Trees Learning from Big Data Streams

Ionel Tudor Calistru[✉], Paul Cotofrei, and Kilian Stoffel

Information Management Institute, University of Neuchatel, Neuchatel, Switzerland
{ionel.calistru,paul.cotofrei,kilian.stoffel}@unine.ch

Abstract. In this paper we introduce PdsCART, a parallel decision tree learning algorithm. There are three characteristics that are important to emphasize and make this algorithm particularly interesting. Firstly, the algorithm we present here can work with streaming data, i.e. one pass over data is sufficient to construct the tree. Secondly, the algorithm is able to process in parallel a larger amount of data stream records and can therefor handle efficiently very large data sets. And thirdly, the algorithm can be implemented in the MapReduce framework. Details about the algorithm and some basic performance results are presented.

Keywords: Bigdata · Business datamining · Streams · MapReduce · Decision trees

1 Introduction

Big Data has become the standard term referring the analysis of very large collections of data. Traditionally, these large collections of data were produced in the context of scientific applications e.g. physics, genomics or meteorology. But once the applications in the domain of business and finance gained in interest, an increasing amount of data was generated and collected in this context, confronting the researchers with the difficulties of dealing with very large data set. Finally, a third domain of great importance in the context of Big Data is the domain of data streams, usually generated by all kinds of sensors, such as mobile devices, remote sensors, radio-frequent identification (RFID), etc.

In this paper we are essentially interested in the intersection of the second and third domain mentioned in the previous paragraph, i.e. data produced in the context of business applications in form of streams. Today, the use of apps on mobile phones has become one of the cornerstones at the interaction between enterprises and their clients. These apps, together with the more traditional web sites, produce data logs and other streams which, in volume, count for the most important raw data sets gathered by the companies. Furthermore, the combination of mobile apps and mobile phones containing sensor, such as GPS, generates data in a strict business setting, having a lot of the characteristics typical for a sensor network.

The MapReduce framework [1] has become the de facto standard for the implementation of processes for analysing very large data sets in parallel, using

© Springer International Publishing Switzerland 2015
W. Abramowicz (Ed.): BIS 2015, LNBIP 208, pp. 3–15, 2015.
DOI: 10.1007/978-3-319-19027-3_1

distributed clusters. Its basic structure consists of two main steps: first a map-step which essentially filters and sorts the data to be analysed, followed by a reduce-step which essentially aggregates the data to be analysed. This simple framework allows for great performance, but it is also quite limited in respect to the algorithms that can be implemented in a straight forward manner. Motivated by some business applications, we are particularly interested in decision tree algorithms. But this category of algorithms is exactly one of those which cannot easily be ported to the MapReduce framework, particularly if some of the specific requirements for business applications have to be respected. In this paper we will propose and analyse an approach allowing to implement decision tree algorithms for big data streams in a MapReduce framework.

The remainder of the papers is organized in the following way. In the next chapter we will present the arguments supporting the choice of decision trees as the data mining algorithm of preference. Then we will describe how decision trees can be used to analyse data streams and how this analysis can be integrated into a MapReduce framework. Then we will present the details regarding the implementation of our parallel decision tree algorithm followed by an analysis of its performance, before to conclude and give some outlooks and future work.

2   Decision Trees for Mining Big Data

One of the most effective and widely used techniques in machine learning today is decision tree learning. These models are popular not only because of their adaptability and accurate prediction capabilities, but also because they can provide classification rules that may be easily interpreted by humans. This is a particularly interesting property in the context of mining business data.

However, decision trees come also with some disadvantages. Traditional decision tree algorithms [2,3] have difficulties when the data does not fit into memory, because they have to recursively read the training data sets to construct split decisions. Furthermore, numerical values need to be sorted in order to find the splitting points of a specific node of the tree. To overcome these time and memory consuming constraints several solutions have been proposed.

2.1   Related Work

One of the techniques used by learning decision trees algorithms is the pre-sorting of attributes values, as in SPRINT [4] or ScalParC [5]. Alternatively, another approach is to approximate the data, instead of sorting it, by using histogram data structures, as in pCLOUDS [6], SPIES [7] and SPDT [8]. To build the histograms, some authors use simply the frequency of data, where others (SPIES and pCLOUDS) use sampling techniques. Another important difference lays in the number of passes through the data needed, e.g. SPIES and SSE (a version of CLOUDS) may need to pass several times over the data during the construction process. Although the pre-sorting approaches are in general more accurate, they may not be suitable for streams of data.

Parallel Decision Trees Algorithms. The challenge of dealing with important amounts of data was largely addressed by several parallel decision tree algorithms such as described in [4–7,9].

In Amado et al. [10] and Srivastava et al. [9] four different types of parallel decision tree algorithms have been described: horizontal, vertical, task and hybrid approaches. In horizontal parallel decision trees, the entire data set is split into subsets which are processed individually. In the vertical case, the set of attributes is partitioned. Task parallelism enables the distribution of decision trees nodes to be processed independently. The fourth type, hybrid parallelism, is a combination of all three already mentioned approaches, i.e. in the first phases of the decision tree growing process, vertical and horizontal parallelism are combined, letting the task parallelism taking over at the end.

An example of a hybrid approach is Google's PLANET [11], a technique that applies horizontal parallelism (by implementing a MapReduce algorithm) at the first few levels of the tree and applies task parallelism to the leaves, as soon as the data fits into memory.

In [8] the authors use horizontal parallelism to build data histograms, which are then merged to take decisions and to build the tree in a breadth-first manner. Other examples of horizontal parallelism for building trees are gradient boosted decision trees (GBDT [12]) or regression trees (GBRT [13]).

Based on SPDT, Li [14] proposes a random forest algorithm (SRF) with a MapReduce implementation (similar to PLANET one): in the map phase the local histograms are computed, while in the reduce phase the global histogram enables the best split decision to be taken. However, in each iteration, when a new level of the tree needs to be build, the complete data set (or a predefined number of samples, if the data is too big) has to be read. This overrules the single-pass constraint and makes it inadequate for a potentially infinite number of records like in a data stream.

Decision Trees for Online Stream Mining. While most of the decision tree algorithms available today are designed to enable the mining of data sets that do not fit into memory, the extremely fast growth - in recent years - of the volume of information that needs to be analysed rises the necessity of new techniques. Ideally, these techniques would be able to continuously process streams of data, without losing any valuable information [15].

The (theoretically) infinite number of records of data streams made the exact determination of the best attribute to split impossible. This lead to the idea of estimation of the best attribute. Due to the fact that this estimation needs to be done in respect to the entire data stream, several techniques supporting the single-pass constraint as well as other data streams particularities [16] have been studied [17–19]. These techniques include Hoeffding's tree algorithm, Very Fast Decision Tree (VFDT) and Concept-adapting Very Fast Decision Tree (CVFDT). The Vertical Hoeffding Tree (VHT) classifier, introduced in [20], utilizes vertical parallelism to extend the VFDT classifier. All these methods were supported mathematically by the Hoeffding's inequality [21].

However, in [22,23] the authors showed that the Hoeffding's inequality is not an adequate probabilistic model for the descriptions of ID3, C4.5 or CART algorithms. They propose a new approach (inspired by [17]), called CART for data streams (dsCART), which applies a Gaussian approximation to establish the best attribute for splitting a tree node. One of the major results of dsCART is that the selected attribute to split on in a considered tree node relative to its data set, is the same, with some high probability, as the attribute chosen by analysing the entire data stream [24].

2.2   The Parallel Approach for Data Streams

Following our interest in online mining of business data, we propose PdsCART, a parallel approach to build the decision trees for inferring and predicting from big data streams. We select the dsCART [24] decision tree algorithm for data stream classification as the basis for our approach.

Our proposed solution is a method to adapt the dsCART algorithm to horizontal parallelism by implementing the MapReduce programming model. While several other horizontal parallel solutions have been mentioned already, like SPDT, PLANET, SRF, GBDT, GBRT etc., to the best of our knowledge, none of them have been applied to a single-pass decision tree for data streams algorithm. More details about our approach are presented in the Implementation section.

3   PdsCART Implementation

In this section we describe PdsCART, our approach to parallelise the dsCART decision tree algorithm. To do this, first we introduce both dsCART[1] and PdsCART decision tree algorithms for data streams, and then we detail our MapReduce implementation. It is important mentioning that we are only interested in showing that very similar learning models may be achieved by treating records in parallel, reducing this way the overall stream processing time.

Before presenting the pseudo-code, the following notes are important:

- for each attribute a^i, the set of attribute values A^i is partitioned into two disjoint subsets A_L^i and A_R^i such that $A^i = A_L^i \cup A_R^i$;
- the choice of A_L^i automatically determines the complementary subset A_R^i.
- the set of all possible partitions of the set A^i is denoted by V_i.
- $\overline{g_q^i}$ is the Gini gain computed for the attribute a^i in the leaf L_q.
- $n_{i,\lambda,q}^k$ is the number of elements from the k-th class in the leaf L_q, for which the value of the attribute a^i is equal to a_λ^i, $\left(a_\lambda^i \in A^i\right)$.
- n_q^k is the number of elements from the k-th class in the leaf L_q
- the tie breaking mechanism (θ) forces the split after some fixed number of elements.
- the stopping condition formula is:

$$\epsilon_{G,K} = z_{1-\alpha}\frac{\sqrt{2Q\left(K\right)}}{\sqrt{n}}, \text{ where } Q\left(K\right) = 5K^2 - 8K + 4 \qquad (1)$$

and $z_{(1-\alpha)}$ is the $(1-\alpha)$-th quantile of the standard normal distribution N(0,1); K is the number of classes, n is the number of samples in the considered tree node.

---

[1] We are following very closely the description of dsCART algorithm done by Leszek Rutkowski, Maciej Jaworski, Lena Pietruczuk and Piotr Duda in [24].

The dsCART Algorithm [24]

Inputs:
S is a sequence of examples,
U is a set of discrete attributes,
α is one minus the desired probability of choosing the correct attribute,
θ is the tie breaking parameter.

Output: dsCART decision tree

Procedure dsCART(S, U, α, θ);
Let dsCART be a single leaf L_0 (the root)
Let $U_0 = U$;
/* Initialize counters in L_0 with 0 */
$n_0 = 0$;

for *each example s in S* **do**
  Sort s into tree leaf L_q;
  for *each attr.* $a^i \in U_q$ **do**
    a^i_λ is the value of s for a^i;
    k is the class of s;
    Increment $n^k_{i,\lambda,q}$;
  end
  Label L_q with the majority class;
  if L_q *has more than one class* **then**
    for *each attr.* $a^i \in U_q$ **do**
      for *each partition of the set* A^i *into* A^i_L, A^i_R **do**
        Get $\overline{g_q}\left(A^i_L\right)$ using $n^k_{i,\lambda,q}$;
      end
      Let $\overline{g^i_q} = \max\limits_{A^i_L \in V_i} \left\{ \overline{g_q}\left(A^i_L\right) \right\}$;
    end
    Let $a^x = \arg \max\limits_{a^i \in U_q} \left\{ \overline{g^i_q} \right\}$;
    Let $a^y = \arg \max\limits_{a^i \in U_q \setminus \{a^x\}} \left\{ \overline{g^i_q} \right\}$;
    Get $\epsilon_{G,K}$ using (1);
    if $\left(\overline{g^x_q} - \overline{g^y_q} > \epsilon_{G,K}\right)$ *or* $(\epsilon_{G,K} < \theta)$ **then**
      Split L_q on a^x;
      for *both branches of the split* **do**
        Add a new leaf L_{last+1};
        Let $U_{last+1} = U_q \setminus \{a^x\}$;
        $n_{last+1} = 0$;
        last = last + 1;
      end
    end
  end
end
return dsCART

The PdsCART Algorithm

Inputs:
S is a sequence of examples,
U is a set of discrete attributes,
α is one minus the desired probability of choosing the correct attribute,
θ is the tie breaking parameter.
R is the number of reccords to process in parralel

Output: PdsCART decision tree

Procedure PdsCART$(S, U, \alpha, \theta, R)$;
Let PdsCART be a single leaf L_0 (the root);
Let $U_0 = U$;
Let $Data$ be the list of examples read from the stream;
Let $TLeaves$ be the list of tree leaves;
$TLeaves.Add(L_0)$;
/* Initialize conters in L_0 with 0 */
$n_0 = 0$;

for *each example s in S* **do**
  $Data.Add(s)$
  if $Data.size == R$ **then**
    $Controller(Data, TLeaves, PdsCART)$
    $Data.clear()$
  end
end
if $Data.size > 0$ **then**
  $Controller(Data, TLeaves, PdsCART)$
  $Data.clear()$
end
return $PdsCART$

Procedure Controller$(data, leaves, tree)$

/* Assign Data To Mappers */
$job = Controller.Assign(data)$;
/* Call Map and Reduce */
$job.Run(tree)$;
/* Collect Histograms for each tree leaf */
$Controller.CollectHistograms()$;

for *each leaf* L_q *in leaves* **do**
  Label L_q with the majority class;
  if L_q *has more than one class* **then**
    for *each attr.* $a^i \in U_q$ **do**
      for *each partition of the set* A^i *into* A^i_L, A^i_R **do**
        Get $\overline{g_q}\left(A^i_L\right)$ using $n^k_{i,\lambda,q}$;
      end
      Let $\overline{g^i_q} = \max\limits_{A^i_L \in V_i} \left\{ \overline{g_q}\left(A^i_L\right) \right\}$;
    end
    Let $a^x = \arg \max\limits_{a^i \in U_q} \left\{ \overline{g^i_q} \right\}$;
    Let $a^y = \arg \max\limits_{a^i \in U_q \setminus \{a^x\}} \left\{ \overline{g^i_q} \right\}$;
    Get $\epsilon_{G,K}$ using (1);
    if $\left(\overline{g^x_q} - \overline{g^y_q} > \epsilon_{G,K}\right)$ *or* $(\epsilon_{G,K} < \theta)$ **then**
      $leaves.Remove(L_q)$;
      Split L_q on a^x;
      for *both branches of the split* **do**
        Add a new leaf L_{last+1};
        Let $U_{last+1} = U_q \setminus \{a^x\}$;
        $n_{last+1} = 0$;
        $leaves.Add(L_{last+1})$;
        last = last + 1;
      end
    end
  end
end

3.1   Preliminary Considerations

Following the dsCART algorithm it is easy to see that the most time consuming phase is to find the best split decision; for each attribute in the current node, we must compute the Gini gains in respect to all possible partitions of the set of attribute values. It is important to keep in mind that all these operations take place for each and every new sample that is read from the data stream.

Secondly, we recall here that in [24] has been proven that the attribute chosen in a considered node, according to its current data, is the same one, with high probability, as the one selected after reading the entire data. This means that no matter when these estimations are made, they chose, with some probability, the same attribute.

All these facts motivated us to compute and check the splitting conditions after reading a variable number of samples, while processing them independently. By choosing a sufficiently high probability (α parameter), our algorithm is able to produce very similar (compared with dsCART) decision trees, with the same level of accuracy but with faster processing times. The parallel approach of PdsCART algorithm is detailed in the following subsection, while a summary of the results using this algorithm are given in the Experiments section.

3.2   The MapReduce Implementation

In our version of the distributed PdsCART decision trees, we apply the MapReduce paradigm by using a horizontal partitioning approach.

The controller process coordinates the tree growing, while the mappers and reducers processes fulfill their standard tasks. Assuming that we have P mappers and we want to consume R records in parallel, the controller will assign to each mapper R/P records to process.

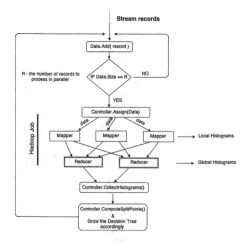

Fig. 1. The PdsCART algorithm logic schema.

In order to keep track of the number of distinct elements for each attribute and class, PdsCART uses some simple data frequency structures, as histograms, which are easy to merge in order compute the Gini gain functions. For each leaf node in the tree, each mapper will build its own local histograms. During the map phase, each input record it is assigned to a leaf node and it is inserted in the corresponding mapper local histogram. The reducers receive from the mappers all the local histograms and merges them into global histograms, one for each leaf node of the tree. The result is serialized into an output file. Given this output file, the controller can then perform a single pass over the leaves global histograms, estimate the best splitting attributes and decide whether to grow the tree in respect to the splitting conditions of each of the tree leaf nodes.

The Map procedure, described in the *Map algorithm*, receives the current version of the tree as well as a set of records. For each record, the tree is traversed in order to find the corresponding leaf node *id*. Based on the node *id*, the mapper updates its dedicated local histogram with the new input record. Once the last record is processed the mapper emits, for each leaf node, the local histograms and their related tree node *id*.

The Reduce procedure, described in *Reduce algorithm*, merges all the local histograms received from the mappers and builds the global ones. The result is exported to the output file.

Map Algorithm	Reduce Algorithm
Inputs:	**Inputs:**
tree : current decision tree;	**(key, values)** pairs emitted by Mappers
records : data assigned to the current mapper	and grouped by the key;
Output: (*key, value*) pairs for each leaf;	**Output: (key, value)** pair
key: the id of the leaf node;	**key**: the id of the leaf node;
value: the corresponding local histogram	**value** : the global histogram - merged

Procedure MAP(*tree, records*)	**Procedure REDUCE**(*key, values*)
HistoMap⟵ local histograms map;	leafId ⟵ key;
while *records.readNextRecord()* **do**	globalHistogram⟵ the global histogram
sample ⟵ SplitRecord(record, delim);	for the current key(leafId);
nodeId ⟵ tree.Traverse(sample);	**for** *each histogram in values* **do**
HistoMap.Add(nodeId, sample);	globalHistogram.MergeHistogram(leafId,
end	histogram);
for *each nodeId in HistoMap* **do**	**end**
histogram ⟵ HistoMap(nodeId);	emit(leafId, globalHistogram);
emit(nodeId, histogram);	
end	

With a single iteration of the output file, the controller can now compute the first and the second best attribute, as well as their related Gini gains, for each and every leaf node in the tree. The PdsCART algorithm can now move forward to compute the splitting conditions, according to the formula (1) and/or θ tie breaking parameter, and to split the leaf nodes if necessary.

In the Experiments section we will show that as, in our approach, the computations of attribute estimation occur quite rarely - only after a larger number of samples are read and processed in parallel from the stream - the overall time of stream processing decreases (compared to dsCART) while the learned decision trees remain similar.

4   Experiments

This section summarizes several results of our experiments. As previously underlined, our parallel approach for decision tree learning from data streams is
designed to achieve similar results as the dsCART algorithm. In fact, in all
our tests, when running with a sufficiently high α value and with the same test
settings (except the number of records processed), we have obtained exactly the
same decision trees with the same level of accuracy as with the dsCART implementation. Due to this fact, we do not need to benchmark the differences in
accuracy between learned models. Instead, we would like to emphasis some theoretical and practical performance gains obtained by implementing our solution,
by processing, in parallel, more than one record at once.

To do so, we first describe our experimental scenarios as well as the details of
the datasets that we have considered, then we present the results of PdsCART
implementation.

4.1   Experimental Scenarios

While it is intuitively that handling a larger number of data records in parallel reduces the processing time, several other aspects have been considered in
order to validate our parallel approach. Some of the aspects that we have taken
into consideration, besides the running time, include: the number of records to
consider per iteration, the number of attributes and the number of bins[2]. Other
aspects, such as the decision tree size, the α parameter and the dependency
relations between all these aspects may be considered for a future work as well.

Table 1. The specifications of the datesets used in our experiments.

#	Data Set	# records	Attributes	Classes	# type
1	D_1	10 thousands	5	2	synthetic
2	D_2	500 thousands	70	5	synthetic
3	D_3	1.5 millions	20	10	synthetic
4	D_4	4 millions	10	5	synthetic
5	D_5	4 millions	15	2	synthetic
6	D^*	4.8 millions	34	20	web data

All these different parameters do not act independently. To evaluate how they
relate to each other, we have considered 5 synthetic and one real world web data
sets. The synthetic ones were generated using MOA (Massive Online Analysis
software environment [25]), while the web one is the KDD CUP 99' data set[3],

---

[2] The standard method of dividing the range of numerical attributes values into bins.
[3] http://archive.ics.uci.edu/ml - for simplicity we have considered only the numerical
attributes.

which was also used to benchmark the dsCART algorithm. Their characteristics are listed in Table 1. In all our tests, the θ tie breaking parameter was not taken into consideration.

4.2   Experiment Results

A first set of results, presented in Table 2, shows the improvement of PdsCART running times, compared with dsCART (first line in the table), while producing exactly the same trees and accuracy.

Table 2. Running time(ms) and accuracy for D_1, D_5, D^* datasets with different # of incoming records processed in parallel. #1 is the dsCART algorithm.

D_1			D_5			D^*		
Records	Accuracy	Time	Records	Accuracy	Time	Records	Accuracy	Time
1	83.11 %	11.44	1	84.94 %	5076.77	1	77 %	9106.35
20	83.11 %	2.10	200	84.94 %	318.68	200	77 %	409.98
40	83.11 %	1.56	400	84.94 %	253.71	400	77 %	297.19
60	83.11 %	1.35	600	84.94 %	215.75	600	77 %	250.04
80	83.11 %	1.23	800	84.94 %	183.67	800	77 %	227.55

Table 2 contains some of the our most important results that leaded to PdsCART algorithm. As we can see in the *Accuracy* column, even if the number of examples taken into consideration for computing the splitting points increases, the accuracy remains the same. This is exactly the property needed to realize the parallel algorithm. Using this characteristic, the execution time of PsdCART can be reduced as can be seen in *Time* column. This results are confirmed by other tests as well (see Fig. 2).

The faster processing times are obtained mainly because of the fewer splitting conditions computations. For example, for D_4 dataset, having 4 millions records, when processing 200 records at time, there are 20 000 splitting computation steps, while when processing 800 records, there are only 5000.

While it might be tempting to choose a very large number of records to be processed in parallel, we have to underline here its side effect as well: the larger the number of records is, the later the algorithm will catch the decision tree changes / splits. Later refers here to the stream records occurrence time and not necessarily to the overall processing time. For example, for business applications that need to have a very responsive prediction model, where the changes in the stream need to be detected and processed immediately, a lower number of records would fit better; for other applications that can afford to have the same prediction model without caring about the amount of stream records consumed, larger number of records may conduct to better processing times.

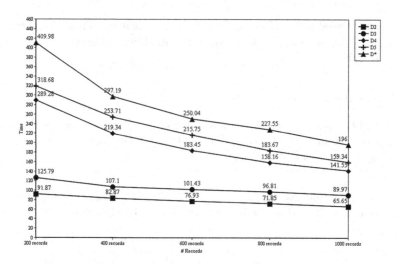

Fig. 2. Running time(ms) per dataset with different # of incoming records processed in parallel.

Detecting the splitting points earlier may facilitate the computations, since the new leaves will have fewer records to take into account for the future splitting conditions computations. This may justify why the differences between processing times with larger number of records (1000 vs. 800) is significantly smaller than with fewer number of records (400 vs. 200).

Table 3. Running time(ms) per dataseta having different # of attributes.

#	Data Set	TreeDepth	TreeNodes	TreeLeaves	# Attributes	Time
1	D_{2a}	2	3	2	2	19.59
2	D_{5a}	5	29	15	5	66.54
3	D_{10a}	10	213	107	10	259.651
4	D_{20a}	20	5501	2197	20	1610.75

However, in Fig. 2 we can see that while D_4 and D_5 synthetic datasets have the same number of records, the running times of D_4 are considerable better than those of D_5, under the same tests settings. This is related this time with another aspect that it is worth mentioning: the number of attributes. Intuitively, more attributes a dataset has, more time it will be needed by the algorithm to process it. Table 3 shows the results obtained after processing 4 synthetic datasets, all of them having 4 millions of records but different number of attributes.

Another result is related to the splitting computing times. When checking or selecting the splitting attributes, the PdsCART algorithm has to consider all possible partitions of the set of attribute values. This is directly related to the

Table 4. Running time(ms) per dataset using histograms with different # of bins.

#	Data Set	2 bins	4 bins	6 bins	8 bins	10 bins
1	D_2	73.28	77.41	82.36	89.40	98.63
2	D_3	79.96	83.70	90.97	100.22	111.77
3	D_4	145.55	153.22	167.28	204.38	229.88
4	D_5	151.89	166.16	190.98	250.27	313.59

number of bins used in the histograms. As we can see in Table 4, the more bins a histogram has, the more time will be needed to evaluate all the partitions. This property may lead to another level of parallelization, where all partitions may be analyzed independently. We are considering this as a solid basis for our future work.

Although these are just a subset of all the experiments conducted, they should prove the potential of our algorithm.

5   Conclusion and Future Work

In this paper we have shown how to implement a decision tree learning algorithm in the MapReduce framework. The first achievement of this algorithm is to be able to produce the decision tree in one single pass over the data. This is crucial in the context of streaming data, were multiple passes over the same data set are very difficult or even impossible. A second important achievement is the performance of the implementation. We were able to show that the algorithm achieves very good results by treating in parallel a larger number of records.

This encouraging results provide a solid basis for the ongoing work. In particular it is necessary to analyze how the algorithm scales with an increasing number of processing units and in which way all other parameters are influencing the behavior of the algorithm. Performance-wise, the outcome is relatively easy to guess. However more work has to be done in order to asses the influence of these parameters regarding the quality of the decision trees. We know that we can achieve similar error rates as the other algorithms (e.g. C4.5). Some other parameters regarding the trees such as size, depth, order of attributes, will have to be further investigated as well.

References

1. Dean, J., Ghemawat, S.: MapReduce: Simplified data processing on large clusters. Commun. ACM **51**(1), 107–113 (2008)
2. Breiman, L., Friedman, J., Olshen, R., Stone, C.: Classification and Regression Trees. Chapman & Hall/CRC, New York (1984)
3. Quinlan, J.R.: C4.5: Programs for Machine Learning. Morgan Kaufmann Publishers Inc., San Francisco (1993)

4. Shafer, C., Agrawal, R., Mehta, M.: SPRINT: a scalable parallel classifier for data mining. In: Proceedings of the 22th International Conference on VLDB, pp. 544–555 (1996)

5. Joshi, M., Karypis, G., Kumar, V.: ScalParC: a new scalable and efficient parallel classification algorithm for mining large datasets. In: Proceedings of the 12th International Parallel Processing Symposium, pp. 573–579 (1998)

6. Sreenivas, M., Alsabti, K., Ranka, S.: Parallel out-of-core divide-and-conquer techniques with applications to classification trees. In: The 10th Symposium on Parallel and Distributed Processing, pp. 555–562 (1999)

7. Jin, R., Agrawal, G.: Communication and memory efficient parallel decision tree construction. In: Proceedings of the 3rd SIAM International Conference on Data Mining (SDM), pp. 119–129 SIAM, (2003)

8. Ben-Haim, Y., Tom-Tov, E.: A streaming parallel decision tree algorithm. J. Mach. Learn. Res. **11**, 849–872 (2010)

9. Srivastava, A., Han, E., Kumar, V., Singh, V.: Parallel formulations of decision-tree classification algorithms. Data Min. Knowl. Discov. **3**(3), 237–261 (1999)

10. Amado, N., Gama, J., Silva, F.: Parallel implementation of decision tree learning algorithms. In: Brazdil, P.B., Jorge, A.M. (eds.) EPIA 2001. LNCS (LNAI), vol. 2258, pp. 6–13. Springer, Heidelberg (2001)

11. Panda, B., Herbach, J., Basu, S., Bayardo, R.: PLANET Massively parallel learning of tree ensembles with MapReduce. In: Proceedings of VLDB-2009 (2009)

12. Ye, J., Chow, J.-H., Chen, J., Zheng, Z.: Stochastic gradient boosted distributed decision trees. In: Proceedings of the 18th ACM Conference on Information and Knowledge Management, pp. 2061–2064 (2009)

13. Tyree, S., Weinberger, K.Q., Agrawal, K., Paykin, J.: Parallel boosted regression trees for web search ranking. In: Proceedings of the 20th International Conference on World Wide Web, pp. 387–396. ACM (2011)

14. Li, B., Chen, X., Li, M.J., Huang, J.Z., Feng, S.: Scalable random forests for massive data. In: Tan, P.-N., Chawla, S., Ho, C.K., Bailey, J. (eds.) PAKDD 2012, Part I. LNCS, vol. 7301, pp. 135–146. Springer, Heidelberg (2012)

15. Rutkowski, L., Jaworski, M., Pietruczuk, L., Duda, P.: A new method for data stream mining based on the misclassification error. IEEE Trans. Neural Netw. Learn. Syst. **26**(5), 1048–1059 (2014)

16. Li, X., Barajas, J.M., Ding, Y.: Collaborative filtering on streaming data with interest-drifting. Intell. Data Anal. **11**(1), 75–87 (2007)

17. Domingos, P., Hulten, G.: Mining high-speed data streams. In: Proceedings of the 6th ACM SIGKDD Conference, pp. 71–80 (2000)

18. Hulten, G., Spencer, L., Domingos, P.: Mining time-changing data streams. In: Proceedings of the Seventh ACM SIGKDD International Conference on Knowledge Discovery and Data Mining, pp. 97–106 (2001)

19. Bifet, A., Holmes, G., Pfahringer, G., Kirkby, R., Gavalda, R.: New ensemble methods for evolving data streams. In: Proceedings of the 15th ACM SIGKDD International Conference Knowledge Discovery and Data Mining (2009)

20. Bifet, A., Holmes, G., Kirkby, R., Pfahringer, B.: DATA STREAM MINING: A Practical Approach. University of Waikato, New Zealand (2011)

21. Hoeffding, W.: Probability inequalities for sums of bounded random variables. J. Am. Stat. Assoc. **58**, 13–30 (1963)

22. Rutkowski, L., Pietruczuk, L., Duda, P., Jaworski, M.: Decision trees for mining data streams based on the McDiarmid's bound. IEEE Trans. Knowl. Data Eng. **25**, 1272–1279 (2013)

23. Rutkowski, L., Jaworski, M., Pietruczuk, L., Duda, P.: Decision trees for mining data streams based on the gaussian approximation. IEEE Trans. Knowl. Data Eng. **26**, 108–119 (2014)
24. Rutkowski, L., Jaworski, M., Pietruczuk, L., Duda, P.: The CART decision tree for mining data streams. Inf. Sci. **266**, 1–15 (2014)
25. Bifet, A., Holmes, G., Kirkby, R., Pfahringer, B.: MOA: massive online analysis. J. Mach. Learn. Res. **11**, 1601–1604 (2010)

Industry 4.0 - Potentials for Creating Smart Products: Empirical Research Results

Rainer Schmidt[2(✉)], Michael Möhring[1], Ralf-Christian Härting[1],
Christopher Reichstein[1], Pascal Neumaier[1], and Philip Jozinović[1]

[1] Business Information Systems, Aalen University, Beethovenstr. 1,
Aalen 73430, Germany
{Michael.Moehring,Ralf.Haertingh,
Christopher.Reichstein}@htw-aalen.de
[2] Business Information Systems, Munich University of Applied Sciences,
Lothstr. 64, Munich 80335, Germany
Rainer.Schmidt@hm.edu

Abstract. Industry 4.0 combines the strengths of traditional industries with cutting edge internet technologies. It embraces a set of technologies enabling smart products integrated into intertwined digital and physical processes. Therefore, many companies face the challenge to assess the diversity of developments and concepts summarized under the term industry 4.0. The paper presents the result of a study on the potential of industry 4.0. The use of current technologies like Big Data or cloud-computing are drivers for the individual potential of use of Industry 4.0. Furthermore mass customization as well as the use of idle data and production time improvement are strong influence factors to the potential of Industry 4.0. On the other hand business process complexity has a negative influence.

Keywords: Industry 4.0 · Cyber physical systems · Empirical research · Business information systems · Study

1 Introduction

Combining the strengths of optimized industrial manufacturing with cutting-edge internet technologies is the core of Industry 4.0 [1]. Therefore it does not surprise that Industry 4.0 is experiencing an increasingly growing attention especially in Europe [1], but also in the United States, coined as Industrial Internet [2]. Industry 4.0 is often compared with proceeding disruptive increases in production [3] such as the industrial revolution(s) initiated by steam, electricity etc. Similar to Industry 4.0 these "revolutions" were initiated not by a single technology, but by the interaction of numbers of technological advances whose quantitative effects created new ways of production [4]. Three disruptive changes of industrial production happened until now [5].

1. At first industrial revolution and the ubiquity of mechanical energy combined with control systems such as the centrifugal force controller enabled huge productivity increase in the textile industry [6].

© Springer International Publishing Switzerland 2015
W. Abramowicz (Ed.): BIS 2015, LNBIP 208, pp. 16–27, 2015.
DOI: 10.1007/978-3-319-19027-3_2

2. The second large bouleversement was the replacement of steam by electricity [7]. Again, the coincidence of a set of technological advances such as transformation of alternating current [8] and advanced means for isolation [9] was necessary.
3. The use of electronics to the automation of production is considered as the third disruptive development. The intelligent control of robots and automated production and their integration provided the breakthrough.

In the same way as these proceeding disruptions, Industry 4.0 will change supply chains, business models and business processes significantly [1]. Therefore, many companies face the challenge to assess the diversity of developments and concepts summarized under the term industry 4.0 and to develop their own corporate strategies [6]. However, as many disruptive developments before, industry 4.0 is also accompanied by hype and overenthusiasm [10, 11]. Therefore, many companies and organizations are exposed to a dilemma: Neither to wait too long with their industry 4.0 implementation nor to start too early and commit fatal errors. There is a lack of research of the potential use of Industry 4.0. Nowadays, it is unclear what important factors influence the potential use of Industry 4.0.

Therefore, this paper aims to provide empirical information on the potentials of use Industry 4.0. In this way it will help academics and practitioners to identify and prioritize their steps towards an Industry 4.0 implementation. This is achieved by identifying those factors with a positive impact on the use of Industry 4.0.

It proceeds as follows. In the next chapter, the concepts of industry 4.0 are defined in detail. The research design and methods are defined, such as the design of the study and the data collection. The results are presented in the following section. Finally a conclusion and an outlook are given.

2 Smart Products, Processes and Technologies

Industry 4.0 is the superposition of several technological developments that embraces both products and processes. Industry 4.0 is related to the so-called Cyber physical systems [12] that describe the merger of digital with physical workflows [13]. In production, this means that the physical production steps are accompanied by computer-based processes. Cyber-physical systems include compute and storage capacity, mechanics and electronics, and are based on the Internet as a communication medium. Another related technology is the so-called Internet of things [14], defined as the ubiquitous access to entities in the internet. The so-called Internet of services [15] pursuits a similar approach with services instead of physical entities. The economic effects of smart products and industry 4.0 are manifold. As same as new products and services will be created [16], also traditional settings will profit from industry 4.0. E.g. the ability to provide more individual or even products that are malleable at the customer site may reduce the number of product returns [17]. In this paper industry 4.0 shall be defined as the embedding of smart products into digital and physical processes. Digital and physical processes interact with each other and cross geographical and organizational borders.

Smart Products. Smart products are products that are capable to do computations, store data, communicate and interact with their environment [18, 19]. Starting from early approaches enabling products to identify themselves via RFID [14] the capabilities of products to provide information on them evolved. Today smart products not only provide their identity but also describe their properties, status and history. Smart products are able to communicate information on their lifecycle. They know not only about the process steps already passed through, but are also able to define future steps. These steps include not only productions steps still to be performed on the unfinished product, but also upcoming maintenance operations. The capability to individually specify its properties can be used for an individual production with varying size. Smart products interact with their physical environment. They are capable to perceive and interact with their environment [6]. E.g. sensors allow to capture physical measures, cameras to get visual information on the product and its environment in real-time. Actors [6] enable products to impact physical entities in their environment without human intervention.

Industry 4.0 implies a huge increase of variety, volume and velocity of data creation [16]. The type and the amount of collected data has grown significantly due to advances in sensor technology and the products contained computer capacities. In the past, selectively measured values were captured. Today, data is collected continuously, creating a continuous stream of data. Also the type of data has evolved. Before, only simple types, such as temperature measurement were collected, now larger data types such as images or even real time videos are used. Due to the significantly higher processing capacity images, sounds and video files can be collected and used for triggering maintenance operations.

Intertwined Digital and Physical Processes Across the Whole Product Lifecycle. The following graphic shows the concepts associated with industry 4.0. They describe an extensive digitization of the value chain and how optimization and flexibility shall be achieved. The steps involved in the creation of value are fully integrated. Through social software [20], customers and suppliers are included in the innovation of the product. In the use phase of the product, the product is networked and connected with cloud services. The product stays connected during its entire life cycle and collected data. Using Big Data [21] a feedback loop into the production phase can be established. Big Data is the interplay of a number of technological advances with changed algorithms and model that are able to process data in an unprecedented volume, velocity and variety [21]. Technologies from the Big Data context such as [22] are able to process the enormous amounts of data and analyze them in near-real-time.

Base Technologies. Industry 4.0 is based on a number of technologies. The most important ones are mobile computing [23], cloud-computing [24] and Big Data [25]. The importance of cloud computing and mobile computing for Industry 4.0 lies not so much in providing scalable compute capacity, but rather in the provision of services, which can be accessed globally via the Internet. So, support services can be easily integrated and used (Fig. 1).

The easy integration of services also promotes the cooperation between the partners along the entire value chain. This has resulted in that relationship and not transactions stand increasingly in the foreground. Cloud computing is also the basis for the creation

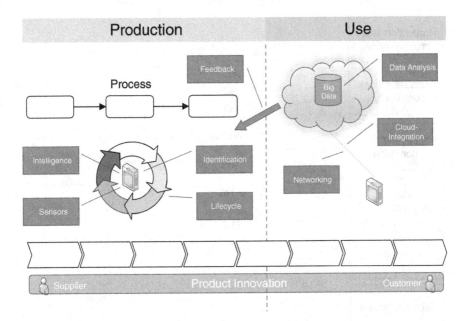

Fig. 1. Industry 4.0 based on [26]

of new business processes and models. Products integrated with cloud computing in the field can provide data that enable a predictive maintenance, and to give information about optimization possibilities in the production.

Already before the rise of cloud-computing and the internet, data was collected during production [27]. However this data remained in the production systems and had to be deleted after some time due to lack of storage capacity. Today the use of integrated networking and integration of products into the Internet data will give far reaching possibilities to collect data [28]. Instead of single data points or short intervals, a continuous stream of data is now available. The huge amounts of data available can now be used to continuously analyze and optimize production. This enables to foster predictive analytics [29].

3   Research Design and Methods

3.1   Design of the Study

To explore the potential use of industry 4.0, we design a quantitative research study. In this section as well as in Fig. 2 we developed our research model. The potential use of Industry 4.0 can be defined as the individual perceived capability of the implementation of Industry 4.0. The design of our study contains six hypotheses shown in Fig. 2.

The importance of production time for supply chain performance is identified in [30]. Its reduction is identified as a potential benefit from Industry 4.0 in [31]. Therefore we created hypothesis 1:

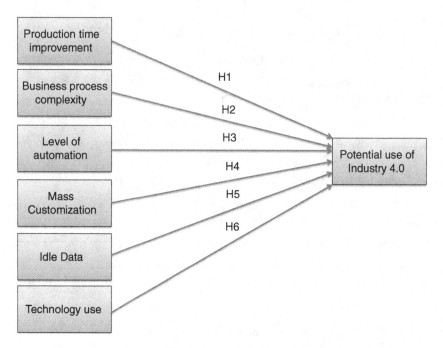

Fig. 2. Research model

H1: An Improvement of the Production Time Positively Influences the Potential Use of Industry 4.0. The complexity of business processes may [32] hamper overall supply-chain performance. However increasing integration and data exchange may overcome these negative effects. As industry 4.0 also fosters integration and data exchange [6], we introduce hypothesis 2:

H2: Complex Business Processes Positively Influences the Potential Use of Industry 4.0. The positive effects of automation in production systems in supply chain performance have already been identified in [33]. Especially the combination with computer integrated manufacturing [34] yields substantial benefits. Therefore, also industry 4.0 should provide these benefits as postulated in [1]. This lead to the creation of hypothesis 3:

H3: A High Level of Automation Positively Influences the Potential Use of Industry 4.0. Mass customization is an important means for competing in consumer-driven markets [35]. Industry 4.0 provides an excellent support for mass customization [1]. Therefore we create hypothesis 4:

H4: Mass Customization Positively Influences the Potential Use of Industry 4.0. The next hypothesis discovers the influence of idle (unused) data as a driver of Industry 4.0. According to Schmidt et al. [36] the amount of idle data has a negative influence on the use of Big Data. Big Data is also one technology driver of Industry 4.0. Therefore, we designed the following hypothesis:

H5: The Amount of Idle Data Negatively Influences the Potential Use of Industry 4.0. The influence of Big Data [16], Cloud-Computing [10], Mobile Computing [37], Internet of Things [14] and Cyber-Physical Systems [13] on industry 4.0 has already been discussed on a theoretical level. However there is still a lack of empirical evidence. Therefore we create hypothesis 6.

H6: Current Technologies Like Big Data, Cloud-Computing, Mobile Computing, Internet of the Things and Cyber-Physical Systems Positively Influences the Potential Use of Industry 4.0. For discovering the special attributes, we use a Likert [38] scale of one to six (1: low to 6: high) for all items (see Table 2). Only the questions about the use of the technology's (e.g. Big Data, Cloud, etc.) are ranged on a scale of one to two (1: use, 0: not in use).

3.2 Research Methods and Data Collection

Our quantitative research study based on a web survey in German speaking countries (Germany, Austria and Switzerland). The study were implemented via the open source tool Limesurvey [39]. To ensure a high quality, we tested our model and study with a pre-test and improve our study on the basis of our results. The main study started in July 2014 and ended in October 2014. We invited several leading experts in the field of information technology as well as manufacturing. These experts were contacted via email, telephone, letter and professional practical journals in Germany, Austria and Switzerland. We collected 592 answers. Because of this very special topic we implemented check-questions (e.g. industry background) to get only experts in the field of Industry 4.0. Only n = 133 answers were collected after data cleaning to ensure a high quality of our study.

More than 42 % of the experts work for enterprises with more than 500 employees. The majority of the asked experts came from the manufacturing sector (54.14 %), followed by the information and communication sector (13.35 %). Further industry sectors are e.g. the energy and facility management sector as well as health care and governmental sector. The experts came from Germany (87.22 %), Austria (5.26 %) and Switzerland (7.52 %). The classification of industry sectors was based on the European Classification of Economic Activities (NACE Rev.2). More than 60 % of the asked enterprises will implement Industry 4.0 or have done an implementation of Industry 4.0.

We used a structural equation modeling (SEM) [40–43] approach to analyze our causal model. Therefore, it is possible to use tools like IBM SPSS AMOS or Smart PLS. Because of our limited data set, we used a Partial Least Squares (PLS) [40] SEM approach. SEM [42] is a method to test the fit of a causal model with empirical data [40, 43]. The smart PLS approach is focusing on a partial least squares regression based on sumscores and significances are calculated via bootstraping.

4 Results

Finally, we analyze our data via the SEM approach by using SmartPLS 3.1 [41] and got the following results:

According to Hypothesis 1 (**An improvement of the production time positively influences the potential use of Industry 4.0**) the results of the SEM shows a positive effect (+0.236) of the production time to the potential use of Industry 4.0. Therefore the hypothesis can be confirmed. One argument for a positive influence of an improvement of the production time might be the fact that the complexity of business processes [32] hinder an overall supply-chain performance. An increasing integration and data exchange may outbalance possible delays created by using Industry 4.0 concepts (Fig. 3).

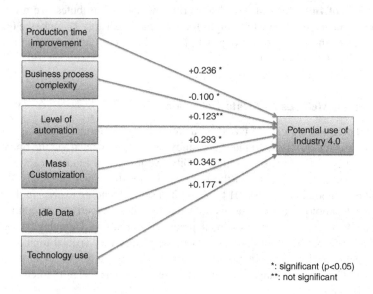

Fig. 3. Structural equation model with coefficients

The influence of the complexity of business processes on the potential use of Industry 4.0 were discussed in Hypothesis 2 (**Complex business processes positively influence the potential use of Industry 4.0**). Based on our model results, we must reject the hypothesis because of a negative value (−0.100). A higher complexity indicates a lower potential use of Industry 4.0. The negative impact of the complexity of business processes can be explained by the theory of transaction costs [44]. Studies based on this approach indicate that new information technologies are able to raise business processes to a higher level of efficiency and to generate economies of scale [45]. But in a situation of high uncertainty respectively high complexity an opposite effect arises. People, who are involved in complex business processes, become uncertain concerning using new technologies like Big Data, Cloud-Computing or Mobile-Computing. That increases the transaction costs, leads to a lack in digital trust and a reduced use of Industry 4.0 technologies.

The level of automation has a positive effect (+0.123) on the potential use of Industry 4.0. But this influence is not significant (p = 0.135 > 0.05). According to this results, hypothesis 3 (**A high level of automation positively influences the potential use of Industry 4.0**) must be rejected.

Furthermore Mass Customization is a very strong driver of Industry 4.0 according to the literature [37]. Our data can confirm this statement following Hypothesis 4 (**Mass Customization positively influences the potential use of Industry 4.0**). The influence of Mass Customization to the potential use of Industry 4.0 is +0.293. Therefore Hypothesis 4 can be confirmed. Industry 4.0 allows an individualized reacting on customer requests with a high degree of self-organization. For example some concepts of Mass Customization are working with intelligent work pieces (e.g. RFID). These pieces contain information regarding relevant process steps and raw materials. Furthermore, consumers who buy products in online shops often return it (especially in the European Union), because they do not fit to their individual preferences [17, 46]. Therefore, mass customization via Industry 4.0 can may reduce product returns by producing more consumer individual products.

The literature shows very interesting results of the impact of idle data to technologies like Big Data [36]. In contrast to [36] we get a positive influence of the amount of idle data (+0.345) to the potential use of Industry 4.0. Following this results, Hypothesis 5 (**The amount of idle data negatively influences the potential use of Industry 4.0**) must be rejected. Therefore, unused data are very important for the implementation of Industry 4.0.

Hypothesis 6 investigates the technology influence on the potential use of Industry 4.0 (**Current technologies like Big Data, Cloud-Computing, Mobile Computing, Internet of the Things and Cyber-Physical Systems positively influences the potential use of Industry 4.0**). Based on our data, Hypothesis 6 can be confirmed (+0.177). Therefore, new technologies are the base and driver of Industry 4.0. Current technologies of Industry 4.0 are also the fundament of new business models. In that context, Industry 4.0 is a disruptive innovation [47]. That means Industry 4.0 and its disruptive technologies have an above-average growth rate and are able to replace conventional technologies shortly.

Tables 1 and 2 contain important quality values of the SEM. Furthermore, the coefficient of determination (R2) is good (0.604 > 0.19) [41] as well as the quality criteria Cronbach Alpha for scales (>0.70) [48].

Table 1. SEM coefficient

SEM-Path	Path coefficient	Significance (P Values)
Automation level – > potential use of Industry 4.0	0.123	0.135
Business process complexity – > potential use of Industry 4.0	-0.100	0.049
Mass-customization – > potential use of Industry 4.0	0.293	0.001
Production time improvement – > potential use of Industry 4.0	0.236	0.001
Technology use – > potential use of Industry 4.0	0.177	0.003
Idle data – > potential use of Industry 4.0	0.345	0.000

Table 2. Cronbach-Alpha and items

Factor and items	Cronbach Alpha
Technology use (Current use of Big Data, Mobile Computing, Internet of the Things, Cyber Physical Systems, Cloud Computing)	**0.834** (5 item)
Production time improvement (use potentials of Industry 4.0 to the reduction of Pre-production time, production time, production wait time, time to market)	**0.831** (4 items)
Business process complexity (Business process complexity of the asked enterprise)	**1** (1 item)
Level of automation (Industry 4.0 impact on the automation level)	**1** (1 item)
Mass customization (Desire for individual products of the customers of the asked enterprise)	**1** (1 item)
Idle data (Amount of idle data in the production sector of the enterprise)	**1** (1 item)

5 Conclusion

Industry 4.0 introduces smart products which capture major data holdings due to significantly increased compute and memory performance, and can evaluate and can identify itself to higher-level systems. Industry 4.0 also intertwines physical and digital processes.

Our empirical research generates some interesting findings of the potential use of Industry 4.0. The use of current technologies like Big Data or Cloud are drivers for the individual potential of use of Industry 4.0. Furthermore Mass Customization as well as the use of idle data and production time improvement are strong influence factors to the potential of Industry 4.0. Business Process Complexity has a negative influence.

Academic research can benefit from our results to get a better understanding of the potential use of Industry 4.0 and can so adopt current approaches in the field of e.g. business intelligence and knowledge management. Managers can use these results for better decision making for e.g. implementing Industry 4.0 and can so save costs.

Our research is limited to the asked Industry 4.0 experts. We only asked experts from German speaking countries and mainly in Germany. One aspect (automation level) is not significant. Furthermore, there may be some further aspects to discover and a qualitative research approach can get some more detailed insights. Therefore, future research can enlarge the number of countries as well as the research approaches (e.g. qualitative interviews) to get a broader view of Industry 4.0. Furthermore some special aspects (e.g. pre-production time and costs) can be a great opportunity for a detailed research in the future.

References

1. Blanchet, M., Rinn, T., Thaden, G., Thieulloy, G.: Industry 4.0. The new industrial revolution. How Europe will succeed. Hg V Roland Berg. Strategy Consult. GmbH Münch. Abgerufen Am 1105 2014 Unter (2014). http://www.Rolandberger.Com/med.0403/Pdf
2. Evans, P.C., Annunziata, M.: Industrial internet: pushing the boundaries of minds and machines. Gen. Electr. **21** (2012)
3. Brynjolfsson, E., McAfee, A.: The Second Machine Age: Work, Progress, and Prosperity in a Time of Brilliant Technologies. W.W. Norton & Company, New York (2014)
4. Brynjolfsson, E., Hofmann, P., Jordan, J.: Cloud computing and electricity: beyond the utility model. Commun. ACM **53**, 32–34 (2010)
5. Dorst, W., EV, B.: Fabrik-und Produktionsprozesse der Industrie 4.0 im Jahr 2020. IM Fachz. Für Inf. Manag. Consult. Nr. **27**, 34–37 (2012)
6. Beckert, S.: Empire of Cotton: A Global History. Knopf, New York (2014)
7. Carr, N.G.: The Big Switch: Rewiring the World, from Edison to Google. WW Norton & Company, New York (2008)
8. Kiebitz, F.: Nikola Tesla zum fünfundsiebzigsten Geburtstage. Naturwissenschaften **19**, 665–666 (1931)
9. Siemens, W.: Ueber telegraphische Leitungen und Apparate. Wissenschaftliche und Technische Arbeiten, pp. 15–29. Springer, Heidelberg (1889)
10. Bauernhansl, T.: Industry 4.0: Challenges and limitations in the production. Keynote. ATKearney Fact (2013)
11. Messe, H.: "Industrie 4.0" muss sich erst beweisen - Industrie - Unternehmen – Handelsblatt. http://www.handelsblatt.com/unternehmen/industrie/hannover-messe-industrie-4-0-muss-sich-erst-beweisen/8044930.html
12. Rajkumar, R., Lee, I.: NSF workshop on cyber-physical systems (2006)
13. Lee, E.A.: Cyber physical systems: design challenges. In: 2008 11th IEEE International Symposium on Object Oriented Real-Time Distributed Computing (ISORC), pp. 363–369. IEEE (2008)
14. Ashton, K.: That "internet of things" thing. RFiD J. **22**, 97–114 (2009)
15. Cardoso, J., Voigt, K., Winkler, M.: Service engineering for the internet of services. Enterp. Inf. Syst. **13**(1), 15–27 (2009)
16. Lee, J., Kao, H.-A., Yang, S.: Service innovation and smart analytics for industry 4.0 and big data environment. Procedia CIRP **16**, 3–8 (2014)
17. Walsh, G., Möhring, M., Koot, C., Schaarschmidt, M.: Preventive product returns management systems-a review and model. In: Proceedings of the 21th European Conference on Information Systems (ECIS), Tel Aviv, Israel (2014)
18. Miche, M., Schreiber, D., Hartmann, M.: Core services for smart products. 3rd European Workshop on Smart Products, pp. 1–4 (2009)
19. Mühlhäuser, M.: Smart products: an introduction. In: Mühlhäuser, M., Ferscha, A., Aitenbichler, E. (eds.) Constructing Ambient Intelligence, pp. 158–164. Springer, Berlin Heidelberg (2008)
20. Nurcan, S., Schmidt, R.: Introduction to the first international workshop on business process management and social software (BPMS2 2008). In: Ardagna, D., Mecella, M., Yang, J., Aalst, W., Mylopoulos, J., Rosemann, M., Shaw, M.J., Szyperski, C. (eds.) Business Process Management Workshops, pp. 647–648. Springer, Berlin Heidelberg (2009)

21. LaValle, S., Lesser, E., Shockley, R., Hopkins, M.S., Kruschwitz, N.: Big data, analytics and the path from insights to value. MIT Sloan Manag. Rev. **52**, 21–32 (2011)
22. White, T.: Hadoop: The Definitive Guide. O'Reilly & Associates, Sebastopol (2015)
23. Forman, G.H., Zahorjan, J.: The challenges of mobile computing. Computer **27**, 38–47 (1994)
24. Mell, P., Grance, T.: The NIST Definition of Cloud Computing. http://csrc.nist.gov/groups/ SNS/cloud-computing/
25. Schmidt, R., Möhring, M.: Strategic alignment of cloud-based architectures for big data. In: Proceedings of the 17th IEEE International Enterprise Distributed Object Computing Conference Workshops (EDOCW), pp. 136–143. Vancouver, Canada (2013)
26. Schmidt, R.: Industrie 4.0 - revolution oder evolution. Wirtsch. Ostwürtt **2013**, 4–7 (2013)
27. Ang, C.L.: Technical planning of factory data communications systems. Comput. Ind. **9**, 93–105 (1987)
28. How big data can improve manufacturing. McKinsey & Company. http://www.mckinsey. com/insights/operations/how_big_data_can_improve_manufacturing
29. Abbott, D.: Applied Predictive Analytics: Principles and Techniques for the Professional Data Analyst. Wiley, New York (2014)
30. De Treville, S., Shapiro, R.D., Hameri, A.-P.: From supply chain to demand chain: the role of lead time reduction in improving demand chain performance. J. Oper. Manag. **21**, 613–627 (2004)
31. Bauer, W., Schlund, S., Marrenbach, D., Ganschar, O.: Industrie 4.0 - Volkswirtschaftliches Potenzial für Deutschland. BITKOM / Fraunhofer (2014)
32. Cardoso, J., Mendling, J., Neumann, G., Reijers, H.A.: A discourse on complexity of process models. In: Eder, J., Dustdar, S. (eds.) BPM Workshops 2006. LNCS, vol. 4103, pp. 117–128. Springer, Heidelberg (2006)
33. Groover Jr., M.P.: Automation, Production Systems and Computer-Aided Manufacturing. Prentice Hall PTR, New Jersey (1980)
34. Groover, M.P.: Automation, Production Systems, and Computer-Integrated Manufacturing. Prentice Hall Press, New Jersey (2007)
35. Gilmore, J.H., Pine 2nd, B.J.: The four faces of mass customization. Harv. Bus. Rev. **75**, 91–101 (1996)
36. Schmidt, R., Möhring, M., Maier, S., Pietsch, J., Härting, R.-C.: Big data as strategic enabler - insights from central european enterprises. In: Abramowicz, W., Kokkinaki, A. (eds.) BIS 2014. LNBIP, vol. 176, pp. 50–60. Springer, Heidelberg (2014)
37. Scheer, A.-W.: Industrie 4.0. Satzweiss.com (2013)
38. Babbie, E.: The practice of social research. Cengage Learning, Boston (2012)
39. Team, T.L.: project: LimeSurvey - the free and open source survey software tool! http:// www.limesurvey.org/de/start
40. Wong, K.K.-K.: Partial least squares structural equation modeling (PLS-SEM) techniques using smartPLS. Mark. Bull. **24**, 1–32 (2013)
41. Ringle, C.M., Wende, S., Will, A.: SmartPLS 2.0 (beta). Hamburg, Germany (2012)
42. Hooper, D., Coughlan, J., Mullen, M.: Structural equation modelling: guidelines for determining model fit. Articles. 2 (2008)
43. Chin, W.W.: The partial least squares approach to structural equation modeling. Mod. Methods Bus. Res. **295**, 295–336 (1998)
44. Williamson, O.E.: The economics of organization: the transaction cost approach. Am. J. Sociol. **87**(3), 548–577 (1981)
45. Wigand, R.T., Picot, A., Reichwald, R.: Information, Organization and Management: Expanding Markets and Corporate Boundaries. Wiley, Chichester (1997)

46. Shulman, J.D., Coughlan, A.T., Savaskan, R.C.: Optimal reverse channel structure for consumer product returns. Mark. Sci. **29**, 1071–1085 (2010)
47. Christensen, C.: The Innovator's Dilemma: When New Technologies Cause Great Firms to Fail. Harvard Business Review Press, Boston (2013)
48. Santos, J.R.A.: Cronbach's alpha: a tool for assessing the reliability of scales. J. Ext. **37**, 1–5 (1999)

Evaluating New Approaches of Big Data Analytics Frameworks

Norman Spangenberg[(✉)], Martin Roth, and Bogdan Franczyk

Information Systems Institute, Leipzig University,
Grimmaische Straße 12, 04109 Leipzig, Germany
{spangenberg,roth,franczyk}@wifa.uni-leipzig.de

Abstract. The big data topic will be one of the leading growth markets in information technology in the next years. One problem in this area is the efficient computation of huge data volumes, especially for complex algorithms in data mining and machine learning tasks. This paper discuss new processing frameworks for big and smart data in distributed environments and presents a benchmark between two frameworks - Apache Flink and Apache Spark - based on a mixed workload with algorithms from different analytic areas with different real-world datasets.

Keywords: Apache Flink · Apache Spark · Big data processing frameworks · Big data analytics · MapReduce

1 Introduction

The hype about the phrase big data and its various Vs - e.g. volume, velocity, variety, veracity - drops but some problems are still out in companies and academia. Whereas the Hadoop system focuses on batch processing web and text data stored in HDFS new demands and applications occurred under the topics smart data and data science [1]. Despite the Hadoop ecosystem has grown and matured for several years it cannot mask the problems of the MapReduce framework [3]. Some of these problems like the inefficient processing of iterations or the slow performance of combining multiple data sources are crucial for data scientists in data mining, machine learning and other algorithms. As well it leads to the turn away of projects from the ecosystem, e.g. the Apache Mahout project. This Hadoop ecosystem extends the functionality of plain MapReduce and HDFS and makes it suitable for many different tasks. But this drawbacks of Hadoop come along with advantages like the HDFS storage system, the linear scalability of applications, a high fault tolerance together with flexibility and simplicity in application development. So this points as well as the problems of real data science tools with huge data volume are drivers for new big data analytics frameworks. This process speeded up with the second version of Hadoop, including the YARN resource manager that allows other programming models than MapReduce [2]. Solely the Apache Foundation has with Flink [4], Giraph [5], Hama [6] and Spark [7] four frameworks that are alternatives for

© Springer International Publishing Switzerland 2015
W. Abramowicz (Ed.): BIS 2015, LNBIP 208, pp. 28–37, 2015.
DOI: 10.1007/978-3-319-19027-3_3

MapReduce. In addition there are projects like AsterixDB [8] or GraphLab [9] which are supported by other communities, the latter even now as a commercial product. While Giraph and Hama are only for suitable iterative and graph applications, AsterixDB is limited to semi-structured data models. In the end there are with Apache Flink, Apache Spark and GraphLab three frameworks that can be summarized as general-purpose big data processing frameworks. This group of processing systems should be able to fulfil the needs for analytical tasks of data scientists. The few comparisons of the aforementioned frameworks have shortcomings and make them unsuitable from a performance perspective to decide which of them is more suitable for big data analytics and data science tasks. For example Elser/Montresor [10] or the AmpLab [11] focusing just on one analytical area, like graph processing or relational queries. Others like Alexandrov et al. [12] or Zaharia et al. [13] include more algorithms but compare the systems only against the Hadoop respectively MapReduce framework. In this paper we present the first direct comparison of Apache Spark and Apache Flink and their applicability for different analytical use cases with real world datasets. The latest versions of the systems were used and benchmarked in a cluster environment. This work is organized as follows: The next section gives a brief technological overview about Apache Flink and Spark. The third section provides the description of the benchmark environment, the workload and a discussion of the results. Finally, a conclusion and recommendations for a decision concerning the frameworks is given.

2   Technological Foundations

This section gives a brief technological overview of the frameworks Apache Spark and Apache Flink. Both extend the number of operators for data processing and make a more efficient use of memory. For detailed explanations of the basic concepts of the frameworks see [12, 14–16] for Apache Flink and [13, 17] for Apache Spark.

2.1   Apache Spark

This framework has its roots in a research group at the UC Berkeley and evolved to an Apache top-level project in 2014. It is implemented in the functional programming language Scala and has further programming interfaces in Java, Python and the declarative query language SparkSQL. Spark rests upon two main concepts: resilient distributed datasets (RDD), which hold the data objects in memory, and transformations or actions that are performed on the datasets in parallel. Resilient distributed datasets are partitioned data structures that are just in a read-only-mode available. RDDs can be computed in several ways. Mainly used are transformations on files in a distributed file system like the Hadoop distributed file system (HDFS), or operations on other RDDs, which result in a new dataset due to the immutable nature of Scala collections [13]. Using materialized checkpoints stored in the distributed file system to achieve

Table 1. Transformations and actions in Apache Spark

Transformations				
map	Join	Union	Distinct	Repartition
mapPartitions	flatMap	Intersection	Pipe	coalesce
cartesian	cogroup	Filter	sample	
sortByKey	groupByKey	reduceByKey	aggregateByKey	
mapPartitionswithIndex		repartitionAndSortWithinPartitions		
Actions				
Reduce	Take	Collect	takeSample	Count
takeOrdered	countByKey	First	Foreach	saveAsTextFile
saveAsSequenceFile		saveAsObjectFile		

fault tolerance is a time consuming task as the datasets in big data applications are very large. Instead, in case of data loss the Spark system recomputes the lost partition with the earlier used transformation function and the persistently stored data in the distributed file system. To avoid large recomputations on datasets with many transformations, the system provides a checkpointing mechanism [17]. The earlier mentioned transformations and actions are the second important component of a Spark application beneath the RDDs. Currently there are 20 transformations and 12 actions usable. The difference between the two kinds of operators is that transformations are functions to generate and manipulate RDDs out of input files or other RDDs, whereas actions can only be used on RDDs to produce a result set in the driver program or write data to the file system [13]. Spark additional operators are for e.g. the grouping, filtering and merging of large datasets (Table 1).

2.2 Apache Flink

The Flink framework [4] has its origins in the Stratosphere project of several German universities. It transfers some basic technologies from relational database systems, e.g. the optimization of execution plans to big data processing frameworks. It is written in Java, but provides APIs for Scala and Python, too. Compared to Spark it has no declarative query language, which would make the implementation of applications easier for data scientists with less programming experience. The Apache Flink framework provides at the moment just 17 functions for data transformation, which are called PACT operators. Many of them have the same functionality like the transformations and actions in Spark with two distinctions. The two iteration operators are unique to Apache Flink and allow the efficient computation of various machine learning and data mining algorithms. While the bulkIteration operator uses the whole dataset in every run, the deltaIteration operator divides the data in a workset and solution set. With this differentiation in each iteration fewer data has to be computed and sent between nodes [14] (Table 2).

Table 2. Operators in Apache Flink

PACT-Operators				
Map	Union	Rebalance	flatMap	Join
mapPartition	Cross	firstN	hashPartition	Reduce
Filter	reduceGroup	project	deltaIteration	bulkIteration
coGroup	Aggregate			

Another feature of the Flink framework is the PACT optimizer, which improves a PACT program with database technologies and transforms it in a job graph that can be executed from the Flink execution engine. After a pre-processing step, a logical optimization creates several semantically equivalent execution plans by reordering the PACT operators based on code analysis and data access conflicts [12,16]. The physical optimization chooses efficient strategies for data transport and operator execution on specific nodes. A cost-based strategy selects the most efficient plan based on criteria like network or storage I/O costs.

3   Comparison of Apache Spark and Apache Flink

The used computational resources were made up of four nodes, each with 16 GB memory, 8 cores and Centos 6.5 as operating system. The test data resides in the Hadoop distributed file system in version 2.4.1. In the benchmarks not only the integrated resource manager of Apache Spark and Flink were tested, additionally the performance with the YARN cluster manager was evaluated. YARN was configured with a minimum allocation size of 3584 MB and a maximum allocation size of 14336 MB for each container, reserving nearly 2 GB memory for operating system purposes on each node. In both frameworks the integrated resource manager settings were the default values or had if possible the same configuration. Additionally to some network configurations in both systems the number of instances or container per node was varied to see which system benefits from a higher degree of intra-node parallelism. In Apache Spark the file spark-env.sh was prepared with different values for driver memory as well as the number of instances, cores and memory per executor. In Apache Flink the same settings have been realized with the configuration of the heap size of the jobmanager, taskmanager and the number Of TaskSlots. The last mentioned property sets how many different operators per instance or container run in parallel. The number of instances in Flink was specified when the system got started.

3.1   Benchmark Workload

Instead of using generated test data, real world datasets were used for the benchmark. Because of the different characteristics of the algorithms different datasets were applied. Table 3 summarizes the amount of data elements in each workload. The workload for the benchmark consisted of four different algorithms

Table 3. Amount of data elements in the test datasets for the specific algorithms

	1	2	3	4	5
WordCount	173,795,924	868,979,785	4,335,048,410	8,189,796,149	16,379,592,298
k-Means	3,335,010	16,675,050	33,350,100	66,700,200	133,400,400
PageRank	718,513	3,592,565	7,185,130	14,370,260	28,740,520
Relational query	3,335,010 + 13,292,124 + 91,788	16,675,050 + 13,292,124 + 91,788	33,350,100 + 13,292,124 + 91,788	66,700,200 + 13,292,124 + 91,788	133,400,400 + 13,292,124 + 91,788

from the topics batch processing in text files, data mining, combining different data sources and graph processing. With this combination of batch processing, iterative algorithms and relational queries the universality of the systems could be tested and the changing analytical tasks of data scientists can be simulated. The algorithms were:

1. WordCount: represents the use-case of classic web, text and log mining, which were the first applications for big data analytics and caused the development of big data processing frameworks like MapReduce and Hadoop at Google or Yahoo. It can be used as a metric to measure the quality of data elements [18]. The WordCount algorithm includes a map function which splits the dataset and converts it in a key-value object of the form (word, 1). The second step is a reduce function that sums up all values of one key. This algorithm can be easily implemented with the map and reduce operators that both frameworks supply. For the benchmark the WordCount example has been executed on Wikipedia datasets of different sizes.

2. K-Means-Clustering: one of the key tasks of data scientists is to discover unknown dependencies and correlations in large datasets. The k-Means-Clustering, first described in [19], is an unsupervised method in data mining to group data elements with a high similarity and separate elements with less similarity. In each iteration it assigns the data points to its nearest cluster center with a map function. Subsequently the data points are grouped by its center with a groupBy operator. At the end of each iteration a new cluster center based on the current data points in the cluster is computed again with a map function. This process ends after a fix number of iterations, in this case 10 iterations. After that in one single step the final cluster center is computed. The used test data were two features extracted from a Twitter dump of April 2014.

3. PageRank: an emerging topic in data analytics are graph algorithms and graph processing applications, which are common for the analysis of technical and social networks or semantic correlations in datasets. One of the most famous algorithm in this area is the PageRank algorithm [20]. It is a key component of the Google search Engine for web data but can also be used for other domains such as the identification of key users in social networks. Its implementation consists of a map and a groupBy stage, which creates an adjacency list. This list stores each vertex and its edges to other vertices.

The iterative part of the algorithm consists of a join, flatMap, groupBy, map and reduce operators that computes a PageRank value for every vertex at the end of the algorithm. The number of iterations is a fixed value of 30. As with the clustering example the test data was extracted from a Twitter dataset, containing linkages between users by a tweet-retweet-relationship.

4. Relational query: Another important feature for a data analyst is to bring together the information from different datasets and sources. The implemented query is similar to the TPC-H queries 3 and 10 [21] and joins instant messages from Twitter with weather station information from several sources of the National Climatic Data Center (NCDC). Therefore the program contains two join operators, as well as a filter operation to prepare the input data and a groupBy operation to aggregate the intermediate data in a final result (Fig. 1).

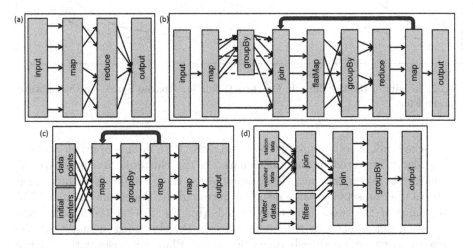

Fig. 1. Visualization of the algorithms: (a) WordCount (b) relational query (c) k-Means clustering (d) PageRank

3.2   Benchmark Results and Related Work

The graphics in Fig. 2 represent the results of the benchmark for each algorithm category. The experiments come to the following conclusions:

1. Batch processing applications like web or text mining are faster in Apache Spark than in Apache Flink. Former evaluations with Hadoop pointed out similar results. In Fan Liang et al. [22] Apache Spark was three times faster than MapReduce, while Alexandrov et al. showed [12] that the Flink ancestor Stratosphere is just 20 works faster, Apache Spark has a ten to twenty percent lesser resource usage than Flink. The reason that we found to explain the gap between both systems is that Flink uses just 70 CPU while Apache Spark had a CPU utilization of more than 90.

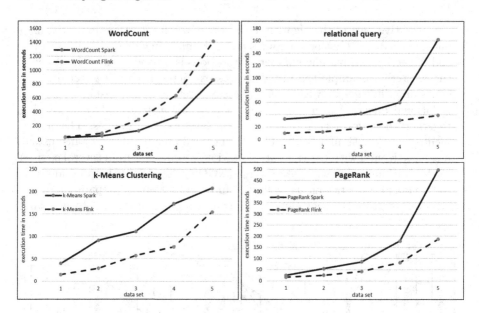

Fig. 2. Results of the benchmark. On the left side for Apache Spark, on the right for Apache Flink

2. Apache Flink is better suitable for the iterative data mining or graph processing algorithms used in this benchmark. In the clustering example Apache Flink is between two and three times faster than Spark. With increasing size of the datasets the same results occurred in the PageRank example. The screening of the two applications unfolds (see Fig. 3) that the time for one iteration differs between the frameworks. In the clustering example the difference in execution time between Spark and Flink decreases with a higher number of iterations. Instead in the PageRank algorithm inverted behavior was noted, because the difference increased with more iterations.

One assumption is that the complexity of the PageRank algorithm compared to the clustering algorithm leads to more important role of the optimizer. So with an increasing complexity it can be expected that Flink has a performance advantage compared with Spark. Another reason for the faster PageRank computation in Apache Flink - which is explained more precise in the next point - is the more efficient join processing, which is performed in every iteration of the algorithm.

3. Relational queries which include merging and processing of different sources run in Apache Flink considerably faster than in Spark. The first-mentioned shows an at least two times faster performance and this factor increases with larger datasets to factor five. One reason for this performance gap could be the optimization of execution plans with database technologies in Apache Flink. Further investigations revealed that in Flink pipelining data to the next operator is more efficient than in Spark (see Fig. 4). Another reason for the better

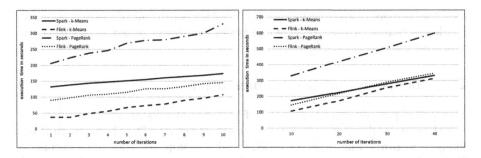

Fig. 3. Duration of applications with increasing number of iterations

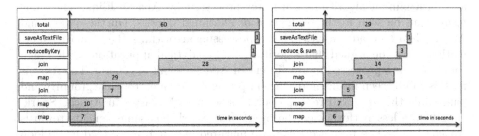

Fig. 4. Join processing in Apache Spark (left) and Flink (right)

performance is a quicker join processing of large datasets in Flink, as already seen in point two. Compared to Apache Hive both systems perform more efficient. Apache Spark is by a factor between two and ten faster, but this value depends heavily on the workload [11]. The Flink framework showed an 8- till 10-times better performance than Hive [12].

4. The choice of the resource manager cant be made on the basis of performance reasons. The results of the benchmarks indicate that neither the YARN nor the integrated resource manager of both systems have a definite better performance. In 47 experiments YARN was faster, in 42 were more efficient. In the remaining experiments both were identical. So a decision about a resource and cluster manager should be made based on administrative reasons, e.g. YARN allows a central resource allocation for more than just one processing framework so it reduces the administrative overhead.

5. Apache Spark works more efficient with less but larger containers. The advantage towards a doubled number of containers is in 80 results in Apache Flink are not that clear. A small tendency to a better performance with a higher number of containers is visible.

This five conclusions about the performance of the frameworks lead to the decision for Apache Flink for data science tasks, because its performance in most of the measured tasks was better than in Apache Spark. An important feature of Apache Spark, which is still missing in Apache Flink, is the interactive shell. This component allows the user the analysis of huge data amounts without the need for developing applications in an IDE. This is just one of the extensions of Spark which make it more mature and better suitable for a fast implementation in several use cases.

4   Conclusion and Recommendations

This paper presented a comparison of two frameworks of big data analytics and their suitability for different algorithm categories. The chosen algorithms simulate features from different tasks, which are important for data scientists. The implemented applications were executed on a 4-node cluster and real world datasets. The benchmark results come to the conclusion that graph processing and data mining algorithms are faster processed by Apache Flink. The same results can be seen in the evaluation of relational queries. On the other hand Apache Spark has advantages in batch processing algorithms. Based on this information, it can be argued that algorithms and analytical applications with a high complexity are processed more efficient in Apache Flink in general. The reasons for this can be seen in the more efficient processing of operators like groupBy and join, while the map and reduce operators seem to be faster in Apache Spark. Despite the less performance the Spark framework has some benefits towards Apache Flink. The framework is more matured, e.g. it is embedded in Hadoop distributions like MapR, Hortonworks or Cloudera. It has a broad community and more supporters, which use it in production environments and ensure the further development of the framework. This aspects lack in the current state of Apache Flink. The larger operator set in Spark provides opportunities for developers to optimize algorithms with more specific and appropriate operators for different tasks. Furthermore it increases the comfort during application development because the adjustment of more general operators isn't necessary anymore. Future works in this area could pick up other frameworks like the Google Cloud Dataflow engine, which is a new competitor next to Flink and Spark but currently only available as a cloud service. Further the future development of the frameworks compared in this paper should be noticed as well as other systems not mentioned in this paper, e.g. GraphLab.

Acknowledgements. The work presented in this paper was funded by the German Federal Ministry of Education and Research under the project Competence Center for Scalable Data Services and Solutions Dresden/Leipzig (BMBF 01IS14014B) and by the German Federal Ministry of Economic Affairs and Energy under the project InnOPlan (BMWI 01MD15002E).

References

1. Gartner. Hype Cycle for emerging technologies (2014). http://www.gartner.com/newsroom/id/2819918
2. Apache Software Foundation. Apache Hadoop NextGen MapReduce (Yarn) (2015). http://hadoop.apache.org/docs/current/hadoop-yarn/hadoop-yarn-site/YARN.html

3. Lee, K.-H., Lee, Y.-J., Choi, H., Chung, Y.D., Moon, B.: Parallel data processing with MapReduce: a survey. SIGMOD Rec. **40**(4), 11–20 (2012)
4. Apache Software Foundation. Apache Flink (2015). http://flink.apache.org/
5. Apache Software Foundation. Apache Giraph (2015). http://giraph.apache.org/
6. Apache Software Foundation. Apache Hama (2015). http://hama.apache.org/
7. Apache Software Foundation. Apache Spark (2015). http://spark.apache.org/
8. University of California Irvine. AsterixDB (2015). https://asterixdb.ics.uci.edu
9. Dato. GraphLab (2015). http://www.graphlab.com/
10. Elser, B., Montresor, A.: An evaluation study of bigdata frameworks for graph processing. In: Hu, X. et al. (eds.) BigData Conference, pp. 60–67. IEEE (2013)
11. AmpLab. BigDataBenchmark (2015). https://amplab.cs.berkeley.edu/benchmark/
12. Alexandrov, A., et al.: The stratosphere platform for big data analytics. VLDB J. **23**(6), 939–964 (2014)
13. Zaharia, M. et al.: Resilient distributed datasets: a fault-tolerant abstraction for in-memory cluster computing. In: Proceedings of the 9th USENIX Conference on Networked Systems Design and Implementation. NSDI 2012, pp. 2–2. USENIX Association, San Jose (2012)
14. Ewen, S., Tzoumas, K., Kaufmann, M., Markl, V.: Spinning fast iterative data flows. Proc. VLDB Endow. **5**(11), 1268–1279 (2012)
15. Warneke, D., Kao, O.: Nephele efficient parallel data processing in the cloud. In: Raicu, I., Foster, I., Zhao, Y. (eds.) MTAGS 2009, p. 110. ACM, New York (2009)
16. Hueske, F., Peters, M., Sax, M.J., Rheinlnder, A., Bergmann, R., Krettek, A., Tzoumas, K.: Opening the black boxes in data flow optimization. Proc. VLDB Endow. **5**(12), 1256–1267 (2012)
17. Zaharia, M., Chowdhury, M., Franklin, M.J., Shenker, S., Stoica, I.: Spark: cluster computing with working sets. In: Proceedings of the 2nd USENIX Conference on Hot Topics in Cloud Computing. USENIX Association (2010)
18. Blumenstock, J.E.: Size matters: word count as a measure of quality on wikipedia. In: Proceedings of the 17th international conference on World Wide Web. ACM (2008)
19. Hartigan, J., Manchek, A.: Algorithm AS 136: a k-means clustering algorithm. In: Applied statistics, pp. 100–108 (1979)
20. Page, L., Brin, S., Motwani, R., Winograd, T.: The PageRank citation ranking: bringing order to the web. http://ilpubs.stanford.edu:8090/422/1/1999-66.pdf
21. Transaction Processing Performance Council: TPC Express Benchmark H Decision Support-Standard Specification (2014). http://www.tpc.org/tpch/default.asp
22. Liang, F., Feng, C., Lu, X., Xu, Z.: Performance benefits of DataMPI: a case study with bigdatabench. In: Zhan, J., Rui, H., Weng, C. (eds.) BPOE 2014. LNCS, vol. 8807, pp. 111–123. Springer, Heidelberg (2014)

Automated Equation Formulation for Causal Loop Diagrams

Marc Drobek[1,2]([⊠]), Wasif Gilani[1], Thomas Molka[1], and Danielle Soban[2]

[1] SAP UK Ltd., Belfast, UK
{marc.drobek,wasif.gilani,thomas.molka}@sap.com
[2] Department of Mechanical and Aerospace Engineering,
Queens University Belfast, Belfast, UK
d.soban@qub.ac.uk

Abstract. The annotation of Business Dynamics models with parameters and equations, to simulate the system under study and further evaluate its simulation output, typically involves a lot of manual work. In this paper we present an approach for automated equation formulation of a given Causal Loop Diagram (CLD) and a set of associated time series with the help of neural network evolution (NEvo). NEvo enables the automated retrieval of surrogate equations for each quantity in the given CLD, hence it produces a fully annotated CLD that can be used for later simulations to predict future KPI development. In the end of the paper, we provide a detailed evaluation of NEvo on a business use-case to demonstrate its single step prediction capabilities.

Keywords: Business dynamics · Causal loop diagrams · Neural networks · Evolutionary algorithms · Big data · Predictive analyses

1 Introduction

The prediction of Key Performance Indicators (KPIs) in large enterprises is one of the major assets for business analysts and decision makers to drive company success. Traditional approaches, such as time series analyses are most common and yield quick results [1]. However, their restriction on small-dimensional dependencies limits the capability to identify actual root causes and main drivers for the particular KPI under study. Especially nowadays, in the Big-Data era, the massive amount of readily available business data raises the question, whether it can be incorporated in the prediction process to pinpoint root causes. The Business Dynamics (BD) domain tries to overcome this small-dimensionality by establishing broader scopes of causal chains and feedback loops [2,3]. In the field of BD, enterprise KPIs are modelled with the help of Causal Loop Diagrams (CLDs) and State & Flow Diagrams (SFDs) that are used to identify causality and feedback between the KPIs, as well as the material/resources flowing through the system. Once the modeller has arrived at a reasonable CLD and SFD, she needs to annotate the model with equations and parameters to create a

© Springer International Publishing Switzerland 2015
W. Abramowicz (Ed.): BIS 2015, LNBIP 208, pp. 38–49, 2015.
DOI: 10.1007/978-3-319-19027-3_4

final simulation model that will eventually produce prediction results. We have already summarised traditional approaches for the process of parameter estimation and equation formulation (PEEF) and concluded that these traditional concepts are cumbersome, resource-intensive and are in general only manually applicable [4]. In this paper, we will demonstrate the automated equation formulation and annotation of a given CLD based on a given historical data set. This is done by training neural networks in combination with evolutionary algorithms. The trained neural network is then annotated to the associated CLD element as a function surrogate (FS) that can replay the historical data or predict its future development. However, the main difference to traditional time series analyses is the incorporation of all target KPI dependencies given in the CLD. Rather than analysing the historical information of an isolated variable (silo-mode), we are exploiting the dependency relationships provided by a CLD and incorporate the historical information of all influencing elements to predict its future development. In the following section, we briefly describe the traditional annotation and simulation process in BD as well as provide a background of neural networks and evolutionary algorithms, and how they can be employed together for time series predictions. Afterwards, we define the research question that is derived from the traditional BD annotation approach and provide a novel approach (NEvo) to tackle this question. We then show the application of NEvo on a simple business use-case and evaluate its results. The paper ends with a conclusion.

2   Background

As has been stated earlier, it is most common in the field of BD to annotate SFDs (for instance their flows and variables) and finally simulate those to produce prediction output [2,5]. The retrieved mathematical equations are usually of a simply nature (as is intended to improve model understanding) and are rarely based on actual system data, but rather are manually derived [4]. Outside the BD domain, approaches have been developed that simplify the creation of function surrogates. For instance, Neural networks (NNs) are well known to reproduce and/or predict time series behaviour, given that the modeller has access to historical training data, knows how to train the NN efficiently, is capable of determining the correct network topology and can identify an adequate number of inputs for the NN [6,7]. Good examples for such applications are, for instance, traditional KPI time series analyses or the prediction of the stock market [8,9]. In both cases, the historical information of the one target KPI is used as training data for the NN, before the NN produces predictions of the target KPI. However, the two latter properties (network topology and NN input) are quite restrictive, since an expert-level background is required to pinpoint the correct topology (feed-forward, recurrent, partial-recurrent, Elman, etc.), the topology configuration (number of input neurons, number of hidden layers, number of hidden layer neurons, hidden layer activation functions) and input variables (number of historical input information) for the problem at hand. Dependent on the target KPI to predict, different network topologies and input variables will yield diverse prediction results. The design and parameterization of NNs is an optimisation problem whose search space is based on the different configurations and its desired optimum is such a configuration that, given a well trained matrix, yields very accurate replay and prediction results for its target KPI. A concept that is particularly well suited to tackle such optimisation problems is that of Evolutionary Algorithms (EAs) [10,11]. Holland, Rechenberg, Goldberg and Schwefel among others, transferred the idea of the biological evolution concept to the field of computer science and mathematical optimisation, thus introducing *evolution strategy*, *evolutionary programming* and *genetic & evolutionary algorithms* [10–13]. Exploiting EAs to improve NNs in all different dimensions (input data, architecture design, weight matrix improvements, learning rule adaptation, etc.) has been extensively researched [14,15].

3   Network Evolution

The scope of this paper is to answer the question, if we can employ the massive available historical business data, which is tracked on an hourly/daily/monthly basis and resides in the companies databases, to support the BD modeller with the annotation process. Furthermore, we are challenging the traditional BD annotation process, by completely avoiding any SFDs, which are so far the pillars of any BD modelling process. We will show, that a given CLD can be automatically annotated, as long as each vertex in the CLD (each KPI) is associated with a historical time series data set. These data sets need to be available in the same time format, e.g., hourly, daily or monthly and are expected to be timely ordered. Ideally, the entire CLD creation was based on those time series data sets (see [16]). However, in the business domain, every KPI depends on a variety of variables that drive the behaviour and future development of the KPI. These causal dependencies (and their associated time series) are reflected within a CLD and are the input for NEvo. The goal is to create an FS for each target KPI which is reflected by an NN, because they are well proven to replay and/or predict time series. In our implementation, we have decided to start with a Feed-Forward network topology, because it can be easily created in an automated fashion. We are handling the challenge of parameterizing the NN topology properties as an optimisation problem that we tackle with the help of an EA. The EA is fed with a parent population that consists of multiple NN individuals. Each individual contains a genotype that represents all topology properties of one NN. Equation 1 defines such a genotype.

$$G = \{\underbrace{g_0, .., g_i,}_{G^{IL},} \quad \underbrace{g_{i+1}, .., g_j,}_{G^{HL},} \quad \underbrace{g_{j+1}, .., g_k}_{G^{AF}}\} \quad (1)$$

A genotype G embodies $k + 1$ genes, each reflecting one particular structural property of an NN topology. In our genotype model, the following structural NN properties are represented:

– **Input Layer:** Subsequence $G^{IL} = \{g_0, .., g_i\}$ represents the number of historical input data points for each dependent variable used to predict the target KPI.

- **Hidden Layers:** Sequence $G^{HL} = \{g_{i+1}, .., g_j\}$ contains all genes that reflect the hidden layers. Gene g_{i+1} defines the number of hidden layers in the neural network and the subsequence $\{g_{j+2}, .., g_j\}$ represents the number of neurons per hidden layer, respectively.
- **Activation Functions:** Subsequence $G^{AF} = \{g_{j+1}, .., g_k\}$ defines the activation function starting from the input layer to the first hidden layer, then from the first hidden layer to the second and so on to the last hidden layer.

The gene sequence G^{IL} reveals two facts about the NN: Firstly, the number of genes $(i+1)$ represents the number of variables that impact the target KPI, including the target KPI itself. Secondly, the length of the historical information used for each dependency: For example, the sequence $\{5, 3, 7\}$ aggregates to an overall input layer length of 15 neurons, which is split into 5 historical values of the first dependency, 3 historical values of the second dependency and 7 historical values of the actual target KPI. However, one of the main questions in creating a well suited NN is the question of how much historical data has to be incorporated when training the network? In order to answer this question we perform a *lag window computation* via *autocorrelation* [17]. Lag windows describe the length of a repeating pattern in a given stationary function that is produced by any type of periodicity. The NN is then used to train the 'pattern' that spans the lag window and supposed to apply it for prediction values it hasn't been trained on. In the business domain, it is common that all KPIs follow a particularly repeating pattern (seasonality), respectively, which is caused by many impacting factors, e.g., customer behaviour, product demand and so on. However, finding a specific pattern in a given time series can be very challenging, since multiple periodicities are usually overlapping one another. Hence, it seems reasonable to expect a diverse set of multiple lag windows for each KPI that can be fed into the creation of an input layer. Because of the large number of different NNs that undergo the evolution process, it is not of high importance which lag window is chosen for which individual, but rather, that all possible lag windows are used at some point. We have therefore decided to randomly choose lag windows, which is depicted via the $U\{X\}$ operator, that represents the uniform random selection of an element in the given set X. The creation of sequence G^{IL} is shown in Algorithm 1, which requires the target KPI, a list of lag windows for all KPIs and a CLD as input. We have shown earlier how to arrive at such CLDs based on a given time series data set [16].

The overall size of the NN mainly depends on the number of hidden layers and neurons per hidden layer stored in G^{HL}. All layers are connected in a feed-forward fashion. We have decided to restrict the overall NN hidden layer structure in its form to a sideways rotated triangle, whose base is the first hidden layer and whose top is the last hidden layer, i.e., each consecutive layer consists of equal or less neurons compared to its predecessor. This is mainly because we find such a structure to beneficially impact the overall prediction results. However, other structures, such as a square or diamond, are also promising and are subject to further research. Algorithm 2 provides the pseudo-code to produce G^{HL} with the required inputs G^{IL}, an interval for the minimum and maximum number of hidden layers and a global minimum of neurons for any hidden layer, to avoid empty hidden layers. Similarly to the random selection in a set $U\{X\}$ in

---

Algorithm 1. Create an input layer sequence for a target KPI $\mathbb{T}$

Procedure: $buildILSequence : G^{IL} = \{g_0, \ldots, g_i\}$

Require: target KPI $\mathbb{T}$; list of lag window sets Ω;
        causal loop diagram CLD

  $G^{IL} \leftarrow [\,]$

  $dependencies \leftarrow$ find dependencies of $\mathbb{T}$ in CLD

  $dependencies \leftarrow$ add $\mathbb{T}$ to $dependencies$

  $index \leftarrow 0$

  for (dependency $\mathbb{D} \in dependencies$) **do**

    $\Omega^{\mathbb{D}} \leftarrow$ find lag window set for $\mathbb{D}$ in Ω

    $G^{IL}[index_{++}] \leftarrow U\{\Omega^{\mathbb{D}}\}$

  end for

  return G^{IL}

Algorithm 1, we are using $U\{[min, max]\}$ as a discrete uniform random selection of a given integral number interval $[min, max]$.

Once the NN structure has been defined, one needs to determine the activation functions for each layer transition, represented in G^{AF}. We have created a set of commonly used activation functions, which includes the following functions: Linear, Sigmoid, Elliot, Tanh, Sin, Log, Gaussian. For each transition, one of these activation functions is randomly choosen. Algorithm 3 shows this procedure. The genotyope G is then simply a union of the three subsequences created with the Algorithms 1, 2 and 3.

G is a representation of an NN and as such easily transformable into any specific NN runtime representation (a real NN). This NN is then trained with the given historical time series and a training algorithm. We have implemented three well established training algorithms for NEvo: Backpropagation, Resilient-Backpropagation and Scaled-Conjugate-Gradient [18, 19]. After a particular training error has been reached, the training is stopped. In some cases, the training error

---

Algorithm 2. Create a hidden layer sequence for a target KPI

Procedure: $buildHLSequence : G^{HL} = \{g_{i+1}, \ldots, g_j\}$

Require: $G^{IL} = \{g_0, \ldots, g_i\}$; hidden layer interval $[hli_{min}, hli_{max}]$;
        minimum number of neurons per layer $LOWER\_BOUND$

  $G^{HL} \leftarrow [\,]$

  $hiddenLayers \leftarrow U\{[hli_{min}, hli_{max}]\}$

  $maxNeurons \leftarrow \sum_{a=0}^{a=i} G^{IL}[a]$

  $idealNeurons \leftarrow \lceil (maxNeurons + 1)/hiddenLayers \rceil$

  $G^{HL}[0] \leftarrow hiddenLayers$

  for $(layer \leftarrow 1; layer \leq hiddenLayers; layer_{++})$ **do**

    $minNeurons \leftarrow max(maxNeurons - idealNeurons, LOWER\_BOUND)$

    $G^{HL}[layer] \leftarrow U\{[minNeurons, maxNeurons]\}$

    $maxNeurons \leftarrow minNeurons$

  end for

  return G^{HL}

Algorithm 3. Create an activation function sequence

Procedure: *buildAFSequence* : $G^{AF} = \{g_{j+1}, \ldots, g_k\}$

Require: G^{HL}; set of activation functions Φ
> $G^{AF} \leftarrow [\ ]$
> $activationFunctionSize \leftarrow size(G^{HL})$
>   **for** $(index \leftarrow 0;\ index < activationFunctionSize\ ;\ index{+}{+})$ **do**
> $G^{AF}[index] \leftarrow U\{\Phi\}$
>   **end for**
>   **return** G^{AF}

---

might be stuck in a local optima that it can't escape from. This effect can be counteracted by introducing an iteration limit, which automatically stops the training after reaching the number of iterations. Once the network has been trained, it has to be evaluated to compute a measure of its fitness.

Fitness Function. The fitness function is a measurement of the quality for an individual in a population. It determines, whether a given individual i_1 is a better solution for the optimisation problem than individual i_2. The definition of an adequate fitness function for the problem at hand is therefore critical. In our case, the fitness function incorporates the following two major metrics:

- The neural network training error f_{TE}.
- The prediction error of the individual f_{PE}.

The training error f_{TE} of an NN is a direct measure of how much the network aligns to the given input data it has been trained on. This value is a direct output of the training function itself and therefore available without any further computation cost. However, since these values are usually very small f_{TE} needs to be scaled appropriately in the fitness function. The prediction error f_{PE} is a measure for evaluating how well a trained NN performs on data it hasn't seen before. In this paper, we have decided to use 80 % of the available historical data (training data) to train the network and the remaining 20 % (prediction data) to compute the prediction error. The prediction error is computed from the deviation of the given timeseries data and the output of the NN for that time period by applying an error function. Literature provides various different error functions that can be applied, e.g., the mean squared error (MSE), root mean squared error (RMSE), mean absolute error (MAE) or mean magnitude of relative error (MMRE). As we will see later in the evaluation section, we have tested NEvo with all those error functions. The fitness function is then a simple aggregation of both, the scaled network training error f_{TE} and the weighted prediction error f_{PE} (2). This function is used to evaluate each individual and compare it with other individuals in a given population, therefore creating evlutionary pressure.

$$f = w_1 * f_{TE} + w_2 * f_{PE} \tag{2}$$

Evolutionary Operators. To improve the overall fitness of a given population according to the above defined fitness function f, we have implemented several evolutionary operators. Each generation is started with a uniform random selection on a given parent population $Pop = \{G_{P_0}, G_{P_1}, \ldots, G_{P_n}\}$. The selection creates pairs of parent individuals that are then recombined with a uniform crossover at gene g_j (number of hidden layers) in order to create two child individuals. Equation (3) shows the signature and definition of this recombination ρ. Hence, the first created child individual contains the input layer subsequence from the first parent and the hidden layer and activation function subsequence from the second parent. The same holds vice versa for the second individual.

$$
\begin{aligned}
\rho : \ & \langle G_{P_i}, G_{P_j} \rangle \to \langle G_{C_k}, G_{C_l} \rangle \\
\rho(\langle G_{P_i}, G_{P_j} \rangle) = \ & \rho(\langle \{G_{P_i}^{IL}, G_{P_i}^{HL}, G_{P_i}^{AF}\}, \{G_{P_j}^{IL}, G_{P_j}^{HL}, G_{P_j}^{AF}\} \rangle) \\
= \ & \langle \{G_{P_i}^{IL}, G_{P_j}^{HL}, G_{P_j}^{AF}\}, \{G_{P_j}^{IL}, G_{P_i}^{HL}, G_{P_i}^{AF}\} \rangle \\
= \ & \langle G_{C_k}, G_{C_l} \rangle
\end{aligned}
\tag{3}
$$

This procedure is repeated until a sufficiently dense child population has been created. Afterwards, a subset population is selected based on a mutation rate. All child individuals in this subset undergo a mutation operation, which alters one gene or a small gene subsequence. We have implemented four different mutations:

- Input neuron mutation: One gene $g_n \in G^{IL}$ is randomly chosen and replaced with a new lag size from the precomputed pool of different lag sizes for the associated input variable.
- Neuron mutation: The number of neurons in one particular hidden layer $g_n \in G^{HL}$ is reset with a new random value in a given interval.
- Hidden layer mutation: The entire hidden layer gene sequence G^{HL} is rebuilt.
- Activation function mutation: One activation function gene $g_n \in G^{AF}$ is replaced with a randomly chosen element from the given pool of activation functions.

The actual mutation, which is applied to a given individual, is randomly selected from the above defined mutation set. Finally, the recombined and mutated child population is filtered with an environment selection, which in our case is a best selection based on the fitness of each child. This selection reduces the size of the child population to the size of the parent population for the next generation. Since the selection is based on the fitness of the child individuals, it also helps to achieve a convergence of the population fitness towards an optimal solution.

Configuration Parameters. NEvo is a complex FS computation algorithm, and as such, subject to a configuration that guides the algorithm and yield different FS results. The following parameters are required for each NEvo run:

- Generations: The number of generations the evolutionary cycle is executed to improve the overall fitness.
- Number of parent individuals: The number of parent individuals each generation is started with.
- Number of child individuals: The number of children created from the parent population with the recombination operator.
- Mutation rate: The percentage of child individuals in a given child population that are subject to the mutation operators.

– Prediction error function: The error function to compute the deviation of the prediction output and the orginal time series. As explained earlier, multiple functions are available for f_{PE}, e.g., RMSE, MSE, MAE, MMRE.
– Training function: The neural network training function, e.g., Backpropagation, Resilient-Backpropagation, Quick-Propagation or Scaled-Conjugate-Gradient.

4 Usecase

The evaluation of NEvo is based on the business use-case of the company Akron Heating (AH), which operates in the highly competitive retail sector [20]. AH's business model is based on selling goods via an online store and supported by various Business Processes (BPs), e.g., the Order process, Consignment fill-up process, Return-Item process and so on. For managing and controlling their business, AH employs various software solutions (ERP, CRM and HRM). These solutions are thoroughly tracking and storing the generated operational and business data, such as event data and aggregated high level data (revenue, sales, market share, cost of goods sold, cashflow and so forth). We have shown in our previous research work how one could automatically create CLDs based on this data, to evaluate causality between these variables and find root causes for bottlenecks [21]. Such automatically created CLDs are created with the help of Granger-Causality and domain specific ontologies, thus aiming to visualize causation rather than correlation [22]. In this paper, we are using an extended CLD that incorporates more strategic KPIs than created in [21]. It has been automatically generated with the same algorithm. An image section of the CLD is shown in Fig. 1. This image section does not contain loops, since it has been simplified from a larger more complex view. However, the entire complex CLD with all loops is the basis for the evaluation and as such, input for NEvo (including all loops). The image section of the CLD shows 18 high-level KPIs, for instance, profit, sales, expenses, number of orders, etc., as well as their causal relationships. Every KPI in the CLD has to be annotated with an accurate FS computed via NEvo. A created FS is 'accurate' if its respective fitness computed via f is minimal. A minimised fitness is a strong indicator, that the FS is capable of replaying the historical time series data, as well as, predict its future development. As we have explained earlier, the usage of NEvo requires time series data for each KPI and its dependencies. In the AH use-case, such time series data is provided for a range of 10 years (2000–2010) and was monitored on a monthly basis. Each time series therefore provides 120 entries, of which 90 entries are used for NEvo (training and prediction data) and 30 entries are used to evaluate NEvo's prediction capabilites (evaluation data). The 90 entries used for NEvo are split in the above stated 80/20 fashion, which means that 72 data points are used as training data and 18 points are used as prediction data. To create a reference frame for later comparison, we have set a default parameter pool for each NEvo run, which looks as follows: 200 generations, 25 parent individuals, 125 child individuals and a mutation rate of 0.7. These parameters have

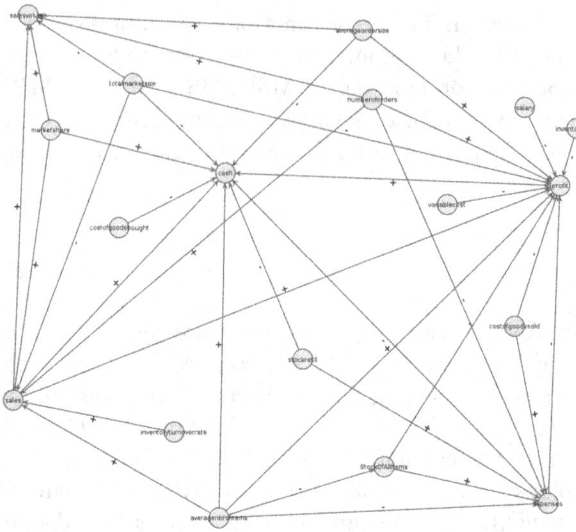

Fig. 1. An image section of a greater CLD that represents interesting dependencies and KPIs in the AH use-case.

been identified to work well throughout several simulation runs. The prediction error function for f_{PE} and the NN training function for f_{TE} are subject to the specific run.

5   Results and Evaluation

The evaluation of NEvo is a comparison of its prediction capabilities with both, a linear regression (LR) and a uniform random estimator (URE). LR produces a linear mean function over the given time series, whereas URE is a uniform random distribution in the interval $[min, max]$, with min and max being the minimum and maximum value of the given time series. We compare the quality of the results for NEvo, LR and URE by using a *cumulative distribution function* (*CDF*) of the relative error. It shows how many y percent of the prediction results are within a relative error range of x percent of the original time series. These results are produced on the evaluation data set, which consists of 30 entries. The NEvo prediction results are created as single step predictions with the evaluation data set as input (rather than previously predicted values). However, none of the evaluation data points have been used to train the NN. Figure 2 shows results for the replay and prediction of the KPIs salesvolume and profit. As we can see in the first picture 2(a), the winning NN nearly reproduces the original training data with an accuracy of a 100 %, but continues less accurate on the remaining prediction data (which was not fed into the NN, but rather used to internally evaluate the NN with the EA). Investigating the evaluation output (green line) clearly shows that only few trends are recognized by NEvo. This observation can be confirmed by analyzing the CDF graph 2(b). The simple LR algorithm

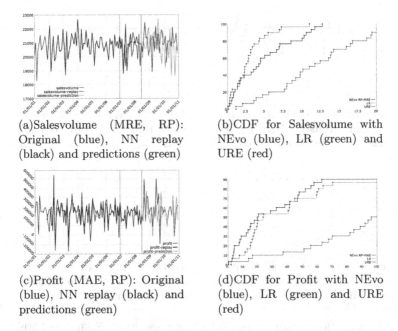

(a)Salesvolume (MRE, RP): Original (blue), NN replay (black) and predictions (green)

(b)CDF for Salesvolume with NEvo (blue), LR (green) and URE (red)

(c)Profit (MAE, RP): Original (blue), NN replay (black) and predictions (green)

(d)CDF for Profit with NEvo (blue), LR (green) and URE (red)

Fig. 2. NEvo results: The left side shows the original time series (blue), the training replay of the winning NN (black) and the predictions of this NN (green). The right side shows the CDF for NEvo (blue), *LR* (green) and *URE* (red) (Color figure online).

outperforms NEvo in this scenario, since 80 % of all prediction results produced by *LR* are better than 5 % of relative error of the original salesvolume time series. This is due to the small overall fluctuation of the salesvolume time series (all values are between a 15 % relative error). However, if we focus more on the results of a KPI with a higher fluctuation, e.g., profit, evidence points to more reliable results predicted with NEvo (see Fig. 2(c) and (d)). The graph 2(c) on the left shows a couple of prediction intervals that are picked up correctly by NEvo, e.g., 03/2009 until 06/2009 or 02/2010 until 06/2010. This is also reflected in the respective *CDF* graph 2(d) on the right, which shows a better performance of NEvo compared to *LR* and *URE*. These sort of prediction results can be found along all given KPIs, if all previously explained error functions and training functions are included (see Fig. 3(a) and (b) for another example of the monthly variable costs of Akron Heating).

An interesting question at this point is that of the 'best' NEvo configuration to guarantee results that are at least of the quality as shown earlier in the profit example. As we can see in the CDF graphs 3(c) and (d), such a configuration might not exist. Both figures are clearly showing that different configurations for different KPIs yield the respective 'best' result. For instance, in case of the profit KPI, *RP_MAE* and *RP_MSE* produce the best results, whereas in case of expenses *RP_RMSE* and *SCG_MSE* are better. Nevertheless, the CDF itself can be employed to automatically determine the 'best' winning NN accurately.

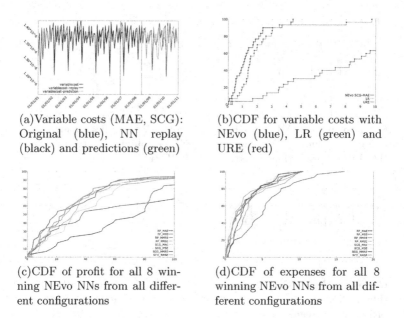

(a)Variable costs (MAE, SCG): Original (blue), NN replay (black) and predictions (green)

(b)CDF for variable costs with NEvo (blue), LR (green) and URE (red)

(c)CDF of profit for all 8 winning NEvo NNs from all different configurations

(d)CDF of expenses for all 8 winning NEvo NNs from all different configurations

Fig. 3. The prediction output for the variable cost and its CDF (3(a) and 3(b)). 3(c) and 3(d) show a comparison of all CDFs for the profit and expenses KPI (Color figure online).

6   Conclusion and Future Work

In this paper we have demonstrated an EA-guided NN function retrieval algorithm that is capable of annotating all variables in a given CLD, thus preparing the CLD itself for a later multi step prediction (simulation). Our future research will be focussed on evaluating if the simulation of CLDs annotated with NEvo can accurately predict the future development of the modelled KPIs for longer time frames (multi step). We have shown that some of the NEvo results are not capable of matching the quality of a simple *LR* algorithm that produces a mean function over the given time series. Some of these results can be explained with a timely shift in the prediction results that has a strong impact on the CDF. However, this clearly shows room for improvement. As stated in the "Network Evolution" section, the NN topology is currently limited to simple feedforward networks. One possible improvement is the implementation of different NN topologies, such as partial-recurrent networks that are capable of maintaining a state. We have also explained earlier that different hidden layer structures are under investigation (square and diamond structures). Furthermore, after analysing the CDF results, one could think of an extension of the EA fitness function by enriching it with a CDF threshold: Whenever y % of the prediction results are above a relative error of x %, an additional penalty is applied. This would surely favour all those individuals that are below such a threshold. It remains to be investigated, whether such improvements yield more accurate prediction results.

References

1. Brockwell, P.J., Davis, R.A.: Time Series: Theory and Methods, 2nd edn. Springer, Heidelberg (2006)
2. Forrester, J.W.: Industrial Dynamics. Currently Available from Pegasus Communications. MIT Press, Cambridge (1961)
3. Sterman, J.D.: Business Dynamics: Systems Thinking and Modeling for a Complex World. McGraw-Hill, New York (2000)
4. Drobek, M., Gilani, W., Soban, D.: Parameter estimation and equation formulation in business dynamics. In: 3rd International Symposium on Business Modeling and Software Design. ScitePress, Noordwijkerhout (2013)
5. Burns, J.R.: Converting signed digraphs to forrester schematics and converting forrester schematics to differential equations. IEEE Trans. Syst. Man Cybern. B Cybern. **10**, 695–707 (1977)
6. Fausett, L.V.: Fundamentals of Neural Networks: Architectures, Algorithms, and Application, 1st edn. Pearson, London (1993)
7. Bishop, C.M.: Neural Networks for Pattern Recognition. Oxford University Press, Oxford (1995)
8. McNelis, P.D.: Neural Networks in Finance: Gaining Predictive Edge in the Market. Academic Press, Orlando (2005)
9. Tudor, N.L.: Intelligent system for time series prediction in stock exchange markets. In: Abramowicz, W., Kokkinaki, A. (eds.) BIS 2014. LNBIP, vol. 176, pp. 122–133. Springer, Heidelberg (2014)
10. Holland, J.H.: Genetic algorithms and the optimal allocations of trials. SIAM J. Comput. **2**, 88–105 (1973)
11. Goldberg, D.E.: Genetic Algorithms in Search. Optimization and Machine Learning. Addison-Wesley Longman Publishing Inc., Boston (1989)
12. Rechenberg, I.: Evolutionsstrategie - Optimierung technischer Systeme nach Prinzipien der biologischen Evolution. Ph.D. thesis (1971)
13. Schwefel, H.P.: Numerische Optimierung von Computer-Modellen, vol. 26th. Birkhaeuser, Basel (1977)
14. Schaffer, J., Whitley, D., Eshelman, L.: Combinations of genetic algorithms and neural networks: a survey of the state of the art. In: COGANN 1992, pp. 1–37. IEEE, Baltimore, MD (1992)
15. Yao, X.: Evolving artificial neural networks. Proc. IEEE **87**, 1423–1447 (1999)
16. Drobek, M., Gilani, W., Soban, D.: A data driven and tool supported CLD creation approach. In: The 32nd International Conference of the System Dynamics Society, pp. 1–20, Delft (2014)
17. Wei, W.W.S.: Time Series Analysis: Univariate and Multivariate Methods, 2nd edn. Pearson, London (2005)
18. Riedmiller, M., Braun, H.: RPROP - a fast adaptive learning algorithm. In: Proceedings of ISCIS VII (1992)
19. Hestenes, M.R., Stiefel, E.: Methods of conjugate gradients for solving linear systems. J. Res. Nat. Bur. Stan. **49**(6), 409–436 (1952)
20. Fritzsche, M., Picht, M., Gilani, W., Spence, I., Brown, J., Kilpatrick, P.: Extending BPM environments of your choice with performance related decision support. In: Dayal, U., Eder, J., Koehler, J., Reijers, H.A. (eds.) BPM 2009. LNCS, vol. 5701, pp. 97–112. Springer, Heidelberg (2009)
21. Drobek, M., Gilani, W., Redlich, D., Molka, T., Soban, D.: On advanced business simulations - converging operational and strategic levels. In: 4th International Symposium on Business Modeling and Software Design. ScitePress, Luxembourg (2014)
22. Granger, C.W.J.: Investigating causal relations by econometric models and cross-spectral methods. Econometrica **37**, 424–438 (1969)

An Empirical Study on Spreadsheet Shortcomings from an Information Systems Perspective

Thomas Reschenhofer[✉] and Florian Matthes

Technical University of Munich, Boltzmannstr. 3, 85748 Garching, Germany
{reschenh,matthes}@in.tum.de

Abstract. The use of spreadsheets to support business processes is widespread in industry. Due to their criticality in the context of those business processes, spreadsheet shortcomings can significantly hamper an organization's business operation. However, it is still unclear, what actual and typical shortcomings of spreadsheets applied as information systems are. Therefore, in this paper we present the results of an empirical study on spreadsheet shortcomings from an information systems perspective. In this sense, we focus particularly on how spreadsheets perform in their respective business context. The result of our work is a list of 20 shortcomings which typically occur in practice.

Keywords: Spreadsheet · Case study · Shortcomings · Business process support · Decision support system

1 Introduction

Spreadsheets are among the most common tools for business-users. They enable business users to get knowledge out of their domain-specific data by analyzing and visualizing it in an end-user-friendly way. Studies have shown that spreadsheets are used in a huge majority of firms in the US and Europe [4,12]. Their area of application is manifold, ranging from financial reporting to workload planning to general administration [15], either as throwaway calculations, or as well-designed business information systems [9]. In addition, spreadsheets not only are regularly used for a variety of purposes, but in many cases they are also critical and important to organizations [6,8]. For these reasons, spreadsheet-related problems can have a considerable impact on an organization's business operation and thus potentially lead to significant financial losses [5,17]. As a consequence, many researchers studied spreadsheet errors in order to propose approaches for either reducing the risk of their occurrence or for identifying, classifying, and fixing spreadsheet errors [16].

However, in addition to their error-proneness, spreadsheets in business suffer from shortcomings with respect to the support of respective business processes. For example, legal regulations like the Sarbanes-Oxley Act [21] in the US or Basel

© Springer International Publishing Switzerland 2015
W. Abramowicz (Ed.): BIS 2015, LNBIP 208, pp. 50–61, 2015.
DOI: 10.1007/978-3-319-19027-3_5

II [3] in the EU stipulate requirements concerning the retention and management of information within organizations. However, those requirements cannot be met by prevalent spreadsheet software, which negatively affects the organization's risk and compliance management [10,14]. Furthermore, those shortcomings also imply a significant lack of efficiency and thus potentially result also in financial losses. While there is a general agreement that spreadsheets in business suffer from this kind of shortcomings [10,14], there is still little research about them, i.e., it is still unclear which shortcomings with respect to the application of spreadsheets as information systems are common in practice.

Therefore, the present work shows the results of an empirical study about spreadsheet shortcomings conducted in two German companies. We discuss types of shortcomings which occur in practice, and why spreadsheets are applied in the respective cases despite suffering from those shortcomings. The contribution of this work is the identification, description, and categorization of 20 shortcomings obtained from nine cases of spreadsheet applications.

The remainder of this paper is organized as follows: Sect. 2 discusses related work in the field of spreadsheet research. Thereafter, Sect. 3 outlines the details of the conducted study, whereas the identified shortcomings are described in Sect. 4. Finally, in Sect. 5 we discuss and conclude the results of the study and this paper's contribution, and propose follow-up research activities.

2   Related Work

Due to their dissemination and popularity in practice, spreadsheets already have been subject of research for some decades [12,16].

In particular spreadsheet errors were investigated extensively in the past. Panko and Halverson [13] outline that spreadsheet errors are widespread, wherefore they propose to classify those errors in order to be able to assess approaches for reducing the risk of errors. Thereby, they differentiate between quantitative and qualitative errors. Quantitative errors are either omission errors, logic errors, or mechanical errors, while qualitative errors are primarily design errors which potentially lead to quantitative errors in the future. Based on this classification, Rajalingham et al. [18] derived a taxonomy as a framework for the systematic classification of spreadsheet errors. This taxonomy also considers the roles of users interacting with the spreadsheet (developers as well as end-users) and furthermore defines different kinds of qualitative spreadsheet errors, e.g., semantic and maintainability-related errors. Therefore, qualitative spreadsheet errors as defined by Panko and Halverson [13] as well as by Rajalingham et al. [18] do not directly lead to incorrect values in spreadsheets, but negatively affect the spreadsheet's usability, maintainability, and related aspects. In this sense qualitative errors are consequences of those shortcomings in the same way as quantitative spreadsheet errors usually are consequences of qualitative spreadsheet errors. The present paper adds a new dimension to the taxonomy as proposed by Rajalingham et al. [18] which also includes causal relationships to qualitative errors.

Grossman et al. [9] urge on perceiving spreadsheets as information systems focusing on business processes instead of personal productivity tools. They highlight the importance of spreadsheet information systems for business, and also show that they are used for a variety of purposes and in a plethora of areas. Consequently, spreadsheets which support business processes also inherit common information system requirements regarding information management, collaboration support, etc. However, the focus of spreadsheets is solving problems [20]. This mismatch of what spreadsheets were primarily designed for and how they are applied in practice is supposed to be a major cause for the shortcomings as described in the present paper. Panko and Port [14] outline that spreadsheet errors are a serious issue, but also point out further concerns which are not related to spreadsheets as such, but to their application for supporting business processes. They explicitly name privacy, security, and compliance as issues raised by spreadsheets. Those issues are induced by the application of spreadsheets in regulated areas like financial reporting, so that they become subject of legal regulations (e.g., Sarbanes-Oxley Act [21]) and thus have to fulfill respective requirements. However, since spreadsheets are not focused on supporting business processes [20], spreadsheet software usually does not address those legal regulations. This is another cause for shortcomings from an information systems perspective as described in this paper.

While Panko and Halverson [13] as well as Rajalingham et al. [18] describe the consequences of spreadsheet shortcomings when applied as information systems, and Grossman et al. [9] as well as Panko and Port [14] outline their causes, there is very little scientific research about which concrete shortcomings actually exist in practice, and in particular there is no work providing an overview over those as targeted by the present paper. For example, Nardi and Miller [11] claim that collaboration support is an essential concern of spreadsheets applied as information systems, and thus they already highlight one concrete shortcoming from an Information Systems perspective. Furthermore, the papers by Rothermel et al. [19] as well as Ayalew and Mittermeir [2] is about testing and debugging spreadsheets, i.e., they address also one common shortcoming of spreadsheets, namely lacking support for Application Lifecycle Management (ALM).

3   Case Studies

In order to identify common shortcomings of spreadsheets with respect to the support of business processes, we conducted nine case studies in two German companies. Thereby, we observed the usage and context of each spreadsheet, and also asked for shortcomings in the respective use cases. We applied the research methodology of exploratory case research in a multiple-case setting [22].

The first cooperating company is a logistics company with more than 100 000 employees worldwide. In the context of this study, we cooperated in particular with an division of this company which provides IT services to the its business units. The second company operates in the financial sector and is part of a big investment and insurance group, also with more than 100 000 employees worldwide. Both companies are internationally active, but located in Germany. In the remainder of this paper, we refer to the former company with "Company 1", and to the latter one with "Company 2". For identifying proper use cases, we asked both companies for spreadsheets they apply as information systems for supporting business processes. We identified four cases in Company 1 and five cases in Company 2. The cases are summarized in Table 1.

For each of the identified use cases, we contacted the responsible designers and users of the respective sheet, and conducted interviews for discussing the context and usage of the respective spreadsheet. On the one hand, we asked them about the life-cycle and design context of the spreadsheet, e.g., "What is the rate of change of the spreadsheet's content and structure?". On the other hand, we were interested in user-related as well as data-related aspects, e.g., "Which users and roles provide input for the spreadsheet or consume its output?". Aside from this context information about the actual usage of the spreadsheet, the interview primarily focused on concrete shortcomings of the respective spreadsheet which its users face in their daily work and hinders them in conducting their business activities. The final part of the interview dealt with why spreadsheets are used in the respective cases despite suffering from the identified shortcomings. In order to gain a better understanding of shortcomings and to have an additional source of evidence, we also asked for the respective spreadsheets themselves as well as—if available—documentations of those spreadsheets and their usage.

4   Shortcomings

By conducting the case study as outlined in Sect. 3, we observed shortcomings for each of the nine cases of spreadsheet information systems. Due to the open-ended nature of the questions about shortcomings in the conducted interviews, we had to interpret and consolidate those shortcomings in order to be able to derive a set of shortcomings which are common across those cases. The result of the consolidation is a list of 20 common shortcomings as listed in Table 2. For the sake of simplicity, we use a positive form for naming the shortcomings, e.g., "Separation of Data, Schema, and Logic" instead of "No Separation of Data, Schema, and Logic".

Furthermore, we categorized those shortcomings into six classes capturing different kinds of aspects of a spreadsheet information system. Those classes are *Readability and Understandability, Extendability, Manageability, Collaboration and Multi-user Support, Data*, and *Processes*. The following section describes each of the identified categories and their shortcomings in detail (order has no relevance).

4.1   Readability and Understandability of the Spreadsheet

This category of shortcomings deals with the readability and understandability of spreadsheets. In particular for users which are not the authors of the spreadsheet it is usually very difficult to understand the design of the spreadsheet. In this context, the design of a spreadsheet refers to its semantic structure and formulas, which in turn constitute a potentially complex network of semantic dependencies between the cells of the spreadsheet.

Table 1. An overview of all observed cases of Companies 1 and 2.

Code	Case Description
Company 1	
1.A	The spreadsheet in this case supports Human Resource (HR) managers to estimate capacities and to determine shortages and surpluses of HRs for a given year. Thereby, HR demands are primarily induced by current and planned projects, which in turn require specific roles and skills. These demands are compared with available HR capacities and visualized as a time-series chart.
1.B	In this case's spreadsheet, IT cost records for the whole company are imported from a financial reporting system and mapped to the responsible IT divisions. The output of this spreadsheet is a chart visualizing the distribution of the company's IT costs over the respective IT divisions.
1.C	The spreadsheet in this case documents all stakeholders of a given project and assesses them by three different dimensions, namely power of the stakeholder, his willingness to engage, and his mindset towards the respective project.
1.D	Similar to case 1.C, this case's spreadsheet documents the risks as well as corresponding mitigation and contingency actions for a given project. Those risks are assessed by their probability of occurrence on the one hand, and by their potential impact on the other hand.
Company 2	
2.A	The spreadsheet in this case enriches financial data imported from a database with manually inputted data and provides the generated information representing returns of investments (ROIs) as an input for another software system.
2.B	Based on data obtained from simulation software, which generates various scenarios, this case's spreadsheet creates a visual report consisting of multiple charts on the one hand, and providing its output for the import by other software systems on the other hand.
2.C	The purpose of the spreadsheet of this case is to calculate cash-flows based on current interests rates and to compare the resulting data with the results from another system.
2.D	In this case's spreadsheet, time-series data obtained from a database is enriched by information which has to be manually extracted from a report from a financial services provider. The resulting data is exported to a database.
2.E	The spreadsheet in this case imports pre-processed scenario-data from a numerical computing software and generates multiple charts summarizing certain aspects of the respective scenarios.

SC01: Formatability and Commentability of Formulas: In prevalent spreadsheet applications, formulas cannot be formatted (e.g., by line-breaks, indents, etc.) or annotated with comments. It is neither possible to visually highlight that certain code elements semantically belong together (e.g., separating those elements from the others through a line-break), nor to add comments allowing the author of the spreadsheet to explain certain parts of the formula.

SC02: Named Cell Addressing: There are three commonly used cell addressing schemes for spreadsheets [1]: A1-style (columns are referenced by letters, rows by numbers), R1C1-style (both columns and rows are referenced by numbers), and named addressing (cells are directly referenced by a unique name). While coordinate-based addressing schemes (which includes both the A1-style and the R1C1-style) are a defining feature of spreadsheets, they negatively affect a formula's readability. In contrast, by using the named addressing schema, the reader of a formula is already able to derive the semantics of referenced cell by the identifier in the formula.

SC03: Transparency of Information Flow: Spreadsheet formulas referencing other cells constitute a network of semantic dependencies between the cells of a spreadsheet and thus represent information flows. In prevalent spreadsheet applications these semantic dependencies are usually not transparent, i.e., it is difficult for the spreadsheet users to get an overview of the information flows between the cells of a spreadsheet. As a consequence, it is hard for the users to determine the provenance of a formula's input on the one hand, and to evaluate the impact of changes of a certain cell or formula on the other hand.

4.2 Extendability

This category is about extending the functionality of spreadsheet software by new components, e.g., to be able to integrate custom data sources, to perform custom calculations, or to display the spreadsheet's data in a tailored visualization.

SC04: Integration of Custom Data Sources: For many use cases of spreadsheet applications, the data from different kinds of data sources has to be integrated. While common data sources like SQL databases and CSV files are integrable by prevalent spreadsheet software, more specialized data sources (e.g., HR information systems as in case 1.A, or object-oriented databases as in case 1.C, c.f., Table 1) are not connectable to spreadsheets. In addition to that, common spreadsheet software does not provide an appropriate infrastructure for extending the set of integrable data sources, so that the integration of additional data source through macros is burdensome.

SC05: Extendability of Computational Expressiveness: Prevalent spreadsheet software already provides a rich set of operations which can be used in formulas, including simple arithmetic operations, conditionals, and advanced statistical methods. Additional functionality can be implemented by macros. However, as the case study has shown, the imperative paradigm of macros is considered as being not suited for the definition of spreadsheet operations by the interviewed spreadsheet designers. Instead, they would prefer a functional programming language. Furthermore, in prevalent spreadsheet software, macros are directly attached to spreadsheets, which hinders the independent development of them.

SC06: Custom Visualizations: In common spreadsheet software, users are able to create charts in order to visualize a spreadsheet's data appropriately with respect to its consumers. Thereby, those spreadsheets provide a certain set of visualization types (e.g., line charts, pie charts, etc.), which are easily configurable by end-users. However, spreadsheets are not providing an infrastructure for extending the set of available visualization types by custom ones.

4.3   Manageability

Shortcomings of this category capture the difficulties when designing and maintaining spreadsheets. In this sense, manageability and flexibility—as one of the defining features of the spreadsheet paradigm—are often contradictory qualities, which is the main reason for the existence of the following shortcomings.

SC07: Managed Evolution of Spreadsheets: After an initial design phase, a spreadsheet's structure still evolves over time, e.g., formulas are changed, or columns are added. In particular when having multiple instances of a spreadsheet, and the intention to merge them in future, the unmanaged evolution of spreadsheets becomes a huge challenge for the respective spreadsheet user.

SC08: Separation of Data, Schema, and Logic: One of the defining features of the spreadsheet paradigm is that data, schema, and logic are not separated from each other, which is the main driver for the flexibility of spreadsheets. The schema of a spreadsheet table is simply determined by the column headers, and its logic is represented by formulas defined in single cells. As a consequence, although the data rows of one data table are semantically of the same type, they have to be defined separately, e.g., row-based formulas and cell-types have to be defined for each individual row separately instead of defining them once.

SC09: Modularity: In many cases of spreadsheet usages (e.g., in cases 1. A and 1.B, c.f., Table 1), certain components (e.g., mapping tables) have to be used multiple times in one spreadsheet, or even in multiple spreadsheets. In order to preserve the Single Source Of Truth principle, those components should be defined only once in such a way that they can be addressed multiple times. As our study revealed, this is already done in many cases. However, on the one hand it is not possible to refer to components of other spreadsheets. And on the other hand the look-up mechanism of prevalent spreadsheet software (e.g., *VLOOKUP* in MS Excel) is cumbersome and error-prone.

SC10: Application Lifecycle Management: Prevalent spreadsheet applications do not support proper Application Lifecycle Management (ALM) for their formulas. In this context, ALM includes in particular testing [19], debugging [2], and central maintenance of spreadsheet formulas.

4.4   Collaboration and Multi-user Support

Due to its early emergence long before the days of collaborative information management, spreadsheet software was designed to be a single-user application. Consequently, prevalent spreadsheet software still suffers from shortcomings related to collaboration and multi-user support as described in the following.

SC11: Collaborative Spreadsheet Design: In prevalent spreadsheet software, components like formulas or visualizations are not shareable, i.e., they cannot be provided to other users so that those are able to reuse the respective functionality. On a related note, spreadsheet software does not maintain the roles of users with respect to those components, e.g., ownerships or responsibilities of certain formulas.

SC12: Element-Based Access Control: Prevalent spreadsheet applications do not support access control on the level of single spreadsheet elements, e.g., rows, cells, and formulas. Thereby, spreadsheet designers are neither able to prevent read access to certain individual elements (e.g., a formula) by certain users, nor to specify that only certain users are able to edit certain elements.

SC13: Element-Based Historization: Similar to SC12, spreadsheet applications do not support element-based historization and tracing of user-activities within the spreadsheet. In this sense, the spreadsheet software is not able to provide information about which cells, formulas, and visualizations were changed at which time by which users, or how a cell's value has evolved over time.

SC14: User-Specific Views: In a collaborative information system, users with different roles require different views on the same data, depending on their specific information demand. However, prevalent spreadsheet applications usually do not support the definition of user-specific or role-specific views. This means, that a spreadsheet provides the same view to each user, regardless of the respective information demand.

4.5   Data

Spreadsheets are used by their end-users in order to process, analyze, and visualize a manageable amount of data of a limited complexity. However, today's business processes for which spreadsheets are applied impose both a considerably bigger size and higher complexity of data to be analyzed. As a consequence, this circumstance leads to the following shortcomings of spreadsheets applied as information systems.

SC15: Complex Data Types: Spreadsheets only support basic cell types, e.g., strings, dates, and numbers. This means that managing linked data and complexobjects (e.g., network-like or hierarchical data structures) is a major challenge in spreadsheets. Furthermore, the result of a formula is also restricted to be of one of those simple types, which impedes the definition of complex calculations and thus makes the analysis of complex data objects even more difficult.

SC16: Scalability: Contrary to the initial design of spreadsheets for the analysis of small-scale data sets, many business processes which are supported by spreadsheets impose the processing of huge data tables. However, prevalent spreadsheet applications either allow just a limited grid size and thus just a limited number of data sets, or calculations based on large-scale data suffer from an enormous loss of performance and even stability.

SC17: Spreadsheet Queries: Analyzing large-scale data sets requires both a query language for defining operations like filters, projections, and aggregations of data entries, and a respective query processing engine which is able to handle a huge data set. Prevalent spreadsheet software usually supports neither of those. One reason for this is the lack of an explicit data schema which could form the basis for a model-driven spreadsheet query language [7].

SC18: Custom Spreadsheet Meta-data: In certain cases of spreadsheet applications, a spreadsheet not only contains usage data, but also meta-data which is not directly related to the data of the spreadsheet, but to the spreadsheet itself. While usually default attributes like the last editor of the spreadsheet or the last modification date are already available and automatically maintained, prevalent spreadsheet software does not support the definition of custom meta-data attributes (e.g., owner of the spreadsheet).

4.6   Processes

Spreadsheet applications heavily rely on manual interaction through end-users and usually lack proper automation and process support. consequently, prevalent spreadsheet information systems suffer from the following shortcomings.

SC19: Support for Automation: Spreadsheets primarily serve as decision support tools [20]. In this sense, based on a manual input and manual definition of calculations, knowledge obtained from the spreadsheet leads to respective actions, which in turn have to be triggered manually by the user. Therefore, applying spreadsheets for decision support requires a significant effort which could be reduced by proper process and data management automation [10].

SC20: Reasoning of Derived Actions: When using spreadsheets as decision support systems, certain actions are triggered based on the knowledge obtained from the analysis of the spreadsheet's data. However, prevalent spreadsheets do not support the automated documentation of derived actions considering the respective context. This would enable the reasoning of derived actions and evaluating them in the respective context at a later point in time.

Table 2. A consolidated list of all obtained shortcomings. Each shortcoming is mapped to the cases in which it occurred. An empty cell means that there was no evidence that the shortcoming occurred in the respective case, i.e., the use case might still suffer from this shortcoming.

Shortcoming		Cases as described in Table 1								
		1.A	1.B	1.C	1.D	2.A	2.B	2.C	2.D	2.E
Readability and Understandability										
SC01	Formatability and Commentability of Formulas	X				X			X	
SC02	Named Cell Addressing	X	X							X
SC03	Transparency of Information Flow			X		X				
Extendability										
SC04	Integration of Custom Data Sources	X	X				X			
SC05	Extendability of Computational Expressiveness					X	X	X	X	X
SC06	Custom Visualizations		X	X		X	X	X	X	X
Manageability										
SC07	Managed Evolution of Spreadsheets					X	X		X	
SC08	Separation of Data, Schema, and Logic					X	X	X	X	X
SC09	Modularity	X	X							
SC10	Application Lifecycle Management					X	X	X	X	X
Collaboration and Multi-user Support										
SC11	Collaborative Spreadsheet Design					X	X	X	X	X
SC12	Element-based Access Control	X		X	X	X	X	X	X	X
SC13	Element-based Historization	X	X	X		X	X	X	X	X
SC14	User-specific Views	X		X						
Data										
SC15	Complex Data Types	X	X			X	X		X	X
SC16	Scalability	X	X				X	X	X	X
SC17	Spreadsheet Queries							X		
SC18	Custom Spreadsheet Meta-data	X								
Processes										
SC19	Support for Automation					X				
SC20	Reasoning of Derived Actions	X		X						

5   Conclusion

In this paper, we described 20 shortcomings of spreadsheets when applied for the support of business processes. Thereby we conducted an empirical study involving nine cases in two companies. We investigated concrete usages of spreadsheets by interviewing the responsible designers and users (c.f., Sect. 3). Based on the results of each single case, we performed a cross-case consolidation and derived common shortcomings of spreadsheet information systems (c.f., Sect. 4).

While the present work already reveals 20 shortcomings spreadsheets typically face when applied for supporting business processes, there might be additional shortcomings we didn't identify, in particular in different contexts and organizations. However, although the list of shortcomings is presumably not complete, it captures most aspects whose relevance were already outlined by related research [9, 10, 13, 14, 18]. Furthermore, in this work we have not done an assessment of shortcomings regarding their frequency or impact. Although several shortcomings occurred in multiple cases of our study, we consider the set of nine cases as too small to be an empirical foundation for such an assessment.

Many of the identified shortcomings are well-known in practice [10]. Therefore, tool vendors already provide extensions to existing spreadsheet software which address certain shortcomings of spreadsheets, e.g., *Slate*[1], *think-cell*[2], and *Google Sheets*[3] address *SC03: Transparency of Information Flow*, *SC06: Custom Visualizations*, and *SC11: Collaborative Spreadsheet Design* respectively. Moreover, many shortcomings are subject of current research activities. For example, Cunha et al. [7] developed model-driven spreadsheet queries and thus address shortcoming *SC17: Spreadsheet Queries*. Nevertheless, there are still shortcomings which are neither addressed by practitioners nor by researches. The reasons why spreadsheets are used in the companies which cooperated in our study despite suffering from those shortcomings are manifold: Previous knowledge in spreadsheet development as well as end-user empowerment were the most common arguments for using them. Also the possibility of prototyping an information system in order to iteratively establish knowledge processes is one of the main drivers for using spreadsheets.

Based on the results of the present paper, in the future we want to focus on certain spreadsheet shortcomings—in particular those of the manageability category. Thereby, we want to propose concepts for an innovative spreadsheet addressing the manageability-related shortcomings. Furthermore, by conducting a quantitative study with the results of the present qualitative study as its foundation, we aim to assess the shortcomings with respect to their frequency of occurrence and impact on the compliance efforts of the respective companies.

---

[1] https://www.useslate.com.

[2] http://www.think-cell.com.

[3] https://docs.google.com/spreadsheets/.

Acknowledgment. This research has been sponsored in part by the German Federal Ministry of Education and Research (BMBF) with grant number TUM: 01IS12057.

References

1. Abraham, R., Burnett, M., Erwig, M.: Spreadsheet Programming. Wiley Encyclopedia of Computer Science and Engineering (2008)
2. Ayalew, Y., Mittermeir, R.: Spreadsheet debugging (2008). arXiv:0801.4280
3. Basel Committee on Banking Supervision: Basel II (2004)
4. Bradley, L., McDaid, K.: Using Bayesian statistical methods to determine the level of error in large spreadsheets. In: Proceedings of the International Conference on Software Engineering, pp. 351–354 (2009)
5. Caulkins, J.P., Morrison, E.L., Weidemann, T.: Spreadsheet errors and decision making: evidence from field interviews. J. Organ. End User Comput. **19**(3), 1–23 (2007)
6. Chan, Y.E., Storey, V.C.: The use of spreadsheets in organizations: determinants and consequences. Inf. Manage. **31**, 119–134 (1996)
7. Cunha, J., Fernandes, J.P., Mendes, J., Pereira, R., Saraiva, J.: Embedding model-driven spreadsheet queries in spreadsheet systems. In: Proceedings of the Symposium on Visual Languages and Human-Centric Computing (2014)
8. Gable, G.G., Yap, C.M., Eng, M.N.: Spreadsheet investment, criticality, and control. In: Hawaii International Conference on System Sciences (1991)
9. Grossman, T.A., Mehrotra, V., Özlük, Ö.: Spreadsheet information systems are essential to business. Working paper (2005)
10. MetricStream GRC Blog: The true cost of spreadsheets in risk and compliance management (2014)
11. Nardi, B.A., Miller, J.R.: An ethnographic study of distributed problem solving in spreadsheet development. In: Proceedings of the Conference on Computer-Supported Cooperative Work, pp. 197–208 (1990)
12. Panko, R.R.: Facing the problem of spreadsheet errors. Decis. Line **37**(5), 8–10 (2006)
13. Panko, R.R., Halverson Jr., R.: Two corpuses of spreadsheet errors. In: Proceedings of the Hawaii International Conference on System Sciences, pp. 1–8 (2000)
14. Panko, R.R., Port, D.N.: End user computing: the dark matter (and dark energy) of corporate IT. J. Organ. End User Comput. **25**(3), 4603–4612 (2012)
15. Pemberton, J.D., Robson, A.J.: Spreadsheets in business. Ind. Manage. Data Syst. **100**(8), 379–388 (2000)
16. Powell, S.G., Baker, K.R., Lawson, B.: A critical review of the literature on spreadsheet errors. Decis. Support Syst. **46**(1), 128–138 (2008)
17. Powell, S.G., Baker, K.R., Lawson, B.: Impact of errors in operational spreadsheets. Decis. Support Syst. **47**(2), 126–132 (2009)
18. Rajalingham, K., Chadwick, D.R., Knight, B.: Classification of Spreadsheet Errors. (2008). arXiv:0805.4224
19. Rothermel, G., Burnett, M., Li, L., Dupuis, C., Sheretov, A.: A methodology for testing spreadsheets. ACM Trans. Software Eng. Methodol. **10**(1), 110–147 (2001)
20. Senn, J.A.: Information Technology: Principles, Practices, and Opportunities, 3rd edn. Pearson Prentice Hall, Upper Saddle River (2004)
21. United States Congress: Sarbanes-Oxley Act (2002)
22. Yin, R.K.: Case Study Research: Design and Methods, 5th edn. Sage Publications, Thousand Oaks (2014)

A Mark Based-Temporal Conceptual Graphs for Enhancing Big Data Management and Attack Scenario Reconstruction

Yacine Djemaiel[✉], Boutheina A. Fessi, and Noureddine Boudriga

Communications Networks and Security Research Laboratory (CN&S),
University of Carthage, Tunis, Tunisia
{ydjemaiel,noure.boudriga2}@gmail.com
boutheina.fessi@isg.rnu.tn

Abstract. The management of big data is mainly affected by the size of the big graph data that represents the huge volumes of data. The size of this structure may increase with the size of data to be handled over the time. Facing this issue, the querying time may be affected and the introduced delay may not be tolerated by running applications. Moreover, the investigation of attacks through the collected massive data could not be ensured using traditional approaches, which do not support big data constraints. In this context, we propose in this paper, a novel temporal conceptual graph to represent the big data and to optimize the size of the derived graph. The proposed scheme built on this novel graph structure enables tracing back of attacks using big data. The efficiency of the proposed scheme for the reconstruction of attack scenarios is illustrated using a case study in addition to a conducted comparative analysis showing how smart big graph data is obtained through the optimization of the graph size.

Keywords: Big data · Smart data · Temporal conceptual graph · Attack scenario · Investigation

1 Introduction

Big data is a new concept appeared in the recent years in all industries and domains (health, government, finance, and technology), due to the increasing advances of information technologies and the development of more complex information systems. It is characterized by five main properties, denoted as 5 V: value, variety, veracity, volume, and velocity. Each feature carries a set of issues that should be resolved to better profit from the flooded information and to extract timely and pertinent information.

Generally, when dealing with big data, the first concern lies in determining how to manage the increasing amount of data to extract the most relevant one. Several researches have been conducted and a number of analytics (such as data mining) are applied but none of them provide a full coverage of big data

© Springer International Publishing Switzerland 2015
W. Abramowicz (Ed.): BIS 2015, LNBIP 208, pp. 62–73, 2015.
DOI: 10.1007/978-3-319-19027-3_6

issues. Recent works have been interested in applying (big) graph to manage and organize complex data. This technique allows parsing data into a set of interrelated clusters. It presents several benefits related to analyzing easily and storing efficiently semi-structured/unstructured information, providing clear phrasing for queries about relationships, and solving new challenges for analysis (including data sizes, heterogeneity, uncertainty, and data quality) [15]. However, the applied technique lacks of dynamicity since the developed graph is static and cannot evolve over time.

Moreover, the reconstruction of attack scenarios using big data is among the topics that is well investigated by the research communities. This process requires a temporal information that is attached in most cases to the performed actions that are related to the handled data by intruders. During the investigation step, a set of missing actions should be identified and an ordering should be established to get a valid scenario that may help investigators identifying the vulnerable components and applying the correct reaction strategy in order to move the monitored system to a secure state. In addition, the interrogation of big data should be performed by optimizing the processing time to ensure an efficient reaction phase. Since current attack strategies are generally distributed and use novel technologies that are potential generators of big data, the interrogation of existing big graph data will introduce an important delay that may not be tolerated for an investigation process.

The objective of this paper is to provide a new model for managing and processing big data to extract smart information. This model is based on dynamic conceptual graph to store data in an optimized manner and to allow the evolution of the data graph over time. The paper, also, details the process of extracting smart data from the proposed model and highlights its efficiency. Moreover, the paper shows the significance of the proposed approach in reconstructing attack scenario.

The paper contributions are four folds. First, we integrate the time feature to the conceptual big data graph to allow more dynamicity and optimization in the data storage and processing. Second, we propose a definition of the smart data concept based on the proposed dynamic big data conceptual graph model. Third, the focus is set on presenting the process of generating smart data from the proposed graph. And fourth, the proposed model is used to enhance the intrusion detection and reaction regarding distributed attacks using big data.

The paper is organized as follows. Section 2 reviews the existing models and presents their weaknesses in coping with big data issues. This section deals with the developed approaches for optimizing the management of the huge amount of data and it details the role of using temporal graphs for Big data. Section 3 describes and discusses the proposed approach, temporal conceptual graph (TCG), for managing the big data and extracting smart information. In this section, we also present the contribution of this approach comparing to the other existing ones. In Sect. 4, we determine the contribution of the use of the proposed approach to reconstruct attack scenario, through providing properties, requirements, and the process to follow for attack reconstruction. In Sect. 5, a case study is given in order to illustrate the capabilities of the proposed model for the reconstruction of attack scenarios in addition to the optimization of the size of generated temporal conceptual graphs compared to traditional conceptual graphs. The last section concludes the work and provides future prospects that should be investigated to overcome the remaining big data challenges.

2   Related Work

The review of the literature shows the existence of various works developed to manage huge amount of data based on different techniques, as it is mentioned in [8], such as map reduce [3], clustering [1], heuristic techniques [5], data mining [6,14]. The paper of Kambtla et al. [11] reviews the different requirements to be considered when adopting big data analytics, including those related to the used hardware platforms for executing analytics applications, and their associated considerations of storage, processing, networking, and energy. The authors also present diverse applications of data analytics, ranging from health and human welfare to computational modeling and simulation.

Recently, many research are directed to cloud computing as a technique for managing big data. In fact, in Hashem et al. paper [9], the importance of cloud computing to perform massive-scale and complex computing is addressed. The paper also discusses the relationship between big data and cloud computing and identifies the research challenges and requirements. In this context, several works [2,3,7,10,16] emphasize the effective usage of cloud computing infrastructure to cope with the storage of complex and various data types and the process power to perform analysis on large datasets. However, the security activity of this infrastructure and of the managed information remains critical tasks to ensure since it is not examined in most cases. The size of the generated huge volumes of data remains an issue that prevents the management of such data in an efficient manner.

Another useful technique that deals with great amount of data is the big graph data. This technique has the intrinsic advantages of basic graphs, including cost storage reduction, enhancement of processing capabilities through easy querying storage data, and scalable data representation that allows representing any type of data in a highly accessible way. Moreover, this technique serves many existing applications, such as health care, economy, social systems, systems biology, power grid, and simulation, since they deal with complex and distributed large datasets [15]. Despite these benefits, Khurana et al. state that existing solutions for graph data management lack for appropriate techniques for temporal annotation, or for storage and retrieval of large scale historical changes on the graph [12].

In this line of research, Djemaiel et al. [4] propose a conceptual graph data management system that manages big data in an efficient manner. The model proposes a management scheme that enables the representation and the retrieval of big data using structured marks. The efficiency of the proposed approach is proved through a case study that describes the data needed to respond to distributed denial of service attacks and how the querying of such data may help to learn unknown attack fragments. Since the proposed approach deals with huge amount of data, the generated conceptual graph will have a great size and its processing and management will take a lot of time.

Another significant concept, temporal graphs (TG), is used to deal with big data management. The concept is introduced in [13] to describe events over periods of time. TG encodes temporal data into graphs. It has the analytical benefits of static graph analysis while retaining the temporal information of the original data. This representation allows the exploration of the dynamic temporal properties of data by using existing graph algorithms. This technique is useful for managing historical or temporal information of large networks at distinct time intervals deriving thus different snapshots and making data analysis tractable [13].

The work of Khurana et al. (2012) is rooted in this context [12]. They developed a graph model, called DeltaGraph, to manage historical data for large evolving information networks with the goal to enable temporal and evolutionary queries and analysis. The model is characterized by extensible, highly tunable, and distributed hierarchical index structure. It allows recording historical information and supporting retrieval of historical graph snapshots for single-site or parallel processing. Even if the effectiveness of the model at managing historical graph information is proved, it is noted that real time massive data could not be managed.

3   A Novel Temporal Conceptual Graph (TCG) Approach

The big data graph is a basic structure that is used to manage big data. The querying of this structure is a frequent task that is needed even to extract knowledge as response to a query or to update the graph content when new data sets are available or deleted. Consequently, the interrogation of such structure may be time consuming and the querying time may not be tolerated by the running applications. In order to deal with big data characteristics, the graph is generated for an instant t and not for all the time period in order to optimize its size and therefore reduces the querying time needed for such structure. Moreover, each element of the big data graph including the nodes that represent in this case the concepts, the data, and the marks in addition to the links that are the predicates and the relations between the concepts are assigned to an instant t. This feature helps during the investigation step since identified actions and data in addition to observations are ordered in time which is required for the reconstruction of the attack scenario. The set of defined predicates, related to the generated data and marks, helps to localize dependent data and therefore performed actions on data that are added to the reconstructed attack scenario.

3.1   The Proposed Model

The proposed model is an extension of the mark based approach for big data management introduced in [4], where a mark based-conceptual graph is used to represent and manage big data in an efficient manner. According to this scheme, the generated graph is a finite, connected, bipartite graph composed of concepts and conceptual relations. The concepts are used to represent the stored data temporarily or permanently in addition to the introduced marks that are represented as structure composed of the different following fields, as detailed in [4]: id_mark, $source$, $target$, $user$, id, $concepts$, $predicates$. This mark holds information related to the identity of the stored data, the source that has generated it, the object that is represented by this data, the user that has generated it in addition to the set of concepts that represent it and the defined predicates that represent the relations between the different concepts.

The proposed scheme introduces the time feature to the aforementioned graph and it becomes a Temporal Conceptual Graph, called TCG, that is defined as follows: $TCG = (t, D(t), M(t), C(t), Map_1(t), Map_2(t), KB(t), Map_3(t))$, with t is the instant where the TCG is generated; $D(t)$ is the set of data handled at an instant t, which may be a file, a video, a stream, a log entry, etc.; $M(t)$ represents the set of defined marks that are available (created or updated) at an instant t and representing the corresponding data; $C(t)$ is the set of concepts that are available (created or updated) at an instant t representing stored big data. A concept that is valid for an instant t may be attached to more than one mark; $Map_1(t)$ is a mapping relation that defines the correspondence between the marks and the data that are available at an instant t. This mapping allows the identification of the needed data as a response to a query fetching big data at an instant t (created, modified, removed, etc.); $Map_2(t)$ is a valid mapping relation for an instant t that defines the relationships between a mark M_{it} and its concepts $C_{i1_t}..C_{in_t}$. It is used for the search and the retrieving process since it holds the association between the mark and the different concepts that represent the same data and that may be used as a criteria in the defined query by the user; $KB(t)$ is the knowledge database that includes the different concepts, the existing relationships between them. The links between the concepts are defined using the set of available predicates; $Map_3(t)$ is a mapping relation that maps a subset (application knowledge) of the $KB(t)$ to $M(t)$. It holds the relationships that may exist between the different concepts belonging to the same mark.

The marking is performed by the application that generates the data based on the $KB(t)$ content associated to the handled data. The set of concepts and predicates are determined by querying the $KB(t)$.

The temporal graph provides a snapshot of the whole graph at an instant t. This feature reduces the size of data that should be handled at an instant t. This graph integrates the set of concepts in addition to the defined predicates, the attached marks, and the related data.

According to the proposed approach, the temporal conceptual graph is generated for an instant t, denoted by TCG_t which includes the following components:

- The set of concepts that represent the stored big data, is referred as: $C_t = \{c_{1t}.. c_{mt}\}$, where m is the total number of concepts that represent the stored big data.
- The set of relations between concepts is called $R_t = \{r_{1t}.. r_{kt}\}$, where k is the total number of relations that are defined for concepts.

Two conceptual graphs TCG_{t_1} and TCG_{t_2} are linked if $TCG_{t_1} \cap TCG_{t_2} \neq \emptyset$, where at least a mark $m_{t_1} \in TCG_{t_1}$ is updated at t_2 ($t_2 > t_1$).

Figure 1 illustrates a sample temporal conceptual graph according to the proposed model, generated for three instants (t_1, t_2 and t_3) in addition to the content of the conceptual graph associated to the previous scheme, as detailed in [4], for the same instants. As it can be noted from Fig. 1, the graph size is largely reduced compared to the first graph observed for the three instants. This decreasing is due to the property of the proposed temporal conceptual graph that includes only the nodes that have been created for the observed instant t or updated from a previous instant t' (where $t' < t$). For instance, $M_{1_{t_2}}$ and $M_{2_{t_2}}$ are the updated marks of $M_{1_{t_1}}$ and $M_{2_{t_1}}$ and they are linked to the $\hat{T}CG_{t_2}$ as it is shown in Fig. 1 using dashed line. These marks are updated since the relation $R_{2_{t_1}}$, that is defined in $M_{1_{t_1}}$ and $M_{2_{t_1}}$, has been modified by adding the concept $C_{4_{t_2}}$ to the relation.

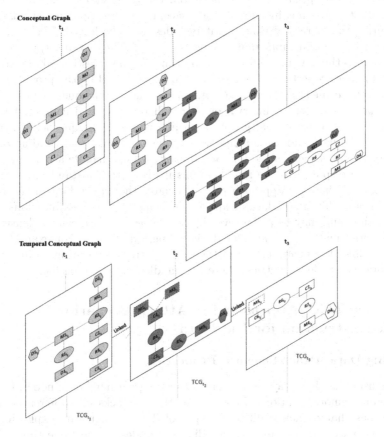

Fig. 1. Illustration of the traditional and novel temporal conceptual graphs

3.2   Smart Big Graph Data

According to the proposed scheme, the graph handled at an instant t does not hold all the data generated from a starting time but only data that has been created at an instant t. According to the figure illustrating the previous model built on conceptual graphs and the proposed novel temporal conceptual graphs, the generated graph at a specific instant, according to the proposed approach, comprises a reduced number of nodes compared to the previous approach illustrated at the top of the figure. According to this property, the obtained graph representing big data at an instant t is considered as a smart big graph data. This feature enables the optimization of the size of the graph without losing knowledge about the stored data in addition to the dependent data through the time. This property is guarantied since the temporal conceptual graphs could be linked. The graph size is reduced since linked graphs include only the updated marks. As illustrated in Fig. 1, the proposed TCG has a largely reduced size compared to conceptual graph. In addition, the resulted TCG for an interval of time $[t_1, t_2]$ does not include all the concepts and relations for all managed data during this period of time but it includes only the linked $TCGs$ that are associated to some instants that belong to $[t_1, t_2]$. The instants t_1 and t_2 may correspond to the starting and the ending times of the investigated attack.

The linking property is also another principle ensured by the provided technique to reduce the size of the TCG. As a consequence, the querying of a TCG for an instant t or for an interval of time will be optimized compared to the interrogation of a conceptual graph according to the technique defined in [4]. The storage of data and the management of marks are also enhanced according to the proposed scheme, since for each instant, we will handle a portion of the data that corresponds to the modifications made at that instant and that will be less in size than the approach that associates the whole data to the mark. Handling only this smart data at an instant helps to trace modifications made through time and may help the investigation of security attacks to identify the actions performed by the intruder during a period of time. For example, this feature enables the identification of the events introduced during an interval of time from the whole log entries where the handled data is a log file.

4   Using TCG Approach for Attack Scenario Reconstruction for Big Data

4.1   Big Data Attack Scenario Properties

Attacks using big data may be performed using a great number of nodes, located at different remote locations. In addition, these attacks use different kind of applications that exchange different types of data at high rate exploiting the 5 V of big data and that may target different nodes at the same time. These features make the intrusion detection process and the investigation activities of such attacks hard to perform. Furthermore, available intrusion detection systems are unable to manage all the generated data and those related to the performed attack, which leads to high set of missing observations and performed malicious actions. Within the high speed and volume of generated data, evidences may be lost since the buffers are configured with a maximal size that makes this data available only for a short period then it is replaced by new events. Moreover, attacks using big data are performed under a large set of access networks, applications, and devices generating various malicious actions at different periods of time and introducing delays for packets transmission through networks. As a consequence, the different actions targeting victim may be separated by varied delays that make the reconstruction of the attack scenarios more difficult.

4.2   Requirements for Attack Reconstruction Using TCG

The reconstruction of the attack scenario using the temporal conceptual graph is performed only if the following conditions are satisfied: (1) At least an event or a malicious action is detected by a deployed monitoring system; (2) The occurrence time of at least an action that belongs to the attack scenario should be reported; (3) The temporal conceptual graphs are generated and maintained for the sites to which belong the victim nodes; and (4) At least logging service should be deployed for each monitored site in order to log performed actions, in addition to the logging services deployed for the network and security resources (e.g. routers and firewalls).

An additional temporal proximity metric is used to identify actions and data related to the detected malicious events that belong to the attack scenario for a specific range of time. This metric defines the dependency on time between temporal graph's nodes. The use of this metric enables the identification of the additional malicious activities that are not identified based on the relationships available at temporal conceptual graphs, due to the missing of links between stored data, the lack of concepts, and the missing knowledge in $KB(t)$. Based on the temporal proximity, the dependencies between data and marks are checked to identify possible relationships between data and therefore finding missing malicious activities belonging to the attack scenario.

4.3   Attack Scenario Reconstruction Process

The reconstruction of the attack scenario starts by fetching the set of linked TCG starting from the detection time, denoted t_d, associated to the reported malicious event as illustrated in Fig. 2. Based on the detection time t_d and the associated reported event (the performed malicious action), the set of data and their associated marks are extracted from the TCG_{t_d} by querying available marks using available marks fields, concepts, and predicates. Based on the collected data $\{D_{t_d}\}$ and their associated marks $\{M_{t_d}\}$, a discovery of the linked $TCGs$ is initiated in backward (for $t < t_d$) to determine the set of $TCGs$ that are linked to the generated TCG_{t_d}. This step aims to identify the set of dependent big data used during the attack scenario in addition to the different instants where the intruders have issued malicious actions. The end of this process in backward enables the identification of the starting time for the attack, denoted

t_i, and the set of dependent data among the different $TCGs$. The same steps are performed in forward (for $t > t_d$) to identify either the end time for the performed attack or to localize the handled data and the attached marks at the current time for ongoing attacks. Based on the identified occurrence times of malicious actions $\{t_k\}$ and the identified data $\{D_{t_k}\}$, where $k \in [1..n]$ and n is the total number of identified instants where intruders have performed malicious actions in relation with the initial reported alert, the next step is to identify the associated performed malicious actions. These actions could be identified directly or through another analysis process. For example, an executed command located in a log file (representing the identified data from the temporal conceptual graph) could be directly identified. An example of data that should be processed to identify the malicious action may be a password file (D_{t_j}). In this case, the discovering process of the related actions starts by checking for instants $t < t_j$ to identify modifications made on D_{t_j}. These modifications are analyzed to find possible actions that may lead to this result. In order to select the most possible action, the TCG_{t_j} is then queried to check the action that allows to get the marks $\{M_{t_j}\}$ as responses.

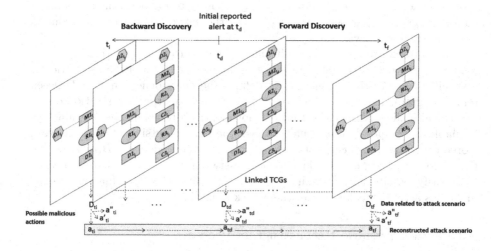

Fig. 2. Attack scenario reconstruction process

5   Case Study

The efficiency of the proposed approach is illustrated through two ways. The former illustrates the reconstruction of the attack scenario based on the collected $TCGs$. The latter is a conducted comparative analysis that illustrates the gain in term of graph size provided by a temporal conceptual graph.

5.1   Reconstruction of the Attack Scenario Capabilities

In order to illustrate the efficiency of the novel temporal conceptual graphs for big data, an enterprise information system is considered as a target for a complex attack using big data to make a set of provided services unavailable. The attack is performed by distributed agents, localized in different area, aiming to generate a distributed denial of service in addition to forging provided services. These attacks have as target the information system services including the enterprise web server, the DNS server, the public FTP server, the storage server, and log server. An alert is reported by a deployed IDS at an instant t_d reporting the download of malicious known backdoor from the web server. Based on this alert, related marks and concepts are identified from TCG_{t_d}. Then, the linked $TCGs$ are fetched in both backward and forward in order to determine respectively the starting time and the current time or end time of the investigated attack. After that, the stored data related to performed attack are collected for the different instants associated to the identified linked $TCGs$. These data includes issued requests, log files, password files, buffers for routers and firewalls. From these data, possible malicious actions are generated including massive spoofed packets and write attempts to log files and password files. These actions are checked against $TCGs$ content to determine those (i.e. actions having observable effects on data) that belong to the reconstructed attack scenario. The identified actions are ordered in time according to the different time instants associated to linked $TCGs$.

5.2   Comparative Analysis Between Mark Based Temporal Graphs and Conceptual Graphs

We consider a conceptual graph where n marks are defined in uniform manner and each mark is attached to m concepts at an instant t_d. Moreover, at each

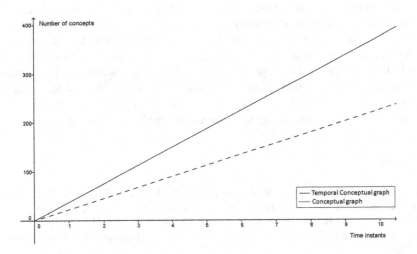

Fig. 3. Evolution of the number of concepts for conceptual and temporal conceptual graphs

instant there is a $\frac{n}{2}$ new marks within their associated concepts and j marks are updated from an instant to another. If we limit the analysis to two instants t_d and $t_{d+\alpha}$, we can notice that for the conceptual graph (as described in [4]), the number of available concepts is $\frac{3}{2}(n \times m)$ at $t_{d+\alpha}$. However, for the proposed temporal conceptual graph, the number of concepts represented is $(\frac{3}{2} \times n - j) \times m$ since for linked marks, the next TCG does not include the associated concepts.

For an increasing number of instants k, the number of concepts added to the conceptual graph will be greater than the new temporal conceptual graph, according to the respective following expressions: $\frac{3}{2}(n \times m \times k)$ and $(\frac{3}{2} \times n - j) \times (m \times k)$. Figure 3 shows a much larger number of concepts in the conceptual graph (solid line) compared to the number of concepts in the temporal conceptual graph, when considering the following values as an example for the defined parameters: $n = 5$, $m = 5$ and $j = 3$. According to this result, the temporal conceptual graph has enabled the definition of a smart big data graph that optimizes enormously the graph size compared to conceptual graphs.

6   Conclusion

In this paper, we have proposed a novel approach for optimizing the management of big data by reducing the size of the big graph data based on the use of a Temporal Conceptual Graph. The support of time for handling big data enables the reduction of the size of the handled big data graph at an instant t in addition to the ability to link generated graphs. This additional feature has enhanced the tracing of performed actions related to generated data through the time and enables the reconstruction of attack scenario using big data. The efficiency of the proposed approach has been illustrated through an analytical study showing the gain in term of size compared to previous scheme using conceptual graphs to manage big data. Furthermore, the capabilities of the proposed approach regarding the generation of complex attack scenarios is also illustrated through a case study in addition to the reducing of the graph size.

Several enhancements will be considered in the near future in order to ensure the detection of malicious activities and the reconstruction of attack scenarios when hiding techniques are deployed by distributed nodes and when the stored data are not well described through concepts. Moreover, the validation of the temporal conceptual graph content is among the other issues that will be investigated to defeat attacks aiming to make these graphs inconsistent.

References

1. Cominetti, O., Matzavinos, A., Samarasinghe, S., Kulasiri, D., Liu, S., Maini, P.K., Erban, R.: Diffuzzy: a fuzzy clustering algorithm for complex datasets. Int. J. Comput. Intell. Bioinform. Syst. Biol. (IJCIBSB) 1(4), 402–417 (2010)
2. David, B.: The promise and peril of big data. Technical report, The Aspen Institute (2010)

3. Dean J., Ghemawat, S.: MapReduce: simplified data processing on large clusters. In: Proceedings of the 6th Conference on Symposium on Opearting Systems Design and Implementation. OSDI 2004, vol. 6, p. 10. USENIX Association, Berkeley (2004)

4. Djemaiel, Y., Essaddi, N., Boudriga, N.: Optimizing big data management using conceptual graphs: a mark-based approach. In: Abramowicz, W., Kokkinaki, A. (eds.) BIS 2014. LNBIP, vol. 176, pp. 1–12. Springer, Heidelberg (2014)

5. Dzemyda, G., Sakalauskas, L.: Large-scale data analysis using heuristic methods. Informatica **22**(1), 1–10 (2011)

6. Fan, W., Bifet, A.: Mining big data: current status, and forecast to the future. SIGKDD Explor. Newsl. **14**(2), 1–5 (2013)

7. Ghit, B., Iosup, A., Epema, D.: Towards an optimized big data processing system. In: 13th IEEE/ACM International Symposium on Cluster, Cloud, and Grid Computing (CCGrid) (2013)

8. Gkoulalas-Divanis, A., Labbi, A. (eds.): Large-Scale Data Analytics. Springer, New york (2014)

9. Hashem, I.A.T., Yaqoob, I., Anuar, N.B., Mokhtar, S., Gani, A., Khan, S.U.: The rise of "big data" on cloud computing: review and open research issues. Inf. Syst. **47**, 98–115 (2015)

10. Ji, C., Li, Y., Qiu, W., Awada, U., Li, K.: Big data processing in cloud computing environments. In: 12th International Symposium on Pervasive Systems, Algorithms and Networks (ISPAN), pp. 17–23, December 2012

11. Kambatla, K., Kollias, G., Kumar, V., Grama, A.: Trends in big data analytics. J. Parallel Distrib. Comput. **74**(7), 2561–2573 (2014). Special Issue on Perspectives on Parallel and Distributed Processing

12. Khurana, U., Deshpande, A.: Efficient snapshot retrieval over historical graph data. CoRR, abs/1207.5777 (2012)

13. Kostakos, V.: Temporal graphs. Physica A: Stat. Mech. Appl. **388**(6), 1007–1023 (2009)

14. Rabbany, R., Zaïane, O.R., ElAtia, S.: Mining large scale data from national educational achievement tests. In: ASSESS Data Mining for Educational Assessment and Feedback Workshop in Conjunction with KDD 2014. New York, 24 August 2014

15. Riedy, J., Bader, D.A., Ediger, D.: Streaming graph analytics for massive graphs. Georgia Institute of Technology, College of computing, 10 July 2012

16. Talia, D.: Clouds for scalable big data analytics. IEEE Comput. **46**, 98–101 (2013)

Semantic Technologies

Ontology-Based Creation of 3D Content in a Service-Oriented Environment

Jakub Flotyński[(✉)] and Krzysztof Walczak

Poznań University of Economics, Niepodległości 10, 61-875 Poznań, Poland
{flotynski,walczak}@kti.ue.poznan.pl

Abstract. In this paper, an environment for semantic creation of interactive 3D content is proposed. The environment leverages the semantic web techniques and the SOA paradigm to enable ontology-based modeling of 3D objects and scenes in distributed web-based environments. The proposed solution combines visual and declarative approaches to 3D content creation. The possible use of various ontologies and knowledge bases permits modeling of content by users with different skills and at different levels of abstraction, thus significantly extending possible use of 3D on the web in various business scenarios.

Keywords: 3D content · Semantic web · 3D web · Ontologies · SOA

1 Introduction

Interactive 3D technologies become increasingly popular in various application domains (and business domains) on the web, such as education, training, tourism, entertainment, social media and cultural heritage. The primary element of any 3D/VR/AR system, apart from the employed interface technologies, is the interactive 3D content presented to users. However, creating interactive 3D content is a much more complex and challenging task than in the case of typical 2D web resources.

The potential of 3D/VR/AR applications on the web can be fully exploited only if there are efficient and flexible methods of content creation and manipulation, which conform to the recent trends in the development of the web. One of such trends is the growing use of semantics – the W3 Consortium initiated the research on semantic web, which aims at the evolutionary development of the current web towards a distributed semantic database linking structured content and documents. The semantic web consists of content described with common ontologies and knowledge bases, which specify the meaning of particular content components at different levels of abstraction. In the case of 3D content, the ontologies used may be either directly related to computer graphics (shapes, textures, materials, elementary animations, etc.) or may be specific to an application domain (cars, exhibits, buildings, animals, etc.). Ontologies and knowledge bases distributed across the web may be accessed by users with different preferences (interests, levels of detail required, etc.) and clients with different

© Springer International Publishing Switzerland 2015
W. Abramowicz (Ed.): BIS 2015, LNBIP 208, pp. 77–89, 2015.
DOI: 10.1007/978-3-319-19027-3_7

capabilities (performance, user interfaces, operating systems, etc.). Furthermore, semantic description of content may take into account not only the knowledge that has been explicitly specified by the content designer (content properties, dependencies and constraints), but also knowledge that has not been explicitly specified, but may be inferred in the reasoning process.

The main contribution of this paper is an architecture for semantic creation of interactive 3D content. The created environment leverages semantic web techniques and a service-oriented architecture. It enables content creation by users distributed across the web, with different skills and equipped with different tools. Content creation is performed by referring to the explicit and implicit meaning of particular content components, which may be specified using different ontologies. The environment leverages a semantic model of interactive 3D content and a method of semantic content creation.

2   Related Works

Several works have been devoted to semantic creation and description of 3D content. In [23], an ontology providing elements and properties that are equivalent to elements and properties specified in X3D has been proposed. Moreover, a set of semantic properties have been proposed to enable description of 3D scenes with domain knowledge. In [31–34], a method of creating VR content based on reusable elements with specific roles has been proposed. The method has been intended to enable 3D content design by non-IT-specialists. In [9], an approach to generating virtual environments upon transformation of domain ontologies has been proposed. In [28], an approach facilitating semantic modeling of content behavior has been proposed. In [13,16,18,19], an approach to building semantic descriptions embedded in 3D web content and a method of harvesting semantic metadata from 3D web content have been proposed. In [22], an ontology-based approach to creating virtual humans as active semantic entities with features, functionality and interaction skills has been proposed.

Several works provide an overview of the use of semantic descriptions of 3D content in artificial intelligence (AI) systems. The idea of semantic description of 3D environments has been summarized in [25]. In [30], a review of the main aspects related to the use of 3D content in connection with the semantic web technologies has been provided. In [26], diverse issues arising from combining AI and virtual environments have been reviewed. In [27], abstract semantic representations of events and actions in AI simulators of 3D environments have been presented. In [24], a technique of integrating knowledge into VR applications has been discussed.

3   The SEMIC Approach

Although several approaches have been proposed for semantic modeling of 3D content (cf. Sect. 2), they have important limitations in the generality and flexibility of 3D content creation. To exploit new technical opportunities provided

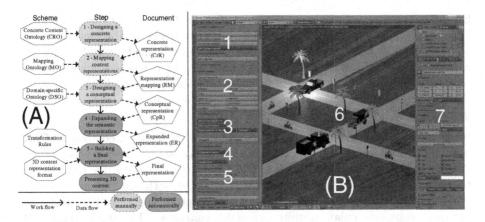

Fig. 1. (A) 3D content creation according to the SEMIC approach. (B) The panels of the SO-SEMIC *Client*: (1) *Concrete Component Panel*, (2) *Concrete Property Panel*, (3) *Class Mapping Panel*, (4) *Property Mapping Panel* and (5) *Conceptual Design Panel* along with the (6) *User Perspective* and (7) other standard Blender tools

by the development of the semantic web, a new approach to modeling 3D content – *Semantic Modeling of Interactive 3D Content* (SEMIC) – has been proposed in [10,20]. SEMIC supports separation of concerns between users with different responsibilities and capabilities. In SEMIC, the creation of 3D content consists of a sequence of partly dependent steps, which use different ontologies and produce knowledge bases (3D content representations) that comply with the ontologies (Fig. 1A). Some of the steps are performed manually—by a content developer, a domain expert and a content consumer, while other steps are performed automatically—by specific software (algorithms). The SEMIC 3D content creation steps are described below.

Step 1—designing a *Concrete Semantic Representation of 3D Content* (CrR) provides elements that are specific to 3D content (concrete content components and properties) and enable representation of domain-specific concepts (classes and properties) in Step 3. Concrete components and properties comply with the *Concrete 3D Content Ontology* (CRO) [11] and are grouped into several partly dependent layers—*geometry layer, structure and space layer, appearance layer, scene layer, animation layer* and *behavior layer*. This step is typically performed by a developer with expertise in 3D modeling, who is equipped with specific tools, e.g., 2D and 3D graphical editors for creating textures and meshes.

Step 2—mapping a CrR to domain-specific semantic concepts enables 3D presentation of domain-specific content representations (created in Step 3). The result of this step is a *Representation Mapping* (RM), which is a knowledge base compliant with the *Mapping Ontology* (MO) [14,17]. Mapping is performed by a developer, once for a particular *Domain-Specific Ontology* (DSO) and a CrR. It enables the reuse of concrete components and properties for forming 3D representations of various domain-specific individuals, which conform to the DSO used.

Step 3—designing a *Conceptual Semantic Representation of 3D Content* (CpR) enables creation of 3D content at an arbitrarily chosen level of abstraction, which is determined by the DSO used. This step can be performed multiple times for a particular DSO, a CrR and an RM. Since this step requires knowledge of particular domain-specific concepts, but not technical 3D modeling concepts, it may be performed by a domain expert, who is equipped with a semantic modeling tool (possibly domain-specific).

Step 4—expanding the semantic representation of 3D content combines a CpR with a CrR using an RM. It produces an *Expanded 3D Content Representation* (ER), which incorporates both the explicit and implicit knowledge (inferred in the reasoning process). An ER represents content at both concrete and conceptual levels of abstraction, i.e. how particular conceptual (high-level) elements are represented by particular concrete (low-level) elements. This step is performed automatically.

Step 5—building a final 3D content representation involves transformation of an ER to its final counterpart, which is encoded using a particular 3D content representation language (e.g., VRML, X3D, ActionScript) [12,15] or based on an internal 3D content representation format implemented by a particular content browser or modeling tool (e.g., Blender, Unity 3D).

4   The SO-SEMIC Environment for 3D Content Creation

The SEMIC approach is independent of particular architectures and technologies, and can be implemented in different ways, e.g., as a standalone or client-server application. The main contribution of this paper is the *Service-Oriented Environment for Semantic Modeling of Interactive 3D Content* (SO-SEMIC). SO-SEMIC implements SEMIC according to the service-oriented architecture (SOA). The use of SOA enables implementation of flexible multi-user content management systems, which separate client-side graphical operations from computationally expensive server-side semantic processing. Also, access to content in distributed mobile environments is possible through the available services.

4.1   System Architecture

The architecture of SO-SEMIC is depicted in Fig. 2. The environment comprises a *Client* and a *Server* in a RESTful SOA architecture. The *Client* is a Python application, which enables visual manipulation of 3D content in the Blender modeling tool as well as transformation of Blender-specific content to its equivalent semantic content representations. The *Server* is a Java application, which enables manipulation of semantic content representations (CrRs and CpRs) using appropriate ontologies (CRO, MO and DSO). Both applications implement multi-layered architectures, which are discussed in detail in the following subsections. SO-SEMIC offers access to new functions of semantic modeling on the web, in contrast to the web-based Blender interface [2], which only provides selected standard functions of Blender.

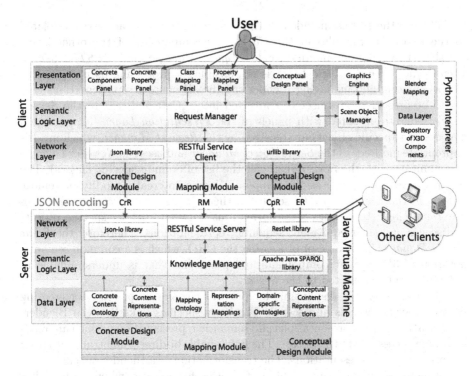

Fig. 2. The architecture of the SO-SEMIC environment. Arrows indicate the flow of information

Client. The multi-layered architecture of the *Client* conforms to the Model-View-Controller (MVC) design pattern [5]. It consists of four layers: the *Semantic Logic Layer* (controller), the *Presentation Layer* (view) as well as the *Network Layer* and the *Data Layer* (model). The *Client* is based on the Blender API [3]. Blender has been selected because it is an advanced, widely-used, open-source environment with extensive documentation, tutorials and several versions available for different operating systems. However, the *Client* could be also developed using other 3D content modeling environments.

The *Presentation Layer* is responsible for handling requests from a user and presenting 3D content in Blender. It employs the Blender *Graphics Engine*, which allows a user to present and manipulate 3D content in the graphical environment, as well as new GUI elements (panels), which extend the standard Blender GUI providing the functionality of the SO-SEMIC environment to a user. The *Graphics Engine* enables access to a number of Blender tools, such as transformations, editors and scene graphs. The *Graphics Engine* is accessible through specific classes and properties of the Blender API. The SO-SEMIC panels include: the *Concrete Component Panel*, the *Concrete Property Panel*, the *Class Mapping Panel*, the *Property Mapping Panel* and the *Conceptual Design Panel* (Fig. 1B). The panels consist of buttons, menus and properties.

Whereas the panels included in the *Presentation Layer* are strictly related to the particular modules of the *Client*, the components of the other layers are shared by different modules, e.g., *Request Manager* and *RESTful Service Client*. The *Semantic Logic Layer* is responsible for processing user requests as well as creating and managing Blender-specific 3D content representations that are to be presented to a user or converted into a CrR and sent to the *Server*. User's requests prepared with panels of the *Presentation Layer* (e.g., create a new domain-specific object) are received by the *Request Manager*, which invokes appropriate methods of the *Scene Object Manager* and uses the *Network Layer* to communicate with the *Server* (e.g., get the list of domain-specific classes). The *Scene Object Manager* uses the *Graphics Engine* to create, modify or remove objects in the Blender-specific scene. Further, the *Scene Object Manager* uses the *Data Layer* to perform transformations between Blender-specific objects and scenes and their equivalent semantic representations. Therefore, the *Scene Object Manager* is capable of processing the CRO, which is a common semantic notation of 3D content components and properties that is understandable to different clients connected to the *Server*.

The *Data Layer* is responsible for providing basic components for creating Blender-specific objects and scenes. The layer specifies links between Blender-specific 3D content representations and semantic representations, which are readable and processable to the *Server*. The links enable bi-directional transformation of both types of representations. The transformation of a Blender-specific representation to its semantic equivalent in a CrR is performed, e.g., when an object has been modified in Blender and it is updated in the corresponding CpR. The transformation of an expanded 3D content representation (ER) is performed every time it is retrieved from the *Server* and should be loaded into Blender.

To provide platform-independence of the *Server*, X3D representations of 3D content are used. The *Repository of X3D Components* is a collection of X3D documents that are Blender-specific equivalents to concrete components (included in a CrR). *X3D Components* are read from the *Repository* and combined by the *Scene Object Manager* into complex Blender-specific objects and scenes. For instance, a texture, which is an *X3D Component* and an equivalent to a `cro:Texture` concrete component, is applied to a material, which is another *X3D Component* and an equivalent to a `cro:Material` concrete component. *X3D Components* are partially independent of their equivalent concrete components, since the properties that are not semantically specified in a concrete component may be arbitrarily set for its equivalent *X3D Component*. For instance, at a high level of abstraction, semantic content creation does not require access to the coordinates of particular vertices of a mesh, which may be specified only in Blender-specific *X3D Components*. Hence, different clients may have independent repositories of *X3D Components*, which have different values of properties that are unspecified for concrete components.

The *Network Layer* is responsible for communication with the *Server*. The *RESTful Client Service* leverages two Python libraries. The *json library* [4] is used to encode and decode semantic 3D content representations in JSON. The *urllib library* [7] is used to create, send and receive HTTP requests (which incorporate encoded semantic representations). As a response to a request, the *RESTful Client Service* typically gets an acknowledgment or an ER, which is further transformed and presented in Blender.

Server. The *Server* manages and provides semantic 3D content representations to clients. The *Server* is independent of particular 3D content representation standards, content browsers and modeling environments. Hence, it could be used with different software clients installed on various devices. The multi-layered architecture of the *Server* consists of three layers: the *Network Layer*, the *Semantic Logic Layer* and the *Data Layer* (Fig. 2).

The *Network Layer* is responsible for communication with clients. The *RESTful Server Service* leverages two Java libraries. The *Restlet* library [6] is used to receive HTTP requests and to create and send HTTP responses to clients. The *json-io* library is used to decode semantic 3D content representations included in the incoming requests and to encode semantic 3D content representations that are sent to clients. The *Semantic Logic Layer* is responsible for processing client requests as well as creating and managing semantic 3D content representations. The *Knowledge Manager* is based on the *Apache Jena SPARQL* library [8] – it can read, modify and perform reasoning on ontologies and content representations. Reasoning on content representations has been implemented using the SPARQL CONSTRUCT clause instead of libraries strictly designed for OWL-based reasoning. Such an approach has allowed to achieve better efficiency through precise inference of only the required facts and skipping the entailments that are not relevant to a particular case. For instance, determining super-classes of 3D cars does not require determining super-classes of 3D traffic lights. The *Data Layer* is responsible for storing semantic representations of 3D content as well as the ontologies to which the representations conform.

Request Processing. User requests to SO-SEMIC are processed by the *Request Manager*. An HTTP request sent by the *Client* to the *Server* is encoded in JSON by the *RESTful Service Client*, transmitted over the network, handled by the *RESTful Service Server*, decoded and conveyed to the *Knowledge Manager*, which processes the requests according to ontologies and content representations obtained from the *Client* or retrieved from the *Data Layer*. Any transformation between a Blender-specific representation and a CrR involves the *Scene Object Manager*, which uses the *Blender Mapping* and *X3D Components*.

4.2   Modeling 3D Content with SO-SEMIC

In this section, the modeling of content with SO-SEMIC (Steps 1-3) is presented.

Step 1—Designing a Concrete 3D Content Representation involves the *Concrete Design Module* with two panels: the *Concrete Component Panel* and the

Concrete Property Panel. The creation of a concrete component requires the selection of a 3D object in the Blender-specific scene (e.g., a mesh, a lamp, a camera) which (or whose component, e.g., color, material, scale) will be the prototype of the new concrete component. A user chooses a CRO class as the base class for the new concrete component and selects the desirable concrete properties to be included in the new concrete component. For instance, a `cro:DiffuseColor` concrete component may be described by three RGB properties (`cro:r`, `cro:g` and `cro:b`).

For a new concrete component, an *X3D Component* may be created to provide a means of representation of the concrete component in the Blender environment. The creation of an *X3D Component* facilitates further modifications of its equivalent concrete component, and enables adding non-semantic properties to the concrete component. Such an approach excludes properties that are indeed necessary for 3D content presentation, but neither need to be semantically accessible nor analyzable (e.g., due to their complexity or large size—vertices of a mesh, sub-objects of a complex structural object) from concrete components, and it speeds up semantic processing of CrRs. If a user creates a concrete component without a *X3D Component*, the use of the concrete component may still imply presentational effects that are determined by the semantic statements of the concrete component. For example, a behavioral component does not need to have geometry nor appearance.

The *Knowledge Manager* adds the received concrete component to the CrR. CrRs are encoded as OWL documents including classes linked with restrictions on properties, like RMs (presented in the next subsection).

An example of a CrR is presented in Fig. 3a. The CrR includes concrete components and concrete properties that will be used for modeling traffic lights: `crr:Post`, `crr:UpperLight`, `crr:CentralLight` and `crr:LowerLight`. While the `crr:Post` is a `cro:StructuralComponent` comprised of a box and a cylinder, the lights are `cro:SpatialComponents` with specific relative positions. Concrete components created in this step of SEMIC are further used for mapping domain-specific concepts in Step 2.

Step 2—Mapping 3D Content Representations involves the *Mapping Module* with two panels—the *Class Mapping Panel* and the *Property Mapping*

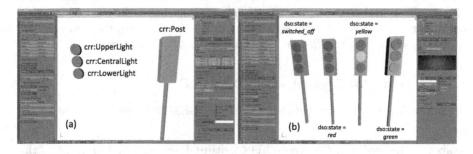

Fig. 3. Examples of a *Concrete 3D Content Representation* (a) and a *Conceptual 3D Content Representation* (b) (Color figure online)

Panel. Due to the mapping, concrete 3D content components and properties determine 3D representations of the domain-specific concepts. SEMIC enables mapping of domain-specific classes and domain-specific properties to concrete classes and concrete properties. Mapping of a domain-specific class requires the specification of the desirable concrete components that will represent the instances of the class. For example, a green tree is a concrete structural component that includes other components—branches and leafs. Like a CrR, an RM is encoded by the *Scene Object Manager*, transferred to the *Knowledge Manager* and stored in the *Data Layer*.

Mapping of a domain-specific property requires the specification of a value of the property and the selection of an already mapped domain-specific class that has to be assigned to an object, if the property of the object is set to the value. An RM is based on OWL restrictions and equivalence of classes. In the presented example of modeling traffic lights, the `dso:GreenTrafficLight` class (Listing 1.1/ lines 1-4), which includes all individuals with the `dso:state` property set to `green` (5-8), is a `cro:StructuralComponent` comprised of three lights (9-13). While the `dso:UpperLight` and the `dso:CentralLight` of a `dso:GreenTraffic Light` individual have no specific materials (14-20), the `dso:LowerLight` has a material with the green color assigned (21-34). The presented RM is generated automatically by the environment as a result of dependencies indicated by a user in the *Mapping Panel*. Traffic lights in the other states (red, yellow and switched off) are mapped in a similar way. Concrete components created in this step of SEMIC may be further leveraged for conceptual modeling of content using domain-specific concepts in Step 3.

Listing 1.1. Mapping of domain-specific concepts to concrete components and properties (the `dso:GreenTrafficLight` class and the `dso:state` property).

```
1    <rdf:Description rdf:about="../dso.owl#GreenTrafficLight">
2      <rdfs:subClassOf rdf:resource="../crr.owl#GreenTrafficLightSC"/>
3      <owl:equivalentClass rdf:resource="../dso.owl#trafficLightState-green"/
       >
4    </rdf:Description>
5    <rdf:Description rdf:about="../dso.owl#trafficLightState-green">
6      <owl:hasValue>green</owl:hasValue>
7      <owl:onProperty rdf:resource="../dso.owl#state"/>
8      <rdf:type rdf:resource="../owl#Restriction"/></rdf:Description>
9    <rdf:Description rdf:about="../crr.owl#GreenTrafficLightSC">
10     <rdfs:subClassOf rdf:resource="../crr.owl#GreenTrafficLightSC-UL"/>
11     <rdfs:subClassOf rdf:resource="../crr.owl#GreenTrafficLightSC-CL"/>
12     <rdfs:subClassOf rdf:resource="../crr.owl#GreenTrafficLightSC-LL"/>
13     <rdfs:subClassOf rdf:resource="../cro.owl#StructuralComponent"/></rdf:
       Description>
14   <rdf:Description rdf:about="../crr.owl#GreenTrafficLightSC-UL">
15     <owl:qualifiedCardinality rdf:datatype="../XMLSchema#integer">1</owl:
       qualifiedCardinality>
16     <owl:onClass rdf:resource="../crr.owl#UpperLight"/>
17     <owl:onProperty rdf:resource="../crm.owl#includes"/>
18     <rdf:type rdf:resource="../owl#Restriction"/>
19   </rdf:Description>
20   <rdf:Description rdf:about="../crr.owl#GreenTrafficLightSC-CL"><!--
       analogously--></rdf:Description>
21   <rdf:Description rdf:about="../crr.owl#GreenTrafficLightSC-LL">
22     <owl:qualifiedCardinality rdf:datatype="../XMLSchema#integer">1</owl:
       qualifiedCardinality>
23     <owl:onClass rdf:resource="../crr.owl#LowerLight"/>
24     <owl:onProperty rdf:resource="../crm.owl#includes"/>
```

```
25      <rdf:type rdf:resource="../owl#Restriction"/></rdf:Description>
26      <rdf:Description rdf:about="../crr.owl#LowerLight">
27      <owl:qualifiedCardinality rdf:datatype="../XMLSchema#integer">1</owl:
            qualifiedCardinality>
28      <owl:onClass rdf:resource="../crr.owl#GreenMaterial"/>
29      <owl:onProperty rdf:resource="../crm.owl#hasMaterial"/>
30      <rdf:type rdf:resource="../owl#Restriction"/></rdf:Description>
31      <rdf:Description rdf:about="../crr.owl#GreenMaterial">
32      <rdfs:subClassOf><owl:onProperty rdf:resource="../cro.owl#g"/>
33      <owl:hasValue>1</owl:hasValue></rdfs:subClassOf><!--g, b - analogously
            -->
34      <rdf:type rdf:resource="../owl#Restriction"/></rdf:Description>
```

Step 3—Designing a Conceptual 3D Content Representation involves the *Conceptual Design Module* with the *Conceptual Design Panel*, which enables modeling of 3D content with domain-specific classes and properties previously mapped to concrete components and properties. The module uses properties of domain-specific objects that are imposed by the RM as well as properties that are imposed by Blender. For instance, the `dso:light` domain-specific property determines the current state of a light source (e.g., its color, texture and intensity), while the position of the light source in the scene is determined by Blender-specific coordinates, which are transformed to concrete properties in the case of modifying the object. Designing a CpR encompasses creating, modifying and removing domain-specific objects in a domain-specific scene.

In the presented example of traffic lights (Fig. 3b), a CpR includes four domain-specific objects (traffic lights) with different values of the `dso:state` property set (`switched_off`, `red`, `yellow` and `green`). Modifying the state of a traffic light with the mapped `dso:state` property requires a user to perform much less actions (the selection of a property and the insertion of a value) in comparison to the use of the standard Blender tools, which require a user to change the colors of two components—a single change requires the selection of the color property and the insertion of its three RGB values.

In contrast to CrRs and RMs, which consist of OWL restrictions on classes, CpRs consist of RDF statements describing individuals of classes by properties.

5   Conclusions and Future Works

The SO-SEMIC environment leverages the semantic web techniques to simplify the 3D content creation process and enable creation of domain-specific objects and scenes without technical knowledge related to 3D graphics. Such simplification of 3D content creation is necessary to enable wider use of interactive 3D content on the web and is a prerequisite element of any sensible business model employing 3D applications. Similarly as writing a formatted text document does not require advanced knowledge about typesetting, font design and printer control commands, creation of 3D content does not have to require knowledge about meshes, normals and light reflection models.

The presented solution has several important advantages in comparison to the available approaches to 3D content creation. First, it combines manual content creation with the use of the semantic web techniques enabling declarative content creation. Second, it permits content creation at different levels of abstraction also taking into account hidden knowledge, which may be inferred and used in the modeling process. Third, the approach permits the reuse of common content elements in multi-user environments, which may be conceptually described with domain-specific ontologies and knowledge bases. Next, due to the conformance to the semantic web approach, the content created with SO-SEMIC is suitable for processing in web repositories. Finally, the SEMIC approach is platform- and standard-independent and it can be developed using diverse 3D content presentation tools, standards and programming languages. The predominance of SEMIC over other approaches to modeling 3D content in terms of functionality and complexity of 3D content representations has been shown in [14,20,21].

In SO-SEMIC, different OWL profiles [1] can be used, determining the level of complexity and decidability in content creation. The possible selection of one of the available profiles makes the OWL-based representations more predictable in terms of the computational time and the semantic problems that can be solved, in comparison to the representations based only on semantic rules.

Possible directions of future research incorporate several aspects. First, the proposed approach can be extended with semantic transformation of declarative descriptions of content behavior. Second, a general model of distributed 3D content including AR/VR services and a method of dynamic discovery of 3D content and semantic web services may be developed, e.g., as an extension of [29]. Third, SO-SEMIC can be evaluated with the focus on efficiency of semantic, manual 3D content creation.

Acknowledgments. This research work has been supported by the Polish National Science Centre (NCN) Grants No. DEC-2012/07/B/ST6/01523 and DEC-2014/12/T/ST6/00039.

References

1. Owl 2 web ontology language profiles (second edition) (2012). http://www.w3.org/TR/owl2-profiles/#Computational_Properties
2. Blenderartists.org (2013). http://blenderartists.org/forum/showthread.php?281307-A-web-UI-for-games-or-realtime-projects
3. Blender api (2015). http://www.blender.org/api/
4. Json encoder (2015). https://docs.python.org/2/library/json.html
5. Microsoft virtual academy (2015). http://www.microsoftvirtualacademy.com/training-courses/introduction-to-asp-net-mvc
6. Restlet api platform (2015). http://restlet.com/
7. urllib (2015). https://docs.python.org/2/library/urllib.html
8. Apache: Apache jena (2015). http://jena.apache.org/
9. De Troyer, O., Kleinermann, F., Pellens, B., Bille, W.: Conceptual modeling for virtual reality. In: Grundy, J., Hartmann, S., Laender, A.H.F., Maciaszek, L., Roddick, J.F. (eds.) Tutorials, Posters, Panels and Industrial Contributions at the 26th International Conference on Conceptual Modeling - ER 2007. CRPIT, vol. 83, pp. 3–18. ACS, Auckland (2007)

10. Flotyński, J., Walczak, K.: Semantic modelling of interactive 3D content. In: Proceedings of the 5th Joint Virtual Reality Conference, Paris, France, 11–13 December 2013
11. Flotyński, J., Walczak, K.: Semantic multi-layered design of interactive 3D presentations. In: Proceedings of the Federated Conference on Computer Science and Information Systems, Kraków, Poland, pp. 541–548. IEEE, 8–11 September 2013
12. Flotyński, J., Walczak, K.: Multi-platform semantic representation of 3D content. In: Proceedings of the 5th Doctoral Conference on Computing, Electrical and Industrial Systems, Lisbon, Portugal, 7–9 April 2014
13. Flotyński, J.: Harvesting of semantic metadata from distributed 3D web content. In: Proceedings of the 6th International Conference on Human System Interaction (HSI), Sopot (Poland), 06–08 June 2013. IEEE (2013)
14. Flotyński, J.: Semantic modelling of interactive 3D content with domain-specific ontologies. Procedia Comput. Sci. 35, 531–540 (2014). 18th International Conference on Knowledge-Based and Intelligent Information & Engineering Systems
15. Flotyński, J., Dalkowski, J., Walczak, K.: Building multi-platform 3D virtual museum exhibitions with flex-vr. In: The 18th International Conference on Virtual Systems and Multimedia, Milan, Italy, pp. 391–398, 2–5 September 2012
16. Flotyński, J., Walczak, K.: Attribute-based semantic descriptions of interactive 3D web content. In: Kiełtyka, L. (ed.) Information Technologies in Organizations - Management and Applications of Multimedia, pp. 111–138. TNOiK, Torun (2013)
17. Flotyński, Jakub, Walczak, Krzysztof: Conceptual semantic representation of 3D content. In: Abramowicz, Witold (ed.) BIS Workshops 2013. LNBIP, vol. 160, pp. 244–257. Springer, Heidelberg (2013)
18. Flotyński, J., Walczak, K.: Describing semantics of 3d web content with rdfa. In: The First International Conference on Building and Exploring Web Based Environments, Sevilla (Spain), 27 January–1 February 2013, pp. 63–68. ThinkMind (2013)
19. Flotyński, J., Walczak, K.: Microformat and microdata schemas for interactive 3D web content. In: Ganzha, M., Maciaszek, L., Paprzycki, M. (eds.) Proceedings of FEDCIS 2013, Kraków, Poland, 8–11 September 2013, vol. 1, pp. 549–556. PTI (2013)
20. Flotyński, J., Walczak, K.: Conceptual knowledge-based modeling of interactive 3D content. Vis. Comput. 35, 1–20 (2014)
21. Flotyński, J., Walczak, K.: Semantic representation of multi-platform 3D content. Comput. Sci. Inf. Sys. 11(4), 1555–1580 (2014)
22. Gutiérrez, M., García-Rojas, A., Thalmann, D., Vexo, F., Moccozet, L., Magnenat-Thalmann, N., Mortara, M., Spagnuolo, M.: An ontology of virtual humans: Incorporating semantics into human shapes. Vis. Comput. 23(3), 207–218 (2007)
23. Kalogerakis, E., Christodoulakis, S., Moumoutzis, N.: Coupling ontologies with graphics content for knowledge driven visualization. In: VR 2006 Proceedings of the IEEE conference on Virtual Reality, Alexandria, Virginia, USA, pp. 43–50, 25–29 March 2006
24. Latoschik, M.E., Biermann, P., Wachsmuth, I.: Knowledge in the loop: semantics representation for multimodal simulative environments. In: Butz, A., Fisher, B., Krüger, A., Olivier, P. (eds.) SG 2005. LNCS, vol. 3638, pp. 25–39. Springer, Heidelberg (2005)
25. Latoschik, M.E., Blach, R.: Semantic modelling for virtual worlds - a novel paradigm for realtime interactive systems? ACM VRST 2008, 17–20 (2008)
26. Luck, M., Aylett, R.: Applying artificial intelligence to virtual reality: Intelligent virtual environments. Appl. Artif. Intell. 14(1), 3–32 (2000)

27. Lugrin, J.L., Cavazza, M.: Making sense of virtual environments: Action represen-
    tation, grounding and common sense. In: Proceedings of the 12th International
    Conference on Intelligent User Interfaces, IUI 2007, pp. 225–234. ACM, New York
    (2007)
28. Pellens, B., Kleinermann, F., De Troyer, O.: A development environment using
    behavior patterns to facilitate building 3D/vr applications. In: Proceedings of the
    Sixth Australasian Conference on Interactive Entertainment, IE 2009, pp. 8:1–8:8.
    ACM (2009)
29. Rumiński, D., Walczak, K.: Dynamic composition of interactive ar scenes with the
    carl language. In: The 5th International Conference on Information, Intelligence,
    Systems and Applications, Chania, Greece, 7–9 July 2014, pp. 329–334. IEEE
    (2014)
30. Spagnuolo, M., Falcidieno, B.: 3D media and the semantic web. IEEE Intell. Syst.
    24(2), 90–96 (2009)
31. Walczak, K.: Flex-vr: configurable 3D web applications. In: Proceedings of the
    Conference on Human System Interactions, pp. 135–140. IEEE (2008)
32. Walczak, K.: Beh-VR: modeling behavior of dynamic virtual reality contents. In:
    Zha, H., Pan, Z., Thwaites, H., Addison, A.C., Forte, M. (eds.) VSMM 2006. LNCS,
    vol. 4270, pp. 40–51. Springer, Heidelberg (2006)
33. Walczak, K., Cellary, W., White, M.: Virtual museum exhibitions. Computer **39**(3),
    93–95 (2006)
34. White, M., Mourkoussis, N., Darcy, J., Petridis, P., Liarokapis, F., Lister, P.F., Wal-
    czak, K., Wojciechowski, R., Cellary, W., Chmielewski, J., Stawniak, M., Wiza, W.,
    Patel, M., Stevenson, J., Manley, J., Giorgini, F., Sayd, P., Gaspard, F.: Arco - an
    architecture for digitization, management and presentation of virtual exhibitions.
    In: Computer Graphics Inter., pp. 622–625. IEEE Computer Society (2004)

Modeling and Querying Sensor Services Using Ontologies

Sana Baccar[1], Wassim Derguech[2]([✉]), Edwards Curry[2], and Mohamed Abid[1]

[1] CES Research Unit, National School of Engineers of Sfax, Sfax, Tunisia
sana.baccar@ceslab.org, mohamed.abid_ces@yahoo.fr
[2] Insight Centre for Data Analytics, National University of Ireland, Galway, Ireland
{wassim.derguech,edward.curry}@insight-centre.org

Abstract. We propose in this paper a service description meta-model for describing services from a functional and non-functional perspectives. The model is inspired from the frame based modeling technique and is serialized in RDF (Resource Description Framework) using Linked Data principles. We apply this model for describing sensor services: modeling sensors and their readings enriched with non-functional properties. We also define a complete architecture for managing sensor data: collection, conversion, enrichment and storage. We tested our prototype using live streams of sensors readings. The paper also reports on the required time and storage size during the management and querying of sensor data.

Keywords: Service modeling · RDF · Linked data · Functional properties · Non-functional properties · Sensors

1 Introduction

Over the last years, there is a considerable interest in the Linked Data [1], which is considered as a best practice for publishing and interconnecting structured data on the Web. Recently, the service computing community tries to benefit from the Linked Data principles to overcome the limitations of the existing Semantic Web Services (SWS) approaches and make services easily interconnected, exchangeable and manageable [2]. Indeed, most of the current SWS approaches add semantics to the service by linking its syntactic operations and messages with some concepts of an external ontology. Although usable in some case, such description is still not able to offer complete information about the implicit knowledge, in addition to the difficulties encountered when it comes to re-use or compose one-to-many parts of the service.

To overcome the previous challenges, a new service description model based on the Linked Data principles is proposed in this paper. Our model describes a service as a structured entity featured via a set of capabilities (functional) and non-functional features. Such description considers a service as an access mechanism to a capability, which is, in its turn, a structured entity that describes what a service can do via an action verb and set of domain-specific attributes.

© Springer International Publishing Switzerland 2015
W. Abramowicz (Ed.): BIS 2015, LNBIP 208, pp. 90–101, 2015.
DOI: 10.1007/978-3-319-19027-3_8

In order to evaluate the applicability of our model, we propose to use a Wireless Sensor Networks (WSNs) environment as a running example throughout this paper. The main reason of choosing such smart environment is its highly configurable and heterogeneous features. In fact, integrating semantics into WSNs is considered as a step toward the understanding, management and use of sensor-based data sources. This challenging integration includes various issues that are derived from the constrained nature of the sensor nodes (variability, unreliability and heterogeneity), their multiple domains and the need for multiple management of sensors through uncoordinated queries. Another aspect required for an effective analysis and decision making over sensor data is the integration of other sources such as enterprise and open data. For example: linking a sensor with its location (data coming from the enterprise) helps to take timely decisions tight to the location of the sensor (e.g., detecting high water usage in a particular location).

This work comes in the context of Waternomics project [3] aiming to create a Linked Water Dataspace [4,5] as an emerging information management approach for collecting, standardizing, enriching and linking water usage data coming from sensors. Applying our proposed model and the proposed sensor data management infrastructure, facilitates to integrate various data sources for effective decision making.

In summary, the contributions of this work are:

- Propose an RDF-based Linked Data model for service description that facilitates service ex-changeability, discovery and composition processes. The model describes various aspects of a service (i.e., functional and non-functional).
- Developing a data integration module for generating RDF-based description of sensor service and interlinking them with enterprise data.
- Validating the proposed service model in a real-time Wireless Sensor Networks (WSNs) by defining sensor services and presenting them as reusable services, extensible and accessible for processing and management.

The remainder of the paper is organized as follows. Section 2 presents our conceptual model for describing services by detailing its different components. Section 3 describes our infrastructure for managing sensor data including: capturing, transforming, enriching, publishing and aggregating. In Sect. 4 we provide qualitative evaluation of our developed system. Before concluding in Sect. 6, we discusses some related works in Sect. 5.

2   Linked Data Model for Describing Web Services

At a high level, we share the same vision of OASIS Reference Model[1] and consider a service as an access mechanism to a *capability* under certain requirements (i.e., NFP properties). In the following, we describe these two concepts in detail.

---

[1] OASIS Reference Model for Service Oriented Architecture 1.0, http://www.oasisopen.org/committees/download.php/19679/soa-rm-cs.pdf.

2.1   Capability

We adopt the definition of capability as defined in [6,7]: "*A capability defines what a program (e.g., a service, a business process, a sensor) does from a functional perspective*". We also adopt their proposed conceptual model that represents capabilities as an action verb and a set of domain related attributes.

Actually, the researchers proposed in [6,7] a new conceptual model that describes functional capabilities as an action verb and set of domain specific attributes captured as an RDF-schema. Such description enables representing capabilities at several levels of abstraction, from the most abstract capability, named capability category, to the most concrete one, named capability offer, by explicitly extracting the relation between capabilities (i.e., specify and extend) and linking between the different levels. Such model deals with capabilities at different abstraction levels in a uniform way. Which takes into account attributes dynamicity, sensitivity and inter-dependency, as well as capability re-usability. This capability meta-model is formalized as an RDF vocabulary[2] which makes it a perfect fit to our work. We do not further elaborate on this meta-model as it has been proved to be good in practice [6,7]. However, our main contribution is centered towards the addition of non-functional properties that are described next (Fig. 1).

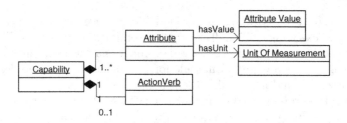

Fig. 1. Capability meta-model

2.2   Non-functional Properties (NFPs)

As there will be many functionally-equivalent capabilities for a specific user request, we propose to consider the non-functional aspect as part of the service description. Such extension leads mainly to filter and select the best offer among these capability offers to better fulfill customer requirements and ensure his satisfaction. Separating the capability from the non functional constraints leads to reuse capabilities under different constraints. Moreover, in their turn, these constraints can be reused by different capabilities.

As it is shown in Fig. 2, the NFP concept contains four types of NFPs: (i) *user-generated* properties, e.g., reputation, (ii) *policy-related* properties e.g., cost

---

[2] Available at: www.deri.ie/cap as accessed on 06/06/2014.

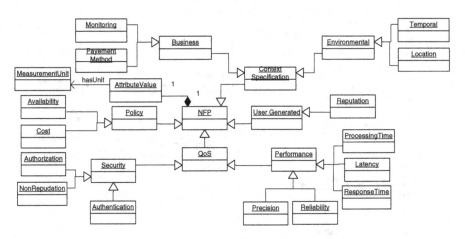

Fig. 2. Non-functional properties meta-model

and availability, (iii) *QoS* properties e.g., security and trust, response time, latency and reliability and (iv) *contextual specifications* e.g., location or business requirements. The NFPs are described by a combination of non-functional (NF) terms (attribute, metrics) and their constraints (attribute values). Such NFP properties will ensure that a decision support model that uses data from a particular service is built with respect to these NFPs.

3 RDF-Data Management and Warehousing

A dataspace, as it is understood in the Waternomics project, is a data integration architecture. It allows integrating data from multiple sources into a single space. A dataspace for Waternomics is not only hosting sensor data but also other relevant data for decision analytics. Relevant data includes: sensor and location meta-data, weather data as well as other relevant data sources identified in the project. In this paper, we are interested in sensor and location meta-data that needs to be standardised, interlinked and published in order to facilitate its reuse internally (i.e., enrichment, aggregations, etc.) or externally (i.e., user or corporate applications). The following sections describe the process of collecting and transforming sensor data to RDF (in Sect. 3.1) then aggregating and storing this RDF data (in Sect. 3.2).

3.1 From Raw Sensor Data to RDF

For this paper, we assume that location meta-data (i.e., data describing the location of sensors) is available in RDF in our dataspace and is constantly updated using a collaborative approach, that we have previously experienced [8]. This location data is actually captured in our model as NFP properties for sensors as in essence they report only the location which is not part of the function of sensors.

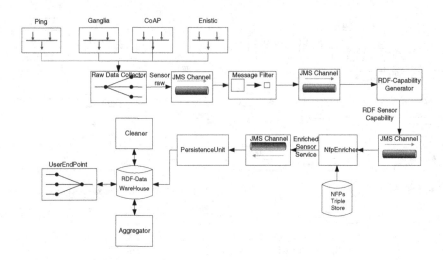

Fig. 3. Data management and warehousing process

In this section, we focus on sensor data management and propose the architecture depicted in Fig. 3 for this purpose. This figure depicts the entire process of collecting, transforming, enriching, publishing and aggregating sensor data.

The main components of the process depicted in Fig. 3 are:

- *Raw Data Collectors* capture incoming data from heterogeneous data sources and publish them in plain text to a Java Message Service (JSM) channel. They are implemented as Universal Datagram Protocol (UDP) listeners, Constrained Application Protocol (CoAP) clients and RESTfull services, which periodically pull sensors data.
- The *Message Filter* is useful to refine and manage the flow of the incoming messages from the Raw Data Collector. It receives the collected messages and filters the irrelevant and unnecessary information in a way that enables the sensed data to be used on-the-fly for further processing.
- The *RDF-Capability transformation generator* identifies possible triples in the event-data, assigns a URI to each identified subject/predicate and represents event-data in RDF. Lines 1 to 10 of Listing 1 show an example of RDF-Capability in N3. In this RDF graph, *sensors:Sensor_0D6F00945B23* is a sensor that has the capability *sensors:SensingCap_d940a313* reporting on its sensed property and its observation time.
- The *Nfp Enricher* is responsible for enriching the service with the necessary non-functional properties retrieved from the Location meta-data store. After extracting the appropriate NFPs, the enricher merges the service capabilities and its adequate NFPs into a single RDF model that represents the entire service description. Listing 1 shows the result of the entire RDF transformation process where lines 12 to 15 show the NFP enrichment part.

Please note that for Listing 1, we are using multiple prefixes for the following:

- sensors: refers to the base URI for identifying sensors and their capabilities.
- snd: stands for Sensor Network Domain, is our customized domain ontology that captures the capability attributes.
- ssn: Semantic Sensor Network Ontology[3]. We use SSN as it is compliant to the W3C and OGC (Open Geospatial Consortium) standard SensorML.
- om: The Ontology of units of Measure and related concepts (OM). It models concepts and relations important to scientific research. It has a strong focus on units and quantities, measurements, and dimensions.

Listing 1. Event-Data converted to RDF

```
1   sensors:Sensor_0D6F00945B23    a    snd:SensingService  ;
2   cap:hasCapability  sensors:SensingCap_d940a313  .
3
4   sensors:SensingCap_d940a313    a    cap:Capability  ;
5   cap:hasActionVerb snd:Sensing  ;
6   snd:hasObservationValue
7   [snd:hasUnit om:cubic_metre_per_second-time  ;
8   ssn:hasValue  ''1.04''];
9   snd:hasObservedProperty   om:Volumetric_flow_rate  ;
10  ssn:endTime ''1396609412296''^xsd:long  .
11
12  sensors:Sensor_0D6F00945B23 a snd:WaterVolumicFlowRateSensor;
13  ssn:featureOfInterest deriRooms:Kitchen  ;
14  ssn:observes om:Volumetric_flow_rate  ;
15  ssn:onPlatform sensors:a209be2b12e686b1b8  .
```

3.2   Data Warehousing

After a successful data transformation into RDF, we propose to store it into a data warehouse for further processing. Components that are manipulating this data store include:

- Data-Persistence Unit: We propose in our implementation the use of Open-Link Virtuos[4] as data store. A dedicated data-persistence unit is continuously receiving enriched sensor data and saves it to the data store.
- Aggregator: Most of the sensors are frequently sending their readings. For example a water flow sensor is sending a flow rate reading each 10 s. It is, however, not necessary to keep every single reading in our data warehouse. In order to avoid handling a large warehouse containing every sensor reading, we propose to aggregate them into a single reading covering a longer period. This operation is done through the aggregator that computes the Sum, Average, Max, Min, etc. of the readings of every sensor. In other words, the aggregator removes all the capabilities of a given sensor that are collected for a predefined period and replaces them by a single aggregated capability.

---

[3] http://www.w3.org/2005/Incubator/ssn/ssnx/ssn.

[4] http://virtuoso.openlinksw.com as accessed on 06/06/2014.

– Data Warehouse Cleaner: Reading and identifying the required data from large
data Warehouse is complex and extensive, both in memory size and time consumption. The cleaner is responsible for limiting the size of the Warehouse and
ensuring a good response time for data access, identification, and management
by removing the old backups and maintaining the recent stored data.

The previously described components have been developed separately, communicate via a JMS server and store the data into the proposed RDF warehouse.
We use ActiveMQ [9] as JMS server. It is a message oriented middleware that
provides simple and easy to use methods to send, receive and handle JMS messages on the JMS channels. It is used in our system as a message exchange server
using publish/subscribe protocols. These JMS messages are then broadcast to
components that subscribe to those topics. The developed system is used for
evaluation purposes reported in the following section.

4   Evaluation

For evaluating the developed system, we used a dedicated server running 64-bit
Windows7 OS, with 4 GB of Ram and an Intel Core i5 (2.66 GHz) CPU. We
use for this evaluation a real-time sensor event collection and publishing them
over ActiveMQ channels. We conducted two quantitative evaluations related to
the required storage size and execution time for capturing, converting, enriching,
storing and retrieving sensor data.

The first experiment that we conducted consists of evaluating the execution
time required for managing sensor data. As it is shown in Fig. 4(a), the required
execution time for executing the entire management process presented previously in Fig. 3 increases linearly to the sensor event frequency. This indicator is
important to consider in managing environments with large number of sensors
and sensor readings, however, due to the fact that we are using a relatively small
environment this issue can not be considered as a problem.

The second evaluation carried out consists of showing the impact of the raw
event data, sensor capability and sensor service on the size of the data warehouse.
As shown in Fig. 4(b), the size of the processed data depends on the size of the
incoming events from each sensor, the frequency of the events and the number
of the non-functional properties. We notice that there is not a big difference in
size after the enrichment step as most of the sensors that we consider share the
same non functional properties that are already stored in the data warehouse
and are not duplicated for each new sensor reading. Furthermore, we notice
a considerable difference when comparing the size of raw sensor readings with
the RDFized data. This is obviously expected as we are generating a complete
description using additional enterprise data during the enrichment step.

As the size of the warehouse is critical either for resource management or
search query processing, we propose the aggregation module that helps to reduce
the size of the data warehouse. Using this aggregator, we reach up to 71 % of
compression rate as shown in Fig. 4(c). Moreover, the difference in size between

the raw sensor readings and the aggregated sensor readings is not big because of the compression rate gained during the aggregation. Even though we have lost the entire readings by aggregating them into a single entry, we gained by having a semantically described and reduced set of sensor readings which also help reducing search queries processing time. Indeed, the proposed approach allows describing heterogeneous sensors uniformly which helps in simplifying the querying process of sensor data.

Finally, to test the performance of our system in data mining and discovery, we define a set of different queries (shown in Table 1), we execute them on different number of triples stored in the data warehouse and we observe the required query execution time. As we can see in Fig. 4(d), the variations in query execution time is due to variation in the number of triples stored in the data warehouse. As an example, the time taken to process a query is between 509 and 537 ms for around 450 RDF-triples; 528 and 546 ms for 1033 triples; 537 and 548 ms for 2376 triples and 556 and 577 ms for 3350 triples. The system takes between 24 and 50 ms to retrieve required data from 2904 triples. We notice that there was a significant difference in the query execution time when considering all sensor services and after the aggregation process is executed.

Table 1. Evaluation queries

Q1	Search per 12 min the average of the water-usage in the kitchen
Q2	Search per minute the sum of the water-usage the showers
Q3	Search the min and the max of the water-usage captured in shawers during the last 10 min

After a random selection and verification of generated capabilities we notice that the system is operating as expected. It is important to notice that the use of a triple store as a data warehouse helps reduce the processing time and complexity of data in contrast to a classical database system. Indeed, OpenLink Virtuoso[5] was hiding the complexity of the data storing process for example when handling duplicate triples.

In conclusion, using our proposed modeling approach in a small smart environment was effective in producing standardized data using RDF. Even though our approach drastically increases the size of raw data, it is still stored in an acceptable size and processed and queried in a reasonable time.

5   Related Work

The work presented in this paper relates to two main research areas: (i) Semantic Web Service Description and (ii) Sensor data integration and management.

---

[5] http://virtuoso.openlinksw.com as accessed on 06/06/2014.

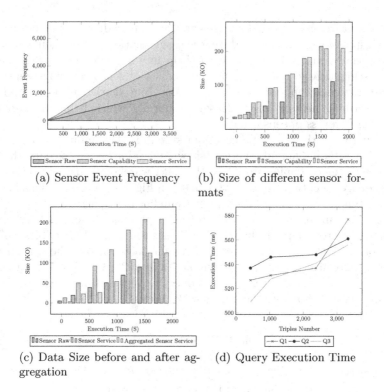

(a) Sensor Event Frequency

(b) Size of different sensor formats

(c) Data Size before and after aggregation

(d) Query Execution Time

Fig. 4. Performance charts

5.1 Semantic Web Service Description

Recently, several ontologies and languages have been proposed to describe Web services using semantics. SA-WSDL[6] [10] uses ontologies to enrich WSDL descriptions and XML Schema of Web services with semantic annotations. The effort of this contribution was put towards modelling services as invocation interfaces which relate mainly to the service implementation. However, in this paper, we consider a service capability as an entity featured via a set of attributes. Using such definition, we share the same vision of OASIS Reference Model that considers a service as an access mechanism to a capability. Within this vision, the invocation interface is only one aspect of the whole service description.

In a more refined fashion, languages such as OWL-S [11], WSMO [12], SWSO[7], provide a semantic description of Web services. These approaches do not go beyond the classical Input, Output, Precondition and Effect paradigm to define services descriptions. These contributions focus more on the machine processing of service descriptions while ignoring end-users intuitive way of describing their needs (i.e., attributes and their values). We adopt in our work

---

[6] http://www.w3.org/TR/sawsdl/.
[7] http://www.w3.org/Submission/SWSF-SWSO/.

the attribute-featured approach for describing services for serving both machine processing and user-centricity.

Sangers et al. [13] propose to give more flexibility to users in searching services while using natural language text and approximate semantic matching. This technique is useful when queries are created without knowing in advance service descriptions. This is not the object in our work, however, our plan is to use the approximate semantic matching of events for the internet of things explored in [14].

5.2   Sensor Data Integration and Management

Managing smart environments requires extensive efforts for collecting, integrating and analysing sensor data. An important step towards an efficient smart environment management is the standardisation of sensor data. Using RDF in such context is a solution that has been also adopted in [15] and [16]. Here, authors specifically address the problem of managing sensor data and propose to transform sensor readings into RDF and generate data cubes on-the-fly from syntactic sensor data. Contrary to our contribution, in these works all the event features are considered as data cube parameters and there is no separation between the sensor capability and its non-functional properties.

Moreover, since event processing requires providing an immediate response over the continuous stream of sensor data, a system that stores generated events and provides fast response to complex queries is needed [17]. To handle such requirements, various database systems have been adopted such as in [18,19]. The proposed systems, have several critical shortcomings that prevent using them directly to process sensor data. In fact, they are too heavyweight and slow, devoting much complexity to handling queries that are based on indexing and query forwarding techniques. To overcome such gaps, we propose in this paper a system that uses an RDF data-warehouse for managing sensor events. Indeed, our data-warehouse is a large store of RDF triples containing both live and historical aggregated data that overcomes the rigidity of a classical database.

Similar to [20], we use SSN in our sensors descriptions. However, [20] go beyond this and propose a naming convention and a data distribution mechanism for a large-scale sensor-based environment.

6   Conclusion

Data integration and processing from sensor services is a challenging issue that requires data integration, standardisation, linking and enrichment. We proposed in this paper a meta model for describing sensors and their produced data using RDF. Our model explicitly differentiates between sensor service functional and non-functional aspects. We defined and implemented an architecture for manipulating sensor data in order to serve as a base for further decision support and data analysis. Using RDF and Linked Data principle in such context helps building a large linked data cloud that makes our data set together with other open sets as a single database.

The proposed system has been tested on a real world scenario showing its applicability and its potential use in data mining and service discovery process. Our ongoing work focuses on extending our system to handle service composition task and support complex queries. However, from sensor service modeling perspective, using capability description in its original form is insufficient when it comes to the composition and planning scenarios. Indeed, we need two main additional attributes that capture the state/change of the world before/after executing the corresponding action. For that, we propose an extension to the introduced capability meta-model (precondition and effect) so that the semantics of capability is captured at two levels: (i) Coarse-grain level: handles the discovery process through a combination of an action verb, domain specific attributes and semantic links between capabilities. (ii) Fine-grain level: handles the composition process via a set of preconditions and effects. The evaluation of the actual meta-model is also part of our future plan.

From an architectural point of view, the main problem with our proposed approach is the fact that it is not designed for large scale environments. As part of the Waternomics project, we are interested in processing real-time data in large environments and this leads us to investigate other architectural styles that help overcome this problem. We are particularly interested in investigating the Lambda Architecture for our future plan. We also plan to extend this architecture for generating additional links between our data store to open data for more knowledge discovery.

Acknowledgments. The research leading to these results has received funding under the European Commission's Seventh Framework Programme from ICT grant agreement WATERNOMICS no. 619660. It is supported in part by Science Foundation Ireland (SFI) under Grant Number SFI/12/RC/2289.

References

1. Bizer, C., Heath, T., Berners-Lee, T.: Linked data - the story so far. Int. J. Semant. Web Inf. Syst. **5**(3), 1–22 (2009)
2. Heath, T., Bizer, C.: Linked Data: Evolving the Web into a Global Data Space. Synthesis Lectures on the Semantic Web. Morgan & Claypool Publishers, San Rafael (2011)
3. Clifford, E., Coakley, D., Curry, E., Degeler, V., Costa, A., Messervey, T., Andel, S.J.V., de Giesen, N.V., Kouroupetroglou, C., Mink, J., Smit, S.: Interactive water services: the waternomics approach. In: 16th International Conference Water Distribution Systems Analysis (WSDA 2014). Elsevier (2014)
4. Curry, E.: System of systems information interoperability using a linked dataspace. In: IEEE 7th International Conference on System of Systems Engineering (SOSE 2012), Genoa, Italy, pp. 101–106. IEEE (2012)
5. Curry, E., Hasan, S., O'Riáin, S.: Enterprise energy management using a linked dataspace for energy intelligence. In: The Second IFIP Conference on Sustainable Internet and ICT for Sustainability (SustainIT 2012), Pisa, Italy. IEEE (2012)

6. Bhiri, S., Derguech, W., Zaremba, M.: Modelling capabilities as attribute-featured entities. In: Cordeiro, J., Krempels, K.-H. (eds.) WEBIST 2012. LNBIP, vol. 140, pp. 70–85. Springer, Heidelberg (2013)

7. Derguech, W., Bhiri, S., Hasan, S., Curry, E.: Using formal concept analysis for organizing and discovering sensor capabilities. Comput. J. **58**(3), 356–367 (2014)

8. Hassan, U., Bassora, M., Vahid, A., O'Riain, S., Curry, E.: A collaborative approach for metadata management for internet of things: linking micro tasks with physical objects. In: Collaborative Computing: Networking, Applications and Worksharing (Collaboratecom), pp. 593–598, October 2013

9. Snyder, B., Bosanac, D., Davies, R.: ActiveMQ in Action. Manning Publications Co., Greenwich (2011)

10. Saquicela, V., Blzquez, L.M.V., Corcho, S.: Adding semantic annotations into (geospatial) restful services. Int. J. Semant. Web Inf. Syst. **8**(2), 51–71 (2012)

11. Martin, D., Paolucci, M., Wagner, M.: Bringing semantic annotations to web services: OWL-S from the SAWSDL perspective. In: Aberer, K., et al. (eds.) ASWC 2007 and ISWC 2007. LNCS, vol. 4825, pp. 340–352. Springer, Heidelberg (2007)

12. Roman, D., de Bruijn, J., Mocan, A., Lausen, H., Domingue, J., Bussler, C.J., Fensel, D.: WWW: WSMO, WSML, and WSMX in a nutshell. In: Mizoguchi, R., Shi, Z.-Z., Giunchiglia, F. (eds.) ASWC 2006. LNCS, vol. 4185, pp. 516–522. Springer, Heidelberg (2006)

13. Sangers, J., Frasincar, F., Hogenboom, F., Chepegin, V.I.: Semantic web service discovery using natural language processing techniques. Expert Syst. Appl. **40**(11), 4660–4671 (2013)

14. Hasan, S., Curry, E.: Approximate semantic matching of events for the internet of things. ACM Trans. Internet Technol. **14**(1), 2:1–2:23 (2014)

15. Mehdi, M., Sahay, R., Derguech, W., Curry, E.: On-the-fly generation of multidimensional data cubes for web of things. In: IDEAS, pp. 28–37 (2013)

16. Lefort, L., Bobruk, J., Haller, A., Taylor, K., Woolf, A.: A linked sensor data cube for a 100 year homogenised daily temperature dataset. In: SSN, pp. 1–16 (2012)

17. Balazinska, M., Deshpande, A., Franklin, M.J., Gibbons, P.B., Gray, J., Hansen, M., Liebhold, M., Nath, S., Szalay, A., Tao, V.: Data management in the worldwide sensor web. IEEE Pervasive Comput. **6**(2), 30–40 (2007)

18. Diao, Y., Ganesan, D., Mathur, G., Shenoy, P.J.: Rethinking data management for storage-centric sensor networks. In: CIDR, pp. 22–31 (2007)

19. Madden, S.R., Franklin, M.J., Hellerstein, J.M., Hong, W.: Tinydb: an acquisitional query processing system for sensor networks. ACM Trans. Database Syst. **30**(1), 122–173 (2005)

20. Barnaghi, P.M., Wang, W., Dong, L., Wang, C.: A linked-data model for semantic sensor streams. In: Green Computing and Communications (GreenCom), 2013 IEEE and Internet of Things (iThings/CPSCom), IEEE International Conference on and IEEE Cyber, Physical and Social Computing, pp. 468–475. IEEE (2013)

Validation and Calibration of Dietary Intake in Chronic Kidney Disease: An Ontological Approach

Yu-Liang Chi[1(✉)], Tsang-Yao Chen[1], and Wan-Ting Tsai[2]

[1] Department of Information Management, Chung Yuan Christian University,
200 Chung-Pei Rd., Chung-Li 32023, Taiwan
{maxchi,polochen}@cycu.edu.tw

[2] Department of Management Information Systems, National Chengchi University,
No.64, Sec. 2, Zhinan Rd., Wenshan District, Taipei City 11605, Taiwan
lhya623@gmail.com

Abstract. This study develops a pilot knowledge-based system (KBS) for addressing validation and calibration of dietary intake in chronic kidney disease (CKD). The system is constructed by using Web Ontology Language (OWL) and Semantic Web Rule Language (SWRL) to demonstrate how a KBS approach can achieve sound problem solving modeling and effective knowledge inference. In terms of experimental evaluation, data from 36 case patients are used for testing. The evaluation results show that, excluding the interference factors and certain non-medical reasons, the system has achieved the research goal of CKD dietary consultation. For future studies, the problem solving scope can be expanded to implement a more comprehensive dietary consultation system.

Keywords: Knowledge-based system · Chronic kidney disease · Ontology · Semantic rules

1 Introduction

The care for patients with chronic kidney disease (CKD) is closely related to the patient's daily diet management. The type, quantity, and nutrient content of the patient's food intake need to be strictly controlled [2]. However, dietary management involves complicated interactions among various factors. This complexity not only reduces the quality of dietary management, but can also consume medical resources if the patients are to have constant dietary consultation. As a result, CKD patients often do not receive enough professional guidance in dietary control, which can lead to disease progression, low life quality, and even malnutrition. Traditional CKD dietary consultation requires dietitians to perform a series of steps, such as collecting patient physical profile and biomedical examination data as baseline information; calculating patient's clinical CKD stage, nutrient baselines, and suggested servings from each food group; comparing the patient's actual diet to the suggested servings for dietary adjustment. These consultation steps involve intensive knowledge from different sources. The dietitian then considers the complicated logical relationships between the patient's conditions and the various knowledge-intensive sources in order to provide dietary guidance.

© Springer International Publishing Switzerland 2015
W. Abramowicz (Ed.): BIS 2015, LNBIP 208, pp. 102–112, 2015.
DOI: 10.1007/978-3-319-19027-3_9

This study employs an ontological engineering approach with focus on the construction of knowledge models and knowledge reasoning with logical rules. Ontology has been widely adopted in various fields of study to model and construct taxonomies for domains of interest [7]. Over time, ontology is seen as a synonym of "conceptual model" [8]. In information science research, especially in the areas of artificial intelligence, ontologies are created with properties and relationships to enable knowledge inference. In knowledge-based systems (KBS), concepts are used not just as terms, but also as computable objects with logical definitions, which enable knowledge for inductive and deductive reasoning [9]. The W3C recommended the OWL (Web Ontology Language) as a formal specification for ontology representation [10]. In terms of development tools, Protégé is a prevalent platform created by the Stanford Center for Biomedical Informatics Research for OWL-based ontology development, OWL-based problem solving modeling, and KBS execution [11]. In addition to conceptual representation of ontology, SWRL (Semantic Web Rule Language) can be used to develop semantic rules in the instance layer of the ontology to enable reasoning using rule inference engines [12, 13].

This study is collaboration between dietitians and knowledge engineers. The first task in knowledge engineering is to decompose and reassemble professional knowledge content in order to analyze the concepts and data required for modeling. Through this process, the nature of the problem, the scope of the knowledge domain, and the logical relationship between concepts are clarified. Then, the task is followed by the construction of a conceptual taxonomy of the domain and by the definition of conceptual properties for inference. Last, semantic rules are developed to create an "instance layer" for knowledge inference. The major knowledge-based elements designed in this process include: (1) using generalized knowledge sources to construct a domain ontology that consists of common constructs, concepts, and instances with super-subordinates and inheritance created using the "*is-a*" hierarchical relationships; (2) to enable problem-solving, the development of an objective-oriented task ontology with a "*has-a*" expression to describe the logical relationship of subsumption and composition aggregation among concepts; and (3) definition of the problem-solving steps in the instance layer with semantic rules to infer implicit knowledge based on known factual knowledge.

2 Problem Analysis

Chronic kidney disease is caused by the biomedical abnormalities in the kidneys [3]. The function of the kidneys is to metabolize nitrogenous waste (such as uric acid) in order to maintain the body's balance of minerals (sodium, potassium, phosphorus, etc.) and to assist in blood pressure control and blood cell production [4]. The generally accepted operational definition of CKD is kidney damage and the kidneys' inability to filter blood as measured by Glomerular Filtration Rate (GFR) [1]. The clinical CKD stages are then defined by plotting the estimated GFR (eGFR) [5, 6] as shown in Table 1. In the calculation of eGFR, the Modification of Diet in Renal Disease (MDRD) equation is widely adopted by organizations such as the United States NKF.

Table 1. Clinical stages of chronic kidney disease.

Stage	Description	GFR
1	Kidney damage with normal or increased GFR	≥ 90
2	Kidney damage with mild decrease in GFR	$60 \sim 89$
3	Moderate decrease in GFR	$30 \sim 59$
4	Severe decrease in GFR	$15 \sim 29$
5	Kidney failure	<15

Common CKD knowledge sources used in clinical dietary consultation include clinical stage definition, stage estimation equation, and nutrient restriction. Nutrient knowledge, as a concept, includes at least food groups and the nutrient composition of each food item. The food group includes the categorization of food items and their recommended daily servings in different conditions. Case patient data work as a trigger to interact with the knowledge sources to create dietary suggestions. To implement the problem scenarios analysis, dietitians and knowledge engineers have participated as consultants for verifying knowledge sources and problem scenarios from the beginning. They formalized the problem solving into two groups of non-logical axioms as follows:

1. Examining suggested food group servings against a patient's food combination: The first step is calculating suggested food group servings by individual conditions. Equation (1) is the general equation for the suggested servings $FS_{(p, i, s)}$ in food groups by case (p), calorie level (i), and CKD stage (s) to find the suggested corresponding servings in each food group. The second step is calculating the case patient's food combination intake in each food group. Equation (2) is by case (p) $FSI_{(p)}$ to find the intake corresponding servings in each food group. The last step is calculating the difference between Eqs. (1) and (2) (i.e. of $FSI_{(p)} - FS_{(p, i, s)}$). Equation (3) examines the differences in food group servings between the case patient's food combination and the suggested values. For example, $f_{(i, grains, servings)}$ denotes the number of grain servings in food item i.

$$FS_{(p, i, stage)} = \left(S_{grains}, S_{meat}, S_{milk}, S_{vegetable}, S_{fruit}, S_{oil\ \&nuts} \right) \qquad (1)$$

$$\left(\begin{array}{c} \sum_{i=1}^{n} (f_{i,grains,servings}), \sum_{i=1}^{n} (f_{i,meat,servings}), \sum_{i=1}^{n} f_{i,vegetable,servings}), \sum_{i=1}^{n} (f_{i,Potassium} * s_i), \\ \sum_{i=1}^{n} (f_{i,fruit,servings}), \sum_{i=1}^{n} (f_{i,oil\&nuts,servings}) \end{array} \right) \qquad (2)$$

$$
Case_{(p,food\_group)} = \begin{pmatrix} (\sum_{i=1}^{n} (f_{i,grains,servings}) - S_{grains}), (\sum_{i=1}^{n} (f_{i,meat,servings}) - S_{meat}), \\ (\sum_{i=1}^{n} (f_{i,vetetable,servings}) - S_{vegetable}), (\sum_{i=1}^{n} f_{i,milk,servings}) - S_{milk}), \\ (\sum_{i=1}^{n} (f_{i,fruit,servings}) - S_{fruit}), (\sum_{i=1}^{n} (f_{i,oil\&nuts,servings}) - S_{oil\&nuts}) \end{pmatrix} \quad (3)
$$

2. Examining suggested key nutrient intakes against a patient's food combination: The first step is calculating suggested key nutrient restrictions $NR_{(p)}$ by individual conditions. CKD patients require sufficient calories, but have to restrict the intake of certain key nutrients: proteins, phosphorus, potassium, and sodium. Equation (4) is a general equation for calculating the patient's suggested key nutrient intakes. The second step is calculating the case patient's food combination intake in each key nutrient. Equation (5) is by case (p) $NI_{(p)}$ to find the intake corresponding amounts in each key nutrient. The last step is calculating the difference between Eqs. (5) and (4) (i.e. of $NI_{(p)} - NR_{(p)}$). Equation (6) examines the differences of key nutrient intake between the patient's food combination and the suggested key nutrient intakes. For example, $f_{(i,Calories)}$ denotes the calorie intake per serving of food item i. The value is then multiplied by the number of intake servings (s_i) to obtain the total calories, and then subtracted by the suggested $N_{Calories}$ to calculate the difference.

$$
NR_p = (N'_{stage,Calorie} \times w_p, N_{Calorie}, N_{Protein}, N_{Phosphorus},
$$
$$
N_{Potassium}, N_{Sodium}) \quad (4)
$$

$$
NI_{(p)} = \begin{pmatrix} \sum_{i=1}^{n} (f_{i,Calorie} * s_i), \sum_{i=1}^{n} (f_{i,Protein} * s_i), \sum_{i=1}^{n} (f_{i,Phosphorus} * s_i), \\ \sum_{i=1}^{n} (f_{i,Potassium} * s_i), \sum_{i=1}^{n} (f_{i,Sodium} * s_i) \end{pmatrix} \quad (5)
$$

$$
Case_{(p,nutrients)} = \begin{pmatrix} (\sum_{i=1}^{n} (f_{i,Calorie} * s_i) - N_{Calorie}), (\sum_{i=1}^{n} (f_{i,Protein} * s_i) - N_{Proein}), \\ \sum_{i=1}^{n} (f_{i,Phosphorus} * s_i) - N_{Phosphorus}), \sum_{i=1}^{n} (f_{i,Potassium} * s_i) - N_{Potassium}), \\ (\sum_{i=1}^{n} (f_{i,Sodium} * s_i) - N_{Sodium}) \end{pmatrix}
$$
$$
(6)
$$

3 OWL-Based KBS

This study uses ontological engineering for knowledge modeling. An ontological knowledge model is an abstract structure of concepts, in which each concept has properties and relationships to represent the knowledge connotations. Knowledge models, when extensively analyzed and defined, are robust and extensive. New factual knowledge can be stored into existing knowledge structures, and the existing logical relationships are inherited for reasoning with no data processing required. Therefore, knowledge systems are suitable for solving knowledge-intensive problems that require logical inference. Based on the guidelines proposed by researchers [8, 14–16], we design three components in the KBS including construction of domain ontology, task ontology and semantic rules.

3.1 Construction of Domain Ontology

Domain ontology consists of a general conceptual structure and instances and uses "*is-a*" relation to establish parent-child hierarchy subordination relations with the terminal concepts being further elaborated by giving instances. Therefore, domain ontology is a typological structure which does not aim for specific problems but rather base on common recognition of the domain and thus can work as definition references to other ontologies and as well as communication terminologies for sharing and reusability. As illustrated in Fig. 1, the knowledge model is edited using the Protégé user interface. The main class hierarchy is presented in the left frame. The class hierarchy is related using super- and sub-class relationships. More restrictions or definitions of classes can be formally defined using description logics expressions. Class members or instances are shown in the middle frame. An instance inherits all properties and restrictions of a class. For example, this frame lists several individual names of the "Nutrient Limitation" concept. Meanwhile, the right frame provides the individual editor for editing details. For example, the instance "restriction_stage2" contains properties such as "*has_stage*" and "*has_Protein_Limitation*".

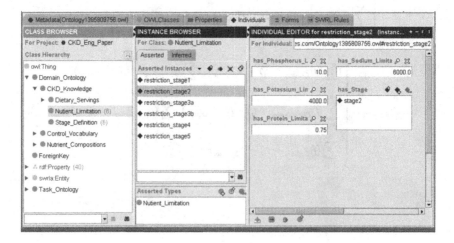

Fig. 1. Develop domain ontology construct and instances with Protégé editor.

This study constructs common terminologies into "Control Vocabulary." From the background and the known factual information of CKD and foods are constructed as initial concepts including "CKD_Knowledge" and "Food_Ingredients". In order to transfer knowledge model into format for information systems, the Protégé ontology editor is used to establish the ontological knowledge framework including classes, properties, and instances.

1. Control Vocabulary: Including 3 sub-concepts such as "Activity Level", "Calorie_Level", and "Food_Groups". Under each concept, the common terminologies are listed to provide reference, indexing, exchange and communication to other concepts and instances.
2. CKD Knowledge: Established 3 sub-concepts by the definition of CKD clinical stages, the key nutrient restrictions in each stage, and the suggested dietary serving in each food group.
3. Food Ingredients: The factual contents of food nutrient composition are adopted from a national government open data platform. However, the due to the difference in data model and the need to integrate with existing models, this study developed a pre-processing procedure for the transformation including data cleaning, relationship mapping. Some protégé plugins are utilized to assist transformation procedure such as DataMaster.

3.2 Construction of Task Ontology

A task ontology aims at solving practical problems. In addition to developing the concepts, the constituent properties of the concepts need to be planned to describe the knowledge framework for describing the problem solving process. To mark the detailed definition of the OWL-based properties, the property content values of known facts or unknown knowledge need to be firstly confirmed to separate asserted property from inferred property. Next, the domain and range of the properties need to be denoted. If a range uses basic data type, it is a data property; whereas if a range uses instances, it is an object property.

The task of this study is to solve the problem of CKD patient diet care, the background knowledge from the aforementioned domain is used to design the conceptual structure of problem solving. Five concepts designed under task ontology are as follows:

1. Personal Profile: including inference properties such as CKD clinical stages, eGFR, calorie requirement.
2. Personal Nutrient Count: including inference properties to obtain suggested amounts of protein, phosphorus, potassium, and sodium.
3. Personal Dietary: including inference properties to obtain the suggested balanced diet servings in food groups.
4. Diet Examination: including inference properties to obtain key nutrients examination and food group examination.
5. Food Selection: including properties for annotating case patient's food combinations.

3.3   Semantic Rules Development

A knowledge base represents factual knowledge in the form of instance, therefore semantic problem solving can be performed by running semantic rules in the instance layer of an ontology to infer implicit knowledge based on known factual knowledge. The ontological inference in the instance layer is performed by semantic rule language (e.g., Semantic Web Rule Language, SWRL) for rule development (Horrocks, Patelschneider, Bechhofer, & Tsarkov, 2005). Protégé platform provides such function through the SWRL Rules tab. In short, SWRL works with OWL-based knowledge bases and is able to perform the inference of implicit knowledge through a rule inference engine (Corsar & Sleeman, 2006).

The analysis of semantic rules starts with the concept in which the property belongs, and then chains the concept to other facts in a step-by-step manner until the objective is achieved. Each step is expressed as an atom and the rule is expressed in the form of "($atom_1$^... ^ $atom_n$) → Consequence" to express the cause-effect relationship. The rules below are implemented equations as designed in Eqs. (1) ~ (3). The equations can be expressed as an SWRL-based rule as follows:

1. Six rules are respectively developed to infer suggested food group servings as designed in Eq. (1). For example, the rule below infers grains servings. Rules for other suggested food group servings can be created in the same manner.

   ```
   Personal_Dietary (?x) ∧ has_Case_Name (?x, ?p) ∧
   has_Stage (?p, ?s) ∧ has_Calorie_Level (?p, ?k) ∧
   Dietary_Servings (?y) ∧ has_Stage (?y, ?s) ∧
   has_Calorie_Level (?y, ?k) ∧ has_Grain_Servings (?y,
   ?ans) → has_Grain_Servings (?x, ?ans)
   ```

2. The rule below is developed to infer the actual food group servings as designed in Eq. (2). The Semantic Query-Enhanced Web Rule Language (SQWRL) is used to help the arithmetic operations of rules.

   ```
   Personal_Dietary (?x) ∧ has_Case_Name (?x, ?x1) ∧
   has_Name (?x1, ?x2) ∧ has_Intake_Food (?x, ?y) ∧
   Food_Selection (?y) ∧ has_Food (?y, ?g) ∧
   has_Food_Group (?y, ?g1) ∧ has_Servings (?y, ?n) °
   sqwrl:makeBag (?s, ?n) ∧ sqwrl:groupBy (?s, ?x, ?g) °
   sqwrl:sum (?ans, ?s) → sqwrl:select (?x2, ?g1, ?ans) ∧
   sqwrl:columnNames ("CSV_Name", "CSV_Food_Group",
   "CSV_Servings")
   ```

3. The rules below examine the differences in food group servings between the case patient's diet and the suggested values as designed in Eq. (3). The rule below infers grains servings. Rules for inferring other food group serving differences can be created in the same manner.

```
Diet_Examination (?x) ∧ has_Case_Name (?x, ?x1) ∧
has_Name (?x1, ?x2) ∧ Personal_Dietary (?y) ∧
has_Case_Name (?y, ?x1) ∧ has_Grain_Servings (?y, ?y1)
∧ Servings (?z) ∧ CSV_NAME (?z, ?x2) ∧
CSV_FOOD_GROUP(?z, "Grain") ∧ CSV_SERVINGS(?z, ?z1) ∧
swrlb:subtract (?ans, ?z1, ?y1) ∧ swrlb:stringConcat
(?ans1, "Inspect Grain    standard value: ", ?y1, "
case: ", ?z1, "  overbalance : ", ?ans, " portion") →
has_Examine-on_servings (?x, ?ans1)
```

4   Experiment

The design of the data flow follows dietary consultation activities. Patients input their basic data and daily actual food combination, and the system answers with inference and estimation from the existing knowledge, similar to dietitians. The system uses Apache Tomcat as the application server to provide a Web platform to connect to the inference services provided by the rules engine (Java Expert System Shell, JESS) available from the Protégé platform. Finally, the user interfaces utilize Java Server Page (JSP) to create Web applications.

4.1   Case Study

This study has designed a patient data collection interface. For example, a case patient may input personal data as: male, height 176, age 65, weight 75 kg, Cr value 1.2, and light activity level. In the Food List block, a user needs to input the daily actual food combination. The food items and servings are added into the My Plate box.

After completing the personal data and daily food combination, the back-end rules engine infers the CKD stage, calorie baseline, suggested food group servings, and suggested key nutrient intakes. This study has devised an interface that presents the four categories that summarize the dietary consultation as described in Eqs. (1) ~ (6) and corresponding rules. Figure 2 demonstrates the results of the rule computation and inference:

1. Calculating suggested food group servings by individual conditions: in the upper left block of Fig. 2.
2. Calculating suggested key nutrient intake by individual conditions: in the upper right block of Fig. 2, the five key nutrient limitations are obtained from executing relevant rules.
3. Examining suggested food group servings against daily diet combination: in the lower left block of Fig. 2, the results are obtained from examining the differences between cases in the food group. For example, the number of servings matches the suggestion in Protein-food, Vegetables, Daily, and Oils. Half serving short in Grains and Fruits.
4. Examining key nutrient intake against daily diet combination: in the lower right block of Fig. 2, the results are obtained from examining the differences in key nutrient

ingestion between cases. For example, in "Protein Restriction", the maximum value is 56.25 g, but the actual intake over 3.45 g. Other examinations on calories, phosphorus, and sodium also show over intake.

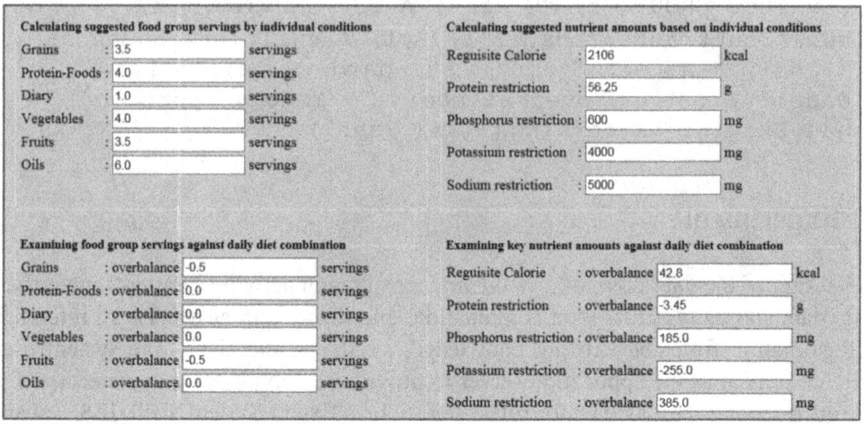

Calculating suggested food group servings by individual conditions				Calculating suggested nutrient amounts based on individual conditions		
Grains	:	3.5	servings	Reguisite Calorie	2106	kcal
Protein-Foods	:	4.0	servings	Protein restriction	56.25	g
Diary	:	1.0	servings	Phosphorus restriction :	600	mg
Vegetables	:	4.0	servings	Potassium restriction :	4000	mg
Fruits	:	3.5	servings	Sodium restriction :	5000	mg
Oils	:	6.0	servings			

Examining food group servings against daily diet combination				Examining key nutrient amounts against daily diet combination		
Grains	: overbalance	-0.5	servings	Reguisite Calorie	: overbalance	42.8	kcal
Protein-Foods	: overbalance	0.0	servings	Protein restriction	: overbalance	-3.45	g
Diary	: overbalance	0.0	servings	Phosphorus restriction	: overbalance	185.0	mg
Vegetables	: overbalance	0.0	servings	Potassium restriction	: overbalance	-255.0	mg
Fruits	: overbalance	-0.5	servings	Sodium restriction	: overbalance	385.0	mg
Oils	: overbalance	0.0	servings			

Fig. 2. Screen snapshot of inference.

4.2 Evaluation

For evaluation purpose, results of the 36 case patients' experiments were sent to the two hospital dietitians for examination and verification of accuracy. The items verified included the evaluation of the case clinical stages, the suggested food servings, the key nutrient restrictions, and the suggested diet combinations. The accuracy is represented as the ratio of the number of identical results (between system outcome and expert examination) over the total number of case results. During the evaluation process, a number of inconsistent results were found between the system outcome and the manual estimation. This finding is similar to a previous study where the system estimation is faster and more accurate than manual estimation [17]. After correcting the manual calculation errors, both sets of evaluation are identical. As the knowledge system-inferred diet suggestions agree with those of the dietitians', it is evidenced that the knowledge system can provide CKD patients with correct dietary consultation.

On the other hand, we compared the inference results from the KBS with the dietitian suggested food combinations. We found noticeable differences in all columns except the CKD Clinical stage. After a joint review of the researchers, it was found that some patients had comorbidity and complication (CC) and few patients had non-medical reasons. These interference factors had caused the differences. However, for the cases without the interference factors, the inference results of the KBS and the dietitians were identical.

5   Conclusion

This study aims to solve the problem of dietary consultation for chronic kidney disease patients. CKD patients are often challenged with dietary management because of the multitude of factors involved: unable to receive frequent and detailed dietary consultation, variation in illness and physical conditions, and lack of nutrient knowledge in daily food intake. This study has contributed to dietary management in the following aspects: (1) building an ontological dietary consultation system with the extensibility to include other chronic diseases in the future; (2) the integration of multiple open knowledge sources into a knowledge model; and (3) the design of task ontology with semantic rules for problem solving. We use CKD as an example domain and develop an ontological knowledge model and a knowledge base system for CKD dietary consultation.

This study uses case information from 36 CKD patients for the case experiment. The resulting dietary suggestions from the experiment are identical to those from the dietitians. This shows that the research objective was accomplished and that this knowledge system is capable of offering good dietary guidance to CKD patients. In the future, with the strengths of open knowledge integration and knowledge base extensibility, this dietary consultation system can be expanded and refined for comorbidity and complication to create a more comprehensive dietary consultation system for chronic disease patients. For future clinical deployment, given the experience from the current research, we suggest to firstly expand the knowledge model to include the closely related knowledge sources of CKD comorbidity and complication, such as specific diabetes mellitus and cardiovascular diseases. Such expansion will enable the KBS to take into account major interaction and interference factors and thus enhance its inference capacity for clinical usage.

Acknowledgments.   The authors would like to thank the National Science Council of the Republic of China, Taiwan for financially supporting this research under Contract No. NSC 102-2410-H-033-036-MY2.

References

1. National Kidney Foundation (NKF). http://www.kidney.org/professionals/kdoqi/pdf/ckd_evaluation_classification_stratification.pdf
2. McCullough, M.L., Feskanich, D., Stampfer, M.J., Giovannucci, E.L., Rimm, E.B., Hu, F.B., et al.: Diet quality and major chronic disease risk in men and women, moving toward improved dietary guidance. Am. J. Clin. Nutr. **76**, 1261–1271 (2002)
3. Levey, A.S., Coresh, J., Balk, E., Kausz, A.T., Levin, A., Steffes, M.W., et al.: National kidney foundation practice guidelines for chronic kidney disease, evaluation, classification, and stratification. Ann. Intern. Med. **139**, 137–147 (2003)
4. Kalista-Richards, M.: The kidney, medical nutrition therapy—yesterday and today. Nutr. Clin. Pract. **26**, 143–150 (2011)
5. Myers, G.L., Miller, W.G., Coresh, J., Fleming, J., Greenberg, N., Greene, T., et al.: Recommendations for improving serum creatinine measurement, a report from the laboratory working group of the national kidney disease education program. Clin. chem. **52**, 5–18 (2006)

6. Stevens, L.A., Coresh, J., Greene, T., Levey, A.S.: Assessing kidney function — measured and estimated glomerular filtration rate. N. Engl. J. Med. **354**, 2473–2483 (2006)
7. Plessers, P., De Troyer, O., Casteleyn, S.: Understanding ontology evolution, a change detection approach. Web Semant. Sci. Serv. Agents World Wide Web. **5**, 39–49 (2007)
8. Welty, C., Guarino, N.: Supporting ontological analysis of taxonomic relationships. Data Knowl. Eng. **39**, 51–74 (2001)
9. García-Castro, R., Gómez-Pérez, A.: Interoperability results for semantic web technologies using OWL as the interchange language. Web Semant. Sci. Serv. Agents World Wide Web. **8**, 278–291 (2010)
10. Horrocks, I., Patel-Schneider, P.F., van Harmelen, F.: From SHIQ and RDF to OWL, the making of a web ontology language. Web Semant. Sci. Serv. Agents World Wide Web. **1**, 7–26 (2003)
11. Gennari, J.H., Musen, M.A., Fergerson, R.W., Grosso, W.E., Crubézy, M., Eriksson, H., et al.: The evolution of Protégé, an environment for knowledge-based systems development. Int. J. Hum. Comput Stud. **58**, 89–123 (2003)
12. Corsar, D., Sleeman, D.: Reusing jesstab rules in Protégé. Knowl.-Based Syst. **19**, 291–297 (2006)
13. Horrocks, I., Patelschneider, P., Bechhofer, S., Tsarkov, D.: OWL rules, a proposal and prototype implementation. Web Semant. Sci. Serv. Agents World Wide Web. **3**, 23–40 (2005)
14. Gómez-Pérez, A., Benjamins, V.R.: Applications of ontologies and problem-solving methods. AI Mag. **20**, 119–122 (2009)
15. Fernández-López, M., Gómez-Pérez, A., Suárez-Figueroa, M.C.: Methodological guidelines for reusing general ontologies. Data Knowl. Eng. **86**, 242–275 (2013)
16. Fürst, F., Trichet, F.: Integrating domain ontologies into knowledge-based systems. In: Proceedings of the Eighteenth International Florida Artificial Intelligence Research Society Conference, pp. 826–827 (2005)
17. Chen, Y., Hsu, C.-Y., Liu, L., Yang, S.: Constructing a nutrition diagnosis expert system. Expert Syst. Appl. **39**, 2132–2156 (2012)

Content Retrieval and Filtering

Evaluation of the Dynamic Construct Competition Miner for an eHealth System

David Redlich[1,2]([⊠]), Mykola Galushka[2], Thomas Molka[2], Wasif Gilani[2],
Gordon Blair[1], and Awais Rashid[1]

[1] Lancaster University, Belfast, UK
mr.redlich@gmail.com, {gordon,marash}@comp.lancs.ac.uk
[2] SAP Research Centre Belfast, Belfast, UK
{mykola.galushka,thomas.molka,wasif.gilani}@sap.com

Abstract. Business processes of some domains are highly dynamic and increasingly complex due to their dependencies on a multitude of services provided by various providers. The quality of services directly impacts the business process's efficiency. A first prerequisite for any optimization initiative requires a better understanding of the deployed business processes. However, the business processes are either not documented at all or are only poorly documented. Since the actual behaviour of the business processes and underlying services can change over time it is required to detect the dynamically changing behaviour in order to carry out correct analyses. This paper presents and evaluates the integration of the Dynamic Construct Competition Miner (DCCM) as process monitor in the TIMBUS architecture. The DCCM discovers business processes and recognizes changes directly from an event stream at run-time. The evaluation is carried out in the context of an industrial use-case from the eHealth domain. We will describe the key aspects of the use-case and the DCCM as well as present the relevant evaluation results.

Keywords: Business process management · Process discovery · Enterprise architecture · Complex event processing · eHealth

1 Introduction

The major objective of the European project TIMBUS is to enable Digital Preservation (DP) of business processes and services [5]. A major innovation of the project is to enable business process centric risk management to help identify critical parts of crucial business processes and services, which need to be preserved in order to ensure long-term availability and business continuity. The TIMBUS risk management process, based on the ISO 31000 standard, starts with first establishing the context of the target system. This means capturing all information about the business processes, including the behavioural information, the execution context, the legal context, the resources needed from the software, hardware and facility level to support the top level business processes, etc. The extracted context information is the basis for different actions: (1) reasoning

© Springer International Publishing Switzerland 2015
W. Abramowicz (Ed.): BIS 2015, LNBIP 208, pp. 115–126, 2015.
DOI: 10.1007/978-3-319-19027-3_10

about whether or not DP is feasible, (2) risk management, and (3) if required, for the execution of digital preservation process. Driven by requirements from three different business processes, coming from the domains of Civil Infrastructure, eScience and eHealth, the individual tools and the overall TIMBUS solution are required to be generic in order to be applicable to the different use-cases.

The first step in the TIMBUS risk management process is the availability of the control-flow information of the target business process and its performance and resource information, which is generally available in the process logs. Two types of Process Extractors have been developed in the project: (1) a statically operating genetic miner that works on the historic business process logs [14] and is suitable for long-life business processes that do not change over time, such as the one from Civil Infrastructure domain. However, due to the characteristics of the static genetic miner (non-deterministic as well as long and unpredictable execution time) it is not applicable for monitoring fast and dynamically changing processes such as the ones from the eHealth domain [3]. For these use-cases it is required to monitor the system for changes in the process which result in changes in the risk assessment and potentially entail a new preservation iteration. This is why the second type of Process Extractor has been included in the project: (2) the Process Monitor that detects changes in the business process during run-time.

In this paper we describe how the Dynamic Constructs Competition Miner (DCCM) has been integrated into the TIMBUS architecture as a Process Monitor and evaluate its suitability in the context of the DrugFusion use-case from the domain of eHealth. In the remainder of the paper we first introduce the Drug-Fusion use-case in Sect. 2, then explain important aspects of the DCCM and its integration into the TIMBUS architecture as a Process Monitor in Sect. 3. This is followed by an evaluation of the DCCM for the DrugFusion use-case in Sect. 4 where we explain findings and shortcomings of the solution. Then in Sect. 5, the paper is concluded by summarizing emerged results and findings.

2   eHealth Use Case (DrugFusion)

Each prescription drug package selling in Europe must contain information about how it works, what is the intended effect, and cautions for its use. Medical practitioners who are authorized to prescribe drugs try to identify the best treatment strategy by assessing a patient's condition and previously prescribed medicines. Such treatment may include a prescription of one or more drugs which need to be taken within the predefined time interval. Sometimes prescribed medications may cause an Adverse Drug Reaction (ADR) [7–9]. The study of ADRs is conducted in the field known as Pharmacovigilance. ADRs describe harms caused by taken medications at a normal dosage during normal use [2,12]. ADRs may occur in the following scenarios: a single dose, a prolonged usage of a drug or a result of combined use of two or more drugs (this scenario is targeted by the DrugFusion system described in this paper). ADRs expression has a different meaning than "side effect", since side effect might also imply that the effects can be beneficial.

A more general term, Adverse Drug Event (ADE) [6,11], refers to any injury caused by the drug (whether drugs were used at normal dosage and/or due to overdose) and any harm associated in such case, i.e. ADRs are a special type of ADEs.

DrugFusion has been developed and made available as a use case by an industrial partner in the TIMBUS project. The main objective was to provide a platform for avoiding ADEs. The high-level description of the DrugFusion process for discovering rules which help to predict ADEs is presented in Fig. 1. The ADE rules discovery process can be split into three distinct phases: *Creation of Dictionaries, Load of Adverse Event Report (AER)* and *Discovery of ADE Rules.* The majority of operations in the second and third phases are implemented using a map-reduce approach and run on a Hadoop cluster.

Creation of Dictionaries Phase. Begins with collecting data for products. A product is an abstraction of the following three types of data: indication, drug, and reaction. The indication data represents patients' diagnosis. The drug data consists of generic names of prescription medications. The reaction data describes undesirable effects on a patient's health, caused by applied treatment. This data is essentially used to create dictionaries for indications, drugs and reactions. The resulting dictionaries are created in parallel and under the supervision of three different groups of experts. Each dictionary is created through a number of iterations. When the scope of each dictionary is finalized, a specialized filter removes information which is not relevant for future lookup operations. The dictionary creation phase is concluded with mapping active ingredients to drugs represented in the dictionary. This process is fully automated, however, if the system cannot make a conclusive mapping decision, it requires an expert to perform the manual assignment.

Load of AER Phase. Begins with downloading drug usage data[1]. The following categories of report data are considered: demographic, indication, drug, and reaction. A loading process performs data cleansing and a replacement of indication, drug and reaction names with unique identifiers obtained from the already created dictionaries. The loaded data of each category is moved to a predefined location to enable loading the joined report. All operations involved in the loading of individual report data can be executed in parallel. The loading of the joined report combines demographic, indication, drug, and reaction information into individual cases using the unique event identifiers. This phase ends with a drug normalization step in which inconsistencies of the drugs' names are resolved. Usually, pharmaceutical companies have their own line of drugs for targeting generic groups of illnesses. These drugs have the same structure but different brand names. Such diversity in drug names causes a significant impact on the complexity of the ADE rules discovery algorithm which is addressed in the normalization step.

---

[1] DrugFusion downloads adverse events report published by United States Food and Drug Administration (http://www.fda.gov) every quarter.

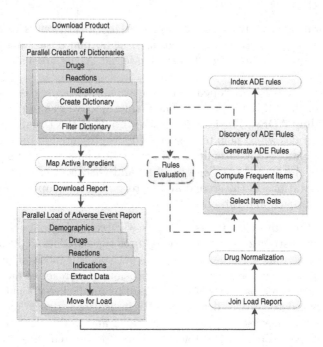

Fig. 1. DrugFusion business process.

Discovery of ADE Rules Phase is implemented by using a specifically modified Apriori Algorithm [15]. It includes: selection of item sets (these item sets combine patients' drug and reaction information), computing of frequent items (these computations run until all items are considered) and generating of ADE rules (where each rule is assigned with confidence and support values, respectively). If some of the obtained rules contradict existing medical observations, a group of experts performs an evaluation of the rules. It might trigger modifications in the items selection algorithm and a relaunch of the discovery process, which can be repeated a number of times. This phase as well as the DrugFusion process is terminated by indexing the obtained rules.

3   Dynamic Process Discovery for Digital Preservation

The goal of the TIMBUS project is to enable the digital preservation of business processes. An important part of this is to be able to make decisions about if, how, and when services/components/modules need to be preserved in order to ensure the execution and continuity of an organization's business functions. Business functions are typically implemented by business processes which are defined as *"...a series or network of value-added activities, performed by their relevant roles or collaborators, to purposefully achieve the common business goal"* [10]. In this paper we focus on the control-flow perspective of a business process, i.e. the activities (steps that represent the execution of actual work) and their execution order defined by control-flow elements, e.g. XOR-Split, XOR-Join, AND-Split, AND-Join.

A number of components have been developed to carry out these analysis on different abstraction levels of the business process and the underlying IT infrastructure, e.g. business process vs. resource level. These analyses operate either in an on-demand or a continuous fashion, depending on requirements of the TIMBUS project use-cases and on the level of automation of the employed methods. In many organizations, the involved business process models are not documented or the documented models deviate from the actual executed process, e.g. for the DrugFusion use case a documented process model is available (see Fig. 1) but does not accurately reflect the reality as will be shown in the evaluation, Sect. 4. The extraction of a business process model from a given event log without the usage of any a-priori information is addressed by *Process Discovery* algorithms [19], e.g. [14,16,18,20].

Of a high importance in the context of the TIMBUS project is the detection of changes in the business process that potentially have implications on the assessment of risks and the infrastructure which is to be preserved. For instance, the process of the DrugFusion use-case shown in Fig. 1 did initially operate without the "Rules Evaluation" carried out by experts in the "Discovery of ADE Rules" phase. The introduction of that step and the associated loop changed the behaviour of the process, thus increasing the severity of risks and introducing additional resources required for carrying out the "Rules Evaluation" activity. To detect these changes in the monitored system, a process monitoring component is required to be part of the TIMBUS solution. As opposed to traditional process discovery algorithms that calculate a process model from an input log in a static fashion, the process monitor has to work in a dynamic fashion, i.e. events are not recorded in a log but directly processed to changes of the system's "state". This event-based processing is an application of Complex Event Processing (CEP), a method to capture and filter low-level events and aggregate them to complex events representing high-level information about the system [13] - in this case information about the business process model.

3.1 Dynamic Constructs Competition Miner

The process monitor in the TIMBUS solution has been implemented with a modified Constructs Competition Miner (CCM) originally proposed in [16]. The original CCM is based on two fundamental steps:

(1) Footprint Creation, in which the footprint consisting of information on the global relations between any two elements of a set of activities is calculated from a given event log. In particular, two different relations are contained in the CCM footprint: (A) *appears before first*, a relative measure that records in how many cases an activity appeared before the first occurrence of another activity, and (B) *appears before*, a measure that records in how many cases an activity appears at some point before another activity (no matter if it was the first occurrence or not). For an example regard the following sequences of activity occurrences: $[DE], [EDE], [ED]$ are three instances of a process execution consisting of the two activities E and D, then appeared E before the first occurrence of D in

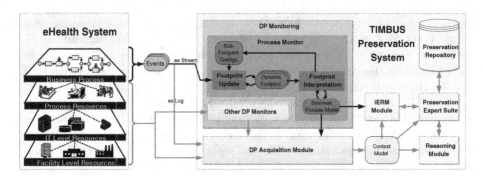

Fig. 2. The high-level view of the TIMBUS DP architecture with focus on the Process Monitor - Rounded corners: Models; Sharp corners: Agents

67% of the cases, and D before the first appearance of E in 33% of the cases (*appears before first* relation), whereas E appears before any D in 67% of the cases, and D before any E in also 67% of the cases[2] (*appears before* relation).

(2) Footprint Interpretation, in which the footprint is interpreted to a business process model construct. The CCM is technically able to identify and build a business process model based on the following constructs: *Sequence, Choice, Parallelism, Loop, Loop over Sequence, Loop over Choice, Loop over Parallelism* and constructs for single activities [16].

These two steps are repeatedly executed in a recursive fashion, with each step identifying a construct and splitting up the set of activities into subsets which are then again analysed the same way. This is carried out until the set of activities cannot be split up any further. The final result of the algorithm is a business process model representing the behaviour recorded in the log. For more details on the functionality as well as the footprint and business process model of the original CCM please see [16].

In [17] modifications to the CCM algorithm have been proposed that enable Dynamic Process Discovery in an event-based fashion as required for the Process Monitoring component in the TIMBUS solution. The result is the Dynamic Constructs Competition Miner (DCCM) in which the two steps, *Footprint Creation* and *Footprint Interpretation*, were completely separated, each with their own respective life cycle. Furthermore, the *Footprint Creation* was altered towards a *Footprint Update* method which is not analysing a complete log but operating on an event-by-event basis, updating the footprint with each event. Due to this a significantly low execution time for the event processing has been achieved and no record of previously occurred events has to be stored. Another feature of the *Footprint Update* is that the influence of older cases on the footprint gradually decreases until it eventually disappears [17], thus supporting the monitoring of dynamically changing processes. Since the *Footprint Interpretation* is

---

[2] For sequence [*EDE*] both relations are true: E appears before D and D appears before E.

relatively cost effective, it is not part of the event occurrence life cycle, but rather executed on-demand or in intervals (e.g. every 10 s). Additionally, other modifications making the original CCM fit for run-time application have been carried out and are described in [17], e.g. introducing a third relation, *direct neighbours*, to speed up the interpretation step. Especially DCCM's features of *robustness* (dealing with exceptional behaviour) and *scalability* (can easily manage 100 s of events/second) qualifies the DCCM for application as Process Monitor in the TIMBUS framework.

3.2   Integration in TIMBUS Project

The Digital Preservation (DP) framework developed within the TIMBUS [4] project provides a unique set of solutions going beyond the scope of existing DP approaches. It covers all aspects of traditional DP system such as preserving a digital content but also addresses enterprise risk analysis and business continuity planning. It covers a wider scope of DP processes, which includes intelligent Enterprise Risk Management (iERM) for automatic identification and prioritization of risks within an enterprise and ability to minimize those risks by taking a specific set of actions including DP. A high-level view of the TIMBUS DP analysis architecture is shown in Fig. 2. It consists of seven main modules: DP Monitoring, DP Acquisition, iERM, Context Model, Preservation Expert Suite, Reasoning, and Preservation repository.

A *Context Model* has been designed and developed in the project [1] that is meant to be populated with the complete context of the business process. It acts as a single data source in the TIMBUS solution for carrying out the risk management, reasoning and the digital preservation process. A number of static context extractors, for example, for software, hardware, and business process, etc., have been developed within the project and are part of the *DP Acquisition Module*. The aim of the DP Acquisition Module is to provide functionality for extracting the contextual information initially or on-demand, and then populating automatically this information into the Context Model. In the iERM Module the Context Model can be imported, annotated with different risk factors, and assessed according to the specified annotations. The assessment is performed in an interaction mode, where a risk expert can alter different model parameters and run simulations to identify the critical subset of resources and business processes, which require preservation. A generated Risk Assessment Report (RAR) is analysed by the Preservation Expert Suite (PES) with the help of the Reasoning module. PES combines a set of tools integrated to provide the centralized control for preservation and redeployment cycles. In the preservation cycle the PES transfers virtualized hardware components from a VM environment into the preservation repository. In the redeployment cycle PES performs a reverse operation where it transfers virtualized hardware components from the preservation repository back into the VM environment.

Since dynamic changes are expected in some of the use-cases, e.g. eHealth, a DP Monitoring Module is integrated in the TIMBUS architecture. It consists of different monitors, each monitoring a different aspect of the business process.

In Fig. 2 the agents and models of the Process Monitor, a module to discover the process behaviour and changes in the process at run-time, are displayed in more detail. The static process extractor employed in the DP Acquisition module could not be utilized for this task due to its non-deterministic behaviour and the non-compliance with run-time requirements (the extractor is based on a genetic algorithm [14]). Instead the Process Monitor was implemented with the Dynamic Constructs Competition Miner as introduced earlier. The Process Monitor module offers a RESTful Webservice API as interface to communicate with the service module. It provides for instance a method onEvent(Event e) which is invoked by the monitored system for each occurring event. Every event triggers a Footprint Update which causes a small alteration of the Dynamic Footprint. The Footprint Update module is highly efficient and takes only a small constant amount of time to execute the update on the footprint (100s of events can be processed per second). The Dynamic Footprint is transformed to a Business Process Model by the Footprint Interpretation. Since this is relatively cost intensive (up to 3 s for very large processes with 100 activities on a normal machine) it is not executed for every event but runs decoupled in a second life cycle, e.g. scheduled for every 10 s. The Footprint Interpretation also formulates special requests (Sub-Footprint Configurations) for the Footprint Update if it is required (see [17]). Past events are discarded after they have been processed. Only the Dynamic Footprint and the current Business Process Model are stored in memory and as a result use only a very limited amount of memory.

The sum of small alterations on the Dynamic Footprint caused by individual events can eventually amount to a change in the Business Process Model, e.g. through introducing a new activity or new behaviour as seen in the evaluation section. The DCCM supports both, disregarding old behaviour and incorporating newly introduced behaviour. To identify a change in the process the Process Monitor always compares the previous Business Process Model with the newly interpreted Business Process Model at every execution of the Footprint Interpretation. If a change occurs the DP Acquisition Module is notified which requests the current Business Process Model with the method BPModel getCurrentModel() provided as a service and updates the Context Model accordingly. This service interface is also directly utilized by the iERM tool which can import the current business process model when performing a run-time risk analysis.

4   Evaluation

In this section we will carry out an evaluation of the DCCM integrated into the TIMBUS solution in the context of the DrugFusion use-case introduced in Sect. 2. In a first step we will examine what the actual business process looks like, i.e. we will hook the Process Monitor into the event stream produced by the use-case. The result after the first recorded 15 process instance executions is displayed in Fig. 3.

The "Creation of Dictionary" phase includes activities from "DownloadNdcRepository" to "MapActiveIngredient". The parallel execution of "Creating a Dictionary" and "Filtering a Dictionary" for Reactions, Drugs, and Indicators

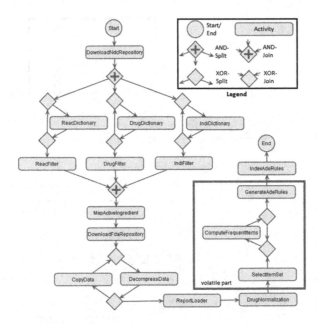

Fig. 3. Originally discovered DrugFusion process

has been identified correctly. However, the dictionary creation can be executed repeatedly until the filtering activity is executed for each type. This shows that the documented process deviates from the actual process execution.

The second phase of "Load Adverse Event Report" ranges in the discovered process from activity "DownloadFdaRepository" to "DrugNormalization". Instead of a parallel behaviour for "Extracting Data" and "Moving Data" for the types Reactions, Drugs, Indicators, and Demographics, a loop over "DecompressData" and "CopyData" (representing the "Extracting Data" and "Moving Data" activities in the original process) is detected. After further investigation it was revealed that "Extract Data" and "Copy Data" are unified activities that will always fire the same event independent of the type of data to be loaded or moved. Because of this underspecification of events "DecompressData" and "CopyData" is recorded four times for each process execution. This revealed a shortcoming of the DCCM: the algorithm has difficulties to detect a loop over a sequence of activities if in every case the sequence is executed more than once. The reason for this behaviour originates from the general concept of the DCCM: it only looks at the global relation between two activities which in this case is indistinguishable from a normal "loop" (instead of "loop over sequence"), i.e. the footprints for the constructs "loop" and "loop over sequence" look exactly the same[3]. Since the DCCM is not able to distinguish these two constructs in this special case the footprint is interpreted as a normal loop. Note, that this

---

[3] Both relations "appears before first" and "appears before" are always true for activities within the loop, e.g. for both sequences $[EDE]$ (normal loop) and $[EDED]$ (loop over sequence) E "appears before" D and vice versa.

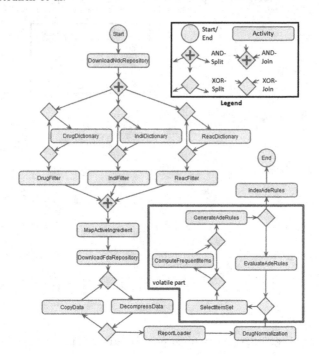

Fig. 4. DrugFusion process with evaluation step

behaviour only occurs if a loop is always looping at least once in every instance execution.

The last phase of the process "Discovery of Adverse Event Rules" is represented in the discovered process by activities from "SelectItemSet" to "GenerateADERules". Here again, the discovered process deviates from the documented one since "ComputeFrequentItems" can in reality be executed multiple times before moving on in the process. Additionally, the activity "IndexAdeRules" has been detected at the end of the process.

In the second part of the evaluation an extra activity "Rules Evaluation" was introduced that would trigger a repeatedly execution of the "Discovery of ADE Rules" phase of the process. The new activity was detected immediately after the first occurrence and the additional loop behaviour shortly afterwards. The business process model discovered 5 instance executions after the change was implemented is shown in Fig. 4. Now, an additional loop over the activities from "SelectItemSet" to "GenerateADERules" can be identified in the discovered process as well as the additional activity "EvaluateADERules" that is executed every time before the "Discovery of ADE Rules" phase is repeated.

After the change was detected the DP Acquisition Module was immediately notified and as a result the Context Model was updated according to the detected change. With the updated Context Model, the iERM module as well as the Reasoning module of the TIMBUS solution were able to perform their analyses on the up-to-date information of the business process.

5 Conclusion

In this paper we presented the integration of the Dynamic Constructs Competition Miner into the TIMBUS solution as a process monitor and evaluated its application in the context of the use-case DrugFusion from the eHealth domain. The main behaviour of the business process of the DrugFusion process as well as its dynamically changing behaviour was discovered. However, due to the unified execution of some of the activities ("DecompressData" and "CopyData"), events were not uniquely mappable to the activities, which in turn resulted in loop behaviour of a constant length. In the light of this, it was identified that the DCCM has difficulties to detect a loop over a sequence of activities if the loop is repeated at least once in every instance. Although not critical for the risk assessment because the general loop behaviour was detected, this is not an optimal result since underspecified event creation and/or loops of constant length may occur in other use-cases as well. In contrast, other process discovery algorithms not based on global relations but on local relations, e.g. direct neighbours, should have less problems to detect the correct loop but do not comply to the run-time requirement. Resolving the issue of detecting a loop of a constant length over a sequence is considered future work for both, the process monitor in the TIMBUS project as well as the general DCCM. Another next step in the context of the TIMBUS project is the monitoring of changes for other aspects of a business process than the control-flow: performance, resources, state of single instances. If this is achieved a continuous automatic risk analysis can be performed.

Acknowledgments. Project partially funded by the European Commission under the 7th Framework Programme for research and technological development and demonstration activities under grant agreement 269940, TIMBUS project (http://timbusproject. net/).

References

1. Antunes, G., Caetano, A., Bakhshandeh, M., Mayer, R., Borbinha, J.: Using ontologies to integrate multiple enterprise architecture domains. In: Abramowicz, W. (ed.) BIS Workshops 2013. LNBIP, vol. 160, pp. 61–72. Springer, Heidelberg (2013)
2. Butt, T., Cox, A., Oyebode, J., Ferner, R.: Internet accounts of serious adverse drug reactions a study of experiences of Stevens-Johnson syndrome and toxic epidermal necrolysis. Drug Saf. **35**(12), 1159–1170 (2012)
3. Galushka, M., Gilani, W.: DrugFusion - retrieval knowledge management for prediction of adverse drug events. In: Abramowicz, W., Kokkinaki, A. (eds.) BIS 2014. LNBIP, vol. 176, pp. 13–24. Springer, Heidelberg (2014)
4. Galushka, M., Taylor, P., Gilani, W., Thomson, J., Strodl, S., Neumann, M.: Digital preservation of business processes with TIMBUS architecture. In: Proceedings of 9th International Conference on Preservation of Digital Objects IPRES2012, pp. 117–125 (2012)
5. Gilani, W., Redlich, D., Galushka, M., Molka, T., Du, Y.: TIMBUS: Digital preservation for timeless business processes and services. In: 23rd Proceedings of eChallenges Conference (e-2013) (2013)

6. Huang, Y., Lin, S., Chiu, C., Yeh, H., Soo, V.: Probability analysis on associations of adverse drug events with drug-drug interactions. In: BIBE 2007, pp. 1308–1312 (2007)

7. Jin, H., Chen, J., He, H., Kelman, C., McAullay, D., O'Keefe, C.: signaling potential adverse drug reactions from administrative health databases. IEEE Trans. Knowl. Data Eng. **22**(6), 839–853 (2010)

8. Jin, H., Chen, J., He, H., Williams, G., Kelman, C., O'Keefe, C.: Mining unexpected temporal associations: applications in detecting adverse drug reactions. IEEE Trans. Inf. Technol. Biomed. **12**(4), 488–500 (2008)

9. Ji, Y., Ying, H., Dews, P., Mansour, A., Tran, J., Miller, R., Massanari, R.M.: A potential causal association mining algorithm for screening adverse drug reactions in postmarketing surveillance. IEEE Trans. Inf. Technol. Biomed. **15**(3), 428–437 (2011)

10. Ko, R.K.L.: A computer scientist's introductory guide to business process management (BPM). Crossroads J., ACM **15**(4), 4 (2009)

11. Koutkias, V., Kilintzis, V., Stalidis, G., Lazou, K., Nis, J., Durand-Texte, L., McNair, P., Beuscart, R., Maglaveras, N.: Knowledge engineering for adverse drug event prevention: on the design and development of a uniform, contextualized and sustainable knowledge-based framework. J. Biomed. Inf. **45**(3), 495–506 (2012)

12. Krska, J., Cox, A.: Adverse drug reactions. Clin. Pharmacol. Ther. **91**, 467–474 (2012)

13. Luckham, D.: The Power of Events: An Introduction to Complex Event Processing. Addison-Wesley Professional, Reading (2002)

14. Molka, T., Redlich, D., Drobek, M., Zeng, X.-J., Gilani, W.: Diversity guided evolutionary mining of hierarchical process models. In: Genetic and Evolutionary Computation Conference (GECCO 2015), ACM (2015) http://dx.doi.org/10.1145/2739480.2754765

15. Rao, S., Gupta, R.: Implementing improved algorithm over APRIORI data mining association rule algorithm. Int. J Comput. Sci. Technol. **1**, 489–493 (2012)

16. Redlich, D., Molka, T., Gilani, W., Blair, G., Rashid, A.: Constructs competition miner: process control-flow discovery of BP-domain constructs. In: Sadiq, S., Soffer, P., Völzer, H. (eds.) BPM 2014. LNCS, vol. 8659, pp. 134–150. Springer, Heidelberg (2014)

17. Redlich, D., Molka, T., Blair, G., Rashid, A., Gilani, W.: Scalable dynamic business process discovery with the constructs competition miner. In: Proceedings of the 4th International Symposium on Data-driven Process Discovery and Analysis (SIMPDA 2014), CEUR 1293, pp. 91–107 (2014)

18. Van Der Aalst, W., Weijters, A., Maruster, L.: Workflow mining: discovering process models from event logs. IEEE Trans. Knowl. Data Eng. **16**(9), 1128–1142 (2004)

19. Van Der Aalst, W.: Process Mining - Discovery Conformance and Enhancement of Business Processes. Springer, Heidelberg (2011)

20. Weijters, A., Van Der Aalst, W., de Medeiros, A.A.: Process Mining with the Heuristics Miner-algorithm. BETA Working Paper Series, WP 166, Eindhoven University of Technology (2006)

Dynamic Reconfiguration of Composite Convergent Services Supported by Multimodal Search

Armando Ordóñez[1], Hugo Ordóñez[2,3]([✉]), Cristhian Figueroa[3,4],
Carlos Cobos[5], and Juan Carlos Corrales[3]

[1] Intelligent Management Systems, Fundación Universitaria de Popayán,
Popayán, Colombia
armando.ordonez@docente.fup.edu.co
[2] Research Laboratory for Software Engineering,
Universidad de San Buenaventura, Cali, Colombia
haordonez@usbcali.edu.co
[3] Telematics Engineering Group, Universidad Del Cauca, Popayán, Colombia
jcorral@unicauca.edu.co
[4] Software Engineering Group, Politecnico di Torino, Turin, Italy
cristhian.figueroa@polito.it
[5] Information Technology Research and Development Group,
Universidad Del Cauca, Popayán, Colombia
ccobos@unicauca.edu.co

Abstract. Composite convergent services integrate a set of functionalities from Web and Telecommunication domains. Due to the big amount of available functionalities, automation of composition process is required in many fields. However, automated composition is not feasible in practice if reconfiguration mechanisms are not considered. This paper presents a novel approach for dynamic reconfiguration of convergent services that replaces malfunctioning regions of composite convergent services considering user preferences. In order to replace the regions of services, a multimodal search is performed. Our contributions are: a model for representing composite convergent services and a region-based algorithm for reconfiguring services supported by multimodal search.

Keywords: Convergent services · Dynamic reconfiguration · Multimodal search

1 Introduction

A composite convergent service (CCS) may be defined as a structured set of services (telecommunication and Web services) that works in a coordinated manner to achieve a common goal [1]. CCS achieve the integration and composition of services offered by IT providers with Telecom operators towards the Web Telecom convergence [2].

W. Abramowicz (Ed.): BIS 2015, LNBIP 208, pp. 127–139, 2015.
DOI: 10.1007/978-3-319-19027-3_11

One example of convergent process is a service that manages environmental early warnings. Environmental manager is in charge of decision making about environmental alarms and crops. In order to do so, it is required information from sensor networks, Telecommunication and Web services that process data and that send information to farmers. One typical requirement of such systems is: to emit an alarm to every farmer within a radius of 2 miles from the river if the river flow is greater than 15 % of average. For solving this request, the sensor data are evaluated and if necessary, an emergency map is generated. This map is created drawing a radius of 2 miles from the sensor. To do so, the system may use geographic services and maps from internet. Finally, the system informs about the alarm to farmers inside the emergency area, in this case the best way to send the information is selected (SMS or call). In both cases, services from Web and Telecommunications work in coordination.

Nowadays, composition of convergent services has been increasingly adopted in telecommunication companies in order to create and offer new CCS that integrate different functionalities of existing convergent services. It helps them to avoid developing new CCS from the scratch and duplication of functionalities. However, composing convergent services is not trivial and is time consuming due to the complexity of the integration of different and heterogeneous messages, operations, and data-types in these services [3]. Due to the latter the automation of this composition has been studied actively. Furthermore, the demanding requirements of the telecommunications environment regarding to performance, availability and accuracy, make it necessary that composed convergent services can replace services (nodes) that may be unavailable or malfunctioning during execution and that can impact negatively in the user experience [4].

Malfunctions or unavailability of convergent services may be originated from diverse factors such as network failures or server overload. In such scenarios, CCS must be fixed or reconfigured in order to continue providing their functionalities. Accordingly, reconfiguration and dynamic composition have been identified as leading challenges in Service Oriented Architectures [5].

Previous works in the literature have addressed the reconfiguration of CCS based mostly in replacement of individual failing services [6]. However, existing proposals leave aside Telecom considerations such as service deployment, real-time requirements or event-based interactions [5]. In addition, replacement of failing services for other ones with different non-functional features cannot maintain initial user constraints [7].

This work proposes a dynamic reconfiguration approach supported by a multimodal model for searching services or sets of services that can be used to reconfigure a CCS where there is a set of services (regions of services) that are malfunctioning or unavailable. The multimodal search is a branch of the information retrieval (IR) that aims for efficiently discovering different kinds of content [8].

The region-based reconfiguration is inspired by the work of Lin et al. [5] which describes a mechanism for selecting a replacement for a region surrounding the failing services instead of replacing the whole CCS. A failing region may be defined as a subset of services from the CCS that needs to be replaced in order to continue the execution of the service without recreating all the CCS. Previously the region-based reconfiguration mechanism was tested in the AUTO platform using automated planning [9–11], in this paper we focus on the multimodal search instead of the automated planning. Multimodal search uses textual and structural information in order to cover more information dimensions for querying during the search [12], this makes that the retrieval of services to be more precise than traditional information retrieval techniques.

The rest of this paper is organized as follows. Section 2 depicts a general overview of AUTO. Section 3 presents the modeling of CCS. Section 4 shows the service adaptation and the multimodal search applied to reconfiguration process. Section 5 presents the experimentation. Finally, Sect. 6 includes the conclusions as well as the relevant related work.

2 The Approach for Dynamic Reconfiguration of Composite Convergent Services

Figure 1, shows the architecture the main modules of AUTO [11]. AUTO, defines sequential phases for service composition namely: creation, synthesis and execution.

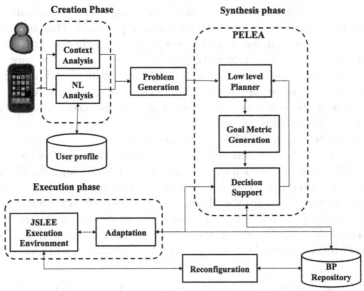

Fig. 1. Architecture of the approach for dynamic convergent service composition

The creation phase accepts user queries (requests) expressed either as voice or as text strings in natural language (NL). The NL Analysis module deals with each query in order to extract a set of significant words, i.e. to delete "stop words", reduce words to their lexical root etc. The Context Analysis module enriches the user queries with information about the user preferences, device capabilities and situational context. The Problem Generation module translates the processed user query into a problem file expressed in the planning language PDDL (Planning Domain Definition Language) [13]. Thus, this problem file contains the goals (user objectives) and metrics (preferences of the user about the composition). Additionally, it can also contain information about the initial state such as location, device information, etc.

The synthesis phase, receives the problem file as input and generates the structure of the composite convergent service CCS. To do so, this phase uses automated planning. In automated planning the generated structures are known as plans. A plan contains a set of services that solves user queries, the obtained plan represents the CCS. This phase is framed in the Planning and Learning Architecture (PELEA) [14]. *PELEA* is an architecture that contains modules for plans generation, monitoring and reconfiguration. Specifically, the low level planner performs the plan generation, the Goal Metric Generator module associates the goals and metrics, obtained from the input problem, with services available for the composition. Finally the decision support performs the monitoring and reconfiguration of the process. The synthesis phase requires that the user receives the plan immediately, therefore the elapsed time spent building the plans must be minimized. Nevertheless, in order to get the plan that represents the best solution, a big computational effort is necessary. On the average case finding the optimal solution requires exploring a very significant part of the search space, which makes such an endeavor impractical. Therefore, the best solution given a time window may be acceptable to begin the execution, and afterwards the planner may refine the plan while the first services are invoked and the associated tasks are performed in the real world. These new plans are potential replacements for the initial CCS, so these CCS are stored in the *BP repository*, which also contains simple convergent services (i.e. those that are not composed with others into a CCS).

The execution phase is mainly performed in the execution environment for telecommunication applications named Java Service Logic Execution Environment (JSLEE). The Adaptation module integrates *PELEA* and JSLEE in order to take the CCS (synthesized plans) and create executable CCS. To do so, the adaptation module associates the planning operators to Java Snippets and generates a CCS using JSLEE Service Building Blocks (SBB). SBBs are the basic components of the JSLEE architecture and are able to call external Web services or Telecom functionalities AUTO monitors CCS execution and repairs plans.

3   Modeling of Services for Multimodal Search

Our approach integrates multimodal search principles into CCS similarity detection during reconfiguration. Multimodal search is a branch of information retrieval (IR) that allows to efficiently discovering different kind of content - such as text, images, and video. Thus, the multimodal model for searching CCS unifies in single search space textual information and the structure of the CCS.

In this case, linguistics represents names and descriptions of the CCS elements (e.g. activities, interfaces, messages, gates, and events), and the structure is defined as codebooks formed of n-structural components ruled by the sequentially of the CCS control-flow (i.e. the union of two or more control-flow components. It has the following advantages: greater speed and precision, diversity in query types, and a good representation of CCS that allows delivery of results that correspond more precisely to user requirements [12].

Our reconfiguration algorithm that is based on the multimodal approach for searching service regions, a CCS must be appropriately represented so it can be dynamically composed and reconfigured. Therefore we have defined a model for representing a CCS as text strings considering users preferences, context conditions and devices capabilities. We have selected some dimensions associated with the devices and the information that can be obtained about the situation of the users using the analysis of their requests.

These dimensions are based on the context analysis module (Fig. 1). It classifies the information according to three dimensions: Device, User Profile and Situation (See Fig. 2). Each one of them obtains the information from different sources and defines the selection of domains. These dimensions allow us to analyze the capability of the system for responding to user preferences.

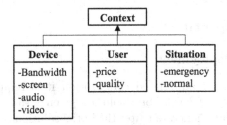

Fig. 2. Dimensions of user Context: Device, User and Situation

The Device dimension gathers the information from the device and the network. To access this information, the device ID or model is consulted in capabilities repositories like the Wireless Universal Resource File (WURFL) and the Composite Capability/Preference Profiles (CC/PP). These repositories store technical specifications of different commercial communication devices.

The User Profile dimension gathers the information using the user ID to look up his/her preferences in the preferences repository in the system. The Situation dimension analyses the Natural Language request identifying specific words such as "urgently" or "emergency". The relationship between these dimensions is established using a preferences function, which assigns weights to each one of them. For instance, considering a user connected that employs a smartphone with video capabilities. This user has registered low cost preference; therefore, cheaper services like SMS or regular voice calls are more likely to be selected because the user price preference has a higher weight in the preferences calculation. On the other hand, if the system detects a situation of emergency, the user preferences from the repository are overridden and the most reliable services are selected instead no matter the cost of the service.

Table 1. Mapping between user context information and service properties

User criteria	Service property	Type
Network	Payload size	Bytes
Device	Payload size	Bytes
Location	Voice, text	Integer
Data subscription	Require subscription	Boolean
Only free services	Cost	Value
Voice subscription	Voice, text	Boolean
Delivery quality	Delivery warranty	Integer

To map the relationship between context and services, the Context Analyzer maps context data to services properties. This information is used in subsequent stages of Service composition, as described further on. Table 1 shows context criteria identified from the request. It additionally, shows how these user criteria are mapped to service properties. These properties are later considered during CCS reconfiguration.

4   Service Adaptation

4.1   Automatic Deployment

As mentioned before, in automated composition it is required that the user to be served immediately, so the best solution given a time window is selected whereas additional plans generated after the initial solution comprise the ranking of alternative plans that are possibly used if required (see Fig. 3). This approach is described in [11]. During the synthesis of the input problem, the best plan that satisfies the user request and preferences in a defined time is selected. The computation and execution of alternative plans is done in background so the user gets an instant response. In order to establish an association between dynamic synthesized plans and JSLEE components, the synthesized CCS are translated into Java components. To do so, the abstract CCS are integrated with execution patterns (conditional, fork, join, ...) defined as Java snippets [11].

The reconfiguration process uses an alarm-based approach. These alarms are described in the JSLEE standard. In case of failure, one or more alarms are activated. The alarms are encrusted in the Java code during the translation between abstract CCS into SBB Components. The black circles in the Fig. 3 represent such alarms.

The alarms encrusted in the executable CCS allow monitoring the execution during the workflow. Monitoring the CCS may lead to three courses of action: if the error is caused by an atomic service, the system selects another service from the implementation services repository and continues on; second, if the problem is caused by the whole CCS and the CCS itself can be modified, a new plan from

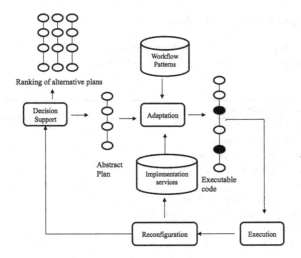

Fig. 3. Reconfiguration schema

the previously generated ranking is selected; finally, if there is a problem in the middle of the executing CCS and no alternatives are available, a new planning CCS is initiated in the Decision Support module to complete the task, starting from the actual state of the world, i.e., a set of values of the variables that define a system in a given moment.

To solve these issues, our approach incorporates a region-based reconfiguration. Next we present the region-based service reconfiguration algorithm.

In the traditional reconfiguration, if the failure is raised in more than one service, it could be necessary to re-plan and/or re-adapt the entire CCS. Besides, it could be necessary to undo all the performed tasks in the previous plan that does not match with the new generated plan. Consequently, the less reconfiguration is performed; the better would be the performance [12]. It was important to clarify that undo actions aren't feasible in some situations when services have been invoked. However, in some scenarios such as early warning management, the services or actions are to configure the sensing time of some devices, or turn on other device. In the latter situations the undo actions may be performed. Additionally, in some scenarios, the reconfiguration of telecommunication services may be done in execution time using techniques such as code injection [15]

In this paper we model the CCS as text strings and perform a novel search algorithm for discovering regions of services based on the multimodal approach. Therefore, it is important to find CCS or fragments thereof that can be reusable for defining new adjustable CCS to meet different requirements of the initial composition. One of the current challenges in this field is precisely the improvement in efficiency in the search for such CCS or fragments that normally is done manually or automated planning, involving heavy demands on time and resources.

4.2 Reconfiguration Algorithm

The Algorithm 1 for convergent service reconfiguration is called when an error is detected during execution, the algorithm is shown below:

Algorithm 1. Reconfiguration Algorithm

Require: faulty CCS si, threshold c
Ensure: replaced sub process $Rf = ri$
 1: Set region $ri = si$; $Rf = \emptyset$, $counter = 0$
 2: **while** $Rf \neq \emptyset$ and $counter < threshold$ c **do**
 3:     $counter + +$
 4:     $ri = \{ri - counter, ri, ri + counter\}$
 5:     Select a created plan ri' that meets QoS of ri
 6:     **if** $ri' \neq \emptyset$ **then**
 7:         add ri' to Rf
 8:     **end if**
 9:     **if** $Rf = \emptyset$ **then**
10:         Search a Rf similar to si using multimodal search
11:         **if** $Rf \neq \emptyset$ **then**
12:             add ri' to Rf
13:         **end if**
14:     **end if**
15: **end while**
16: **return** Rf

Let us suppose that a plan is generated and a CCS si presents malfunctions. Then, a replacement for si in the repository must be found. The new CCS should contain the same non-functional properties as to the initial constraints of the plan defined by the user's preferences.

If such a CCS cannot be found, it is necessary to try to replace the surrounding services, so we increase the counter and expand the region. To do so, we include the previous and the next service. Replacing a set of services together gives us more flexibility as long as the replacing services can meet the combined constraints needed. In this way, the reconfiguration region can be extended or reduced in order to include services that accomplish with the aforementioned constraints. It is worth noting that if a region includes many services for replacement, selecting alternative plans from the ranking is probably a better option.

Given p faulty services and a reconfiguration threshold c ($0 < c < 1$) on the maximum number of services to be repaired, the algorithm first starts a loop (lines 2–15) to repeatedly expand the region. Inside the loop, the algorithm first tries to find a replacement region ri' for the failing service and adds it to the response Rf (lines 5–8). If the required plan does not exist, the algorithm search for similar CCS stored in the repository in order to recompose the region ri. The CCS are found using a multimodal model search approach (lines 9–13), which receives as input a text string representing the faulty services to be replaced.

4.3   Multimodal Search for Convergent Services

The multimodal search of CCS is intended to search for text string representing a convergent service or a fragment thereof. This model transforms the CCS into a matrix for two components: a linguistic component and a structural component. The linguistic component contains all the convergent services and gates that compose the CCS, and the structural component contains pairs of sequential nodes (service-service or service-gate) maintaining the order they appear in the CCS. Next the formation of the linguistic and the structural components is detailed.

Formation of the Linguistic Component: suppose that we have a repository T of CCS: $T = \{CCS_1, CCS_2, \ldots, CCS_i, \ldots, CCS_l\}$, where l is the total number of CCSs that it contains. The first step of the algorithm is to read each CCS_i and to represent it as a tree A. Then the algorithm takes each A_i, extracts the textual characteristics (activity name, activity type, and description) and forms a vector $Vtc_i = \{tc_{i,1}, tc_{i,1}, \ldots, tc_{i,l}, \ldots, tc_{i,L}\}$, where L is the number of textual characteristics $tc_{i,l}$ found in A_i. For each vector Vtc_i, which represents a CCS_i, a row of a matrix of textual components MC is constructed. This row contains the linguistic component for the CCS stored in the repository. In this matrix, i represents each CCS and l a textual characteristic for each of them.

Formation of the Structural Component: a codebook Cd is a set of N structural components describing one or more nodes of the CCS in the form of text strings. The set of codebooks formed from the whole repository is called the codebook component matrix. This matrix is formed by taking each tree A_i, which all contain a vector of codebooks $Vcb_i = \sum_{k=1}^{p} Cd_{i,k}$. Therefore, the codebook component matrix MCd_{ik} is formed where i represents the current CCS and k represents its correspondent codebook.

Indexing Process: the linguistic and codebook components are weighted to create a multimodal search index MI composed of the matrix of the linguistic component MC and the codebook component matrix MCd, i.e. $MI = \{MCd \cup MC\}$. The index also saves the reference to the physical file of each of the models stored in the repository. These models correspond to CCS stored in the BP Repository.

- Weighting: Next, the indexing layer applies a weighting scheme of terms, similar to that proposed in information retrieval (IR), via the document representation vector model to form the term document matrix. This weighting scheme is based on Eq. 1, initially proposed by Salton [16].

$$W_{i,j} = \frac{F_{i,j}}{max(F_i)} \times \log\left(\frac{I}{I_j + 1}\right) \tag{1}$$

In Eq. 1, $F_{i,j}$ is the observed frequency of a component j in CCS_i, the component j may be a linguistic (l) or codebook component k. $max(F_i)$ is the greatest frequency observed in CCS_i, I is the number of CCSs in the collection. Each cell $W_{i,j}$ of the multimodal index matrix reflects the weight of a specific element j of a CCS_i, compared with all the elements of the BP in the repository.

- Index: The multimodal index is stored in a physical file within the file system of the operating system, in which each CCS is indexed through a pre-processing mechanism. This mechanism consists in converting all the terms of the linguistic and codebook component matrices to lowercase and removing stop words, accents and special characters. Subsequently, the stemming technique (Porter algorithm [16]) is applied, which enables each of the matrix elements to be transformed into their lexical root (e.g. the terms "fishing" and "fished" become the root, "fish") and physical file of each CCS to be stored in the repository.

Search Similar CCS: when a CCS fails and there are not other services or CCS that accomplish their IOPE and QoS requirements, the multimodal search process is called. It receives the malfunctioning CCS and forms the linguistic and the structural component. Next, a conceptual rating (score) is expressed expressed by Eq. 2 (based on the *Lucene practical scoring function*[1]) is applied in order to return a ranking list of results containing a set of convergent services with the highest conceptual punctuation within the multimodal index.

$$score\,(q,d) = coord\,(q,d) \times \sum_{t \in q} \Big(tf\,(t \in d) + idf\,(t)^2 \times norm\,(t,d) \Big) \quad (2)$$

In Eq. 2 t is a term of query q; d is the current CCS with q; $tf(t \in d)$ is the term frequency defined as the number of times the term t appears in d so that CCS are ranked according to the values of term frequency scores; $idf(t)$ is the number of convergent services in which term t appears (inverse frequency); $coord(q,d)$ is the scoring factor based on the number of query terms found in the queried CCS (those CCS containing the most query terms are scored highest); $norm(t,d)$ is the weighting factor in the indexing, taken from $w_{i,j}$ in the multimodal index

Eventually, the goal of the planning algorithm is to create a new plan. Furthermore, the planning algorithm considers user preferences trough the cost function that assigns a cost value to each generated plan. This process is described in detail elsewhere [10].

The algorithm continues and, if the repaired regions include too many nodes (as defined by c), the whole plan will be replaced from the existing ranking of plans.

---

[1] http://lucene.apache.org/core/3_6_2/api/core/org/apache/lucene/search/Similarity.html.

5   Experimentation

The experimentation was focused on computing the performance of the proposed approach which is based on the execution time evaluated in two parts: the first part verifies the performance of the proposed approach in a telecommunications environment and the second part verifies the performance of the multimodal search using test plans with different number of nodes.

For first part, an hypothetical composite convergent service was defined by running in it a Telecom Server that invokes some telecom features: send SMS, invoke a Web Service and performs a voice call. The composite service uses a set of services in a sequence. The traditional method reconfigures the whole plan from scratch, to do so; this method must undo the tasks performed before to the error occurs. Only then the new plan can be started. If the error is presented at the beginning, then there is no need for undo any task. Conversely, if the error is present at the last node it is necessary to remake the entire plan or select other plan from the ranking. For representing undo tasks in our experiment, we performed a call establishments and web service invocations. As would be expected, time increases linearly with the number of tasks that must be undone; the higher the node of the error, the higher is the time consumed to undone the previous tasks (continuous line in Fig. 4).

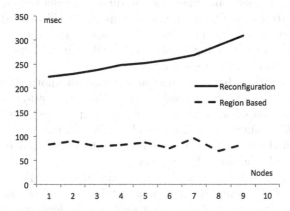

Fig. 4. Performance comparison of traditional vs. Region based reconfiguration using automated planning

Figure 4 shows the performance of the reconfiguration algorithm using automated planning. Figure 4 shows a test a plan with 10 nodes. It is assumed than an error occurred at different nodes of the path from 1st to 10th node (X axis in Fig. 4).

As can be seen in Fig. 4, the algorithm for region-based reconfiguration using automated planning has a better performance that the traditional reconfiguration processes. Next, the aim of the experimentation is to test if the multimodal search presents better results that the region based reconfiguration using automated planning.

Table 2. Performance of the multimodal search for CCS with 10, 20 and 30 nodes

10 Nodes runtime (ms)	20 Nodes runtime (ms)	30 Nodes runtime (ms)
17	40	60

For the second part, the performance of the multimodal search was evaluated finding CCS with 10, 20, and 30 nodes taking into account that most of the CCS contain only few nodes. Table 2 shows, that the Multimodal search approach performing a simple text extraction algorithm (complexity $O(n2)$) takes just a few milliseconds to retrieve the results (17–60 ms) in the CCS with 10–30 nodes. Therefore the multimodal algorithm is suitable to the process of reconfiguration of CCS as it shows a low computational complexity.

6   Conclusions

Convergent composition requires that CCS can be efficiently recovered from failures. This work presents the results of our ongoing work towards the definition of a mechanism for planning based reconfiguration in convergent domains. Furthermore, we present an algorithm based on multimodal model to find services that can be used to reconfigure troublesome regions of a CCS instead of remaking the whole CCS or selecting other plan. "Replanning" the whole CCS or selecting other plan from the ranking involves a big effort in undoing the actions of services previous to the failing service.

Multimodal search is based on text and structural information; due to the latter the quality of the CCS retrieval is higher than using automated planning. Besides, the inclusion of structural information helps to get better plans that sequential plans obtained from traditional planners. Finally the experimental results show that the performance of the multimodal search may reduce in many orders of magnitude the time of execution.

The future work will be focused in using the algorithm for performing the reconfiguration of different failing regions at the same time. Equally we are interested in perform further testing that evaluates performance and quality of the approach in Cloud based platforms for convergent services measuring real user experience.

References

1. Object Management Group: Uml profile for advanced and integrated telecommunication services (TelcoML). Standard, OMG, August 2013. http://www.omg.org/spec/TelcoML/1.0/
2. Ambra, T.: Description and composition of services towards the web-telecom convergence. In: Lomuscio, A.R., Nepal, S., Patrizi, F., Benatallah, B., Brandić, I. (eds.) ICSOC 2013. LNCS, vol. 8377, pp. 578–584. Springer, Heidelberg (2014)

3. Wang, D., Yang, Y., Mi, Z.: A genetic-based approach to web service composition in geo-distributed cloud environment q. Comput. Electr. Eng. (2014)
4. Jula, A., Sundararajan, E., Othman, Z.: Cloud computing service composition: a systematic literature review. Expert Syst. Appl. **41**(8), 3809–3824 (2014)
5. Lin, K.J., Zhang, J., Zhai, Y., Xu, B.: The design and implementation of service process reconfiguration with end-to-end QOS constraints in SOA. Serv. Oriented Comput. Appl. **4**(3), 157–168 (2010)
6. Pernici, D.A.B.: Adaptive service composition in flexible processes. IEEE Trans. Softw. Eng. **33**, 369–384 (2007)
7. Kaldeli, E., Lazovik, A., Aiello, M.: Continual planning with sensing for web service composition. In: AAAI, pp. 1198–1203 (2011)
8. Ordóñez, H., Corrales, J.C., Cobos, C.: Multisearchbp-entorno para busqueda y agrupacion de modelos de procesos de negocio. Polibits **49**, 29–38 (2014)
9. Ordóñez, A., Corrales, J.C., Falcarin, P.: Natural language processing based services composition for environmental management. In: 2012 7th International Conference on System of Systems Engineering (SoSE), pp. 497–502. IEEE (2012)
10. Ordonez, A., Alcázar, V., Borrajo, D., Falcarin, P., Corrales, J.C.: An automated user-centered planning framework for decision support in environmental early warnings. In: Pavón, J., Duque-Méndez, N.D., Fuentes-Fernández, R. (eds.) IBERAMIA 2012. LNCS, vol. 7637, pp. 591–600. Springer, Heidelberg (2012)
11. Ordóñez, A., Alcázar, V., Corrales, J.C., Falcarin, P.: Automated context aware composition of advanced telecom services for environmental early warnings. Expert Syst. Appl. **41**(13), 5907–5916 (2014)
12. Ordoñez, H., Corrales, J.C., Cobos, C.: Business processes retrieval based on multimodal search and lingo clustering algorithm. IEEE Latin Am. Trans. **13**(9), 40–48 (2015)
13. Gerevini, A.E., Haslum, P., Long, D., Saetti, A., Dimopoulos, Y.: Deterministic planning in the fifth international planning competition: PDDL3 and experimental evaluation of the planners. Artif. Intell. **173**(56), 619–668 (2009). Advances in Automated Plan Generation
14. Guzmán, C., Alcázar, V., Prior, D., Onaindia, E., Borrajo, D., Fdez-Olivares, J., Quintero, E.: PELEA: a domain-independent architecture for planning, execution and learning. In: Proceedings of the ICAPS, vol. 12, pp. 38–45 (2012)
15. Adrada, D., Salazar, E., Rojas, J., Corrales, J.C.: Automatic code instrumentation for converged service monitoring and fault detection. In: 2014 28th International Conference on Advanced Information Networking and Applications Workshops (WAINA), pp. 708–713. IEEE (2014)
16. Manning, C.D., Raghavan, P., Schütze, H.: Introduction to Information Retrieval, vol. 1. Cambridge University Press, Cambridge (2008)

On Extracting Information
from Semi-structured Deep Web Documents

Patricia Jiménez and Rafael Corchuelo[⊠]

ETSI Informática, Avda. Reina Mercedes, s/n., 41012 Sevilla, Spain
{patriciajimenez,corchu}@us.es

Abstract. Some software agents need information that is provided by
some web sites, which is difficult if they lack a query API. Information
extractors are intended to extract the information of interest automatically and offer it in a structured format. Unfortunately, most of them rely
on ad-hoc techniques, which make them fade away as the Web evolves.
In this paper, we present a proposal that relies on an open catalogue of
features that allows to adapt it easily; we have also devised an optimisation that allows it to be very efficient. Our experimental results prove
that our proposal outperforms other state-of-the-art proposals.

Keywords: Information extraction · Semi-structured deep-web data
sources

1 Introduction

Since the Web is currently the wealthiest source of data, many authors have
worked on proposals whose goal is to help engineers create information extractors
as automatically as possible.

Our focus is on learning a set of rules that can be used to extract the data
of interest (positive examples) and discard the spurious data (negative examples) from semi-structured deep-web documents. There are many proposals that
address this problem, but most of them are ad-hoc, that is, they rely on specific-
purpose machine-learning techniques that were specifically tailored to the problem of extracting semi-structured web data [7,29,35]; many of them are even
specific to a kind of layout, e.g., lists, tables, or search engine results [1,24].
This makes it difficult to adapt them as the Web evolves, since the features
of the documents on which they rely and the techniques used to analyse them
are built into the proposals. Consequently, extracting web data and structuring
them has become quite an active research field for years, since existing proposals
fade away quickly as the Web evolves. Only a few researchers have tried to use
open catalogues of features to design their information extractors. Such open
catalogues are appealing insofar they ease adapting the proposals as the Web
evolves. Instead of devising a new algorithm to deal with the evolutions of the
Web, one can focus on the features that capture the essence of such evolutions.

© Springer International Publishing Switzerland 2015
W. Abramowicz (Ed.): BIS 2015, LNBIP 208, pp. 140–151, 2015.
DOI: 10.1007/978-3-319-19027-3_12

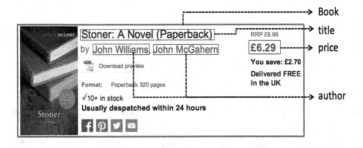

Fig. 1. Sample training document.

The few existing proposals in this category include SRV [13], Irmak and Suel's proposal [18], L-Wrappers [5], and Fernández-Villamor et al.'s proposal [12].

In this paper, we present a proposal to learn rules that are based on an open catalogue of attributive and relational features. By using relational features, our system is able to learn very expressive rules, since they consist of conditions that rely on first-order predicates with which we can model arbitrary properties of a web source. By using an open catalogue of features, our system can evolve as the Web does, which makes it very flexible and adaptable to changes. It follows a top-down covering approach, that is, it starts with the most general rule, and it iteratively adds conditions that are based on the features of the catalogue until the rule does not match any negative examples. The process finishes when every positive example is matched. Otherwise, the process continues to learn new rules. Our proposal also provides mechanisms not to produce very complex or too specific rules. To avoid the former, it includes a version of the Minimum Description Length principle; to avoid the latter, it tries not to include the most promising conditions only, but also conditions that can spread the search space and produce good rules in the forthcoming steps. Our proposal has to search through a typically large search space and the cost of evaluating candidate conditions is high if there are many positive and negative examples. To tackle this problem, we have incorporated a simple technique to discard some negative examples that has proven to work very well in practice.

The rest of the paper is organised as follows: Sect. 2 describes our proposal; Sect. 3 reports on our experimental analysis; Sect. 4 discusses on the related work; finally, Sect. 5 concludes the paper.

2   Description of Our Proposal

2.1   Training Sets

Our proposal works on a training set that is composed of positive examples, negative examples, and feature instantiations. To assemble it, we recommend downloading at least six documents that account for as much variability as possible, e.g., permuted or missing attributes, alternate formats, and so on.

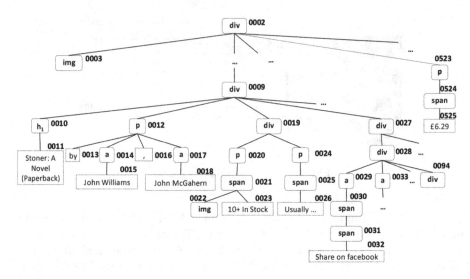

Fig. 2. A sample DOM tree.

Table 1. Partial catalogue of features. (W = word, N = DOM node, V = value.)

FEATURE NAME	FEATURE NAME	FEATURE NAME	FEATURE NAME
firstBigram(N,W,W)	numberOfBlanks(N,V)	numberOfIntegers(N,V)	endsWithPunctuation(N)
parent(N,N)	numberOfUppercaseBigrams(N,V)	isURL(N)	isAlphaNum(N)
lastSibling(N,N)	height(N,V)	width(N,V)	isNumber(N)
coordinates(N,V,V)	isUppercase(N)	backgroundColor(N,V)	numberOfUppercaseTokens(N,V)
firstSibling(N,N)	isCurrency(N)	isLowerCase(N)	isEmail(N)
tagName(N,V)	numberOfAlphaNum(N,V)	beginsWithPunctuation(N)	lineHeight(N,V)
children(N,N)	numberOfLowercaseTrigrams(N,V)	isCapitalised(N)	isBlank(N)
lastBigram(N,W,W)	numberOfChildren(N,V)	fontSize(N,V)	endsWithParenthesis(N)
firstWord(N,W)	numberOfSiblings(N,V)	isDate(N)	containsCurrencySymbol(N)
siblingIndex(N,V)	isNumber(N)	isbnNode(N)	verticalAlign(N,V)
lastWord(N,W)	containsBracketedNumber(N)	beginsWithParenthesis(N)	textAlign(N,V)
beginsWithNumber(N)	lowerCaseTokenAmount(N,V)	display(N,V)	borderBottomWidth(N,V)
nextSibling(N,N)	numberOfLetters(N,V)	fontWeight(N,V)	borderLeftWidth(N,V)
textLength(N,V)	numberOfCapital(N,V)	containsBracketedAlphaNum(N)	containsQuestionMark(N)
textNode(N)	numberOfDigits(N,V)	isYear(N)	isPhone(N)
depth(N,V)	numberOfUpperCaseTrigrams(N,V)	numberOfLowercaseBigrams(N,V)	...
numberOfTokens(N,V)	textDecoration(N,V)	marginBottom(N,V)	
sibling(N,N)	hasNotBlanks(N)	borderRightColor(N,V)	

Figure 1 shows a very simple document that a user has annotated, that is, he or she has specified which the positive examples are and has labelled them; Fig. 2 shows its corresponding DOM tree view. This document must be transformed into a training set by computing the features of every node. Table 1 shows a partial view of our open catalogue, which includes lexical, HTML, rendering or semantic features, to mention a few categories, and Table 2 shows an excerpt of our sample training set.

Table 2. An excerpt of a training set.

% Positive Examples	% Attributive Features	% Relational Features
author(0014).	numberOfTokens(0010, 10).	parent(0010, 0009).
author(0017).	fontFamily(0010, "Arial").	children(0010, 0011).
	fontSize(0010,"20px").	sibling(0010,0012).
% Negative Examples	colour(0010,"#212121").	sibling(0010,0019).
		sibling(0010,0027).
not(author(0001)).	firstWord(0013,"by").	...
...		children(0012, 0013).
not(author(0009)).	...	children(0012,0014).
...	tagName(0014, a).	children(0012,0016).
not(author(0013)).	numberOfDigits(0014,0).	children(0012,0017).
not(author(0015)).	numberOfChildren(0014, 1).	...
not(author(0016)).	numberOfTokens(0014,3).	parent(0014, 0012).
...	...	previousSibling(0014, 0013).
not(author(0018)).	tagName(0017, a).	firstSibling(0014, 0013).
not(author(0019)).	numberOfDigits(0017,0).	lastSibling(0014, 0017).
not(author(0020)).	numberOfChildren(0017, 1).	children(0014, 0015).
...	numberOfTokens(0017,3).	...
not(author(0523)).	...	parent(0017, 0012).
not(author(0524)).	containsCurrencySymbol(0525).	previousSibling(0017, 0016).
not(author(0525)).	isTextNode(0525).	firstSibling(0017, 0013).
	numberOfDigits(0525, 3).	children(0017, 0018).
	isPrice(0525).	...
	...	

```
1: procedure main(docs)
2:     result ← ∅
3:     for each different label in the annotations of docs do
4:         trainingSet ← CreateTrainingSet(docs, label)
5:         trainingSet ← ReduceNegatives(trainingSet)
6:         ruleset ← LearnRuleSet(label, trainingSet)
7:         result ← result ∪ {label ↦ ruleSet}
8:     end for
9:     return result
10: end procedure
```

Fig. 3. The main procedure.

2.2 The Main Procedure

Figure 3 shows our main procedure. It works on a collection of documents and iterates through the labels in the user annotations. In each iteration, it creates a training set and pre-processes it in order to reduce the number of negative examples, which speeds up the learning process since there are typically many such examples. By repeated experimentation we found that the best alternative was to remove 40 % of the negative examples randomly. Finally, this procedure returns a map in which each of the labels in the input documents is associated with a rule set that is specifically tailored to identifying positive examples of the corresponding types.

Learning a Rule Set: Figure 4 shows our procedure to learn a rule set. It works on a training set and a label, and it returns a set of rules. It first creates an empty set of rules and proceeds to create a single rule in each iteration.

```
 1: procedure LearnRuleSet(label, trainingSet)
 2:     ruleSet ← ∅
 3:     repeat
 4:         rule ← LearnRule(label, trainingSet)
 5:         if rule ≠ nil then
 6:             (ruleSet, trainingSet) ← Update(ruleSet, trainingSet, rule)
 7:         end if
 8:     until rule = nil ∨ trainingSet has no positive examples
 9:     ruleSet ← PostProcess(ruleSet)
10:     return ruleSet
11: end procedure
```

Fig. 4. Procedure to learn a rule set.

If it succeeds, then the rule set is updated with the new rule, and the training set is subtracted the positive examples that it matches. The loop finishes when no new rule can be learnt or no positive examples remain in the training set. Finally, the rule set is post-processed to remove subsumed rules and useless conditions.

Learning a Rule: Figure 5 presents the procedure to learn a rule. It works on a label and a training set, and it starts with an empty rule. Then, it generates a set of possible refinements. Each refinement is a condition that consists in an instantiated feature or a built-in comparator. A refinement of the first type is a feature defined in the catalogue in Fig. 1, but the arguments are instantiated with variables of the corresponding types. The variables can be new variables that expand the search space or bound variables that have been introduced in the head of the rule or in previous conditions of the body of the rule. The only constraint is that there must exist a bound variable in every possible refinement. A built-in comparator is a binary condition that compares two variables or a variable and a constant, as long as they both are of the same type. The types with which a built-in comparator works can be numeric or discrete. We use five built-in comparators, namely: $=$, $>$, $>=$, $<$, and $<=$. When the variables to compare are discrete, only the comparator $=$ is available.

In each iteration it attempts to refine the rule by adding conditions, so that it matches as many positive examples as possible and as few negative examples as possible. To do that, it implements an outer loop that iterates until no more refinements can be added or until the rule being learnt is a solution or too complex; in every iteration of the inner loop, it first computes a set of possible refinements for the current rule, that is, a set of conditions that might be added to the rule; then it evaluates each condition, that is, it computes a score that allows to compare its goodness to others, and it checks whether it is bad enough to be pruned immediately. Refinements that are not pruned are stored in a candidate set together with their scores. A candidate rule is the result of adding a candidate refinement to the current rule. The inner loop is executed as long as the current rule can be refined and it is not a solution; a rule is a solution when it matches some remaining positive examples and no negative ones.

```
 1: procedure LearnRule(label, trainingSet)
 2:     rule ← CreateEmptyRule(label)
 3:     repeat
 4:         stop ← false
 5:         candidates ← ∅
 6:         refinements ← GenerateRefinements(rule)
 7:         for each refinement in refinements while ¬stop do
 8:             (score, pruned) ← Evaluate(rule, refinement, trainingSet)
 9:             if ¬pruned then
10:                 candidate ← (refinement, score)
11:                 candidates ← candidates ∪ {candidate}
12:                 stop ← IsSolution(candidate, trainingSet)
13:             end if
14:         end for
15:         if candidates = ∅ then
16:             rule ← nil
17:         else
18:             bestCandidates ← FindBestCandidates(rule, candidates)
19:             rule ← add refinements in bestCandidates to rule
20:         end if
21:     until rule = nil ∨ IsSolution(rule, trainingSet) ∨ IsTooComplex(rule, trainingSet)
22:     if IsTooComplex(rule, trainingSet) then
23:         rule ← nil
24:     end if
25:     return rule
26: end procedure
```

Fig. 5. Procedure to learn a single rule.

Once the set of candidates is computed, the best ones are selected and added to the rule being constructed. When the outer loop finishes, the algorithm checks if the candidate rule is too complex, in which case, it returns *nil* to indicate that no good rule was found.

2.3 Output Rules

Next, we illustrate the rules that our proposal learnt to extract the data in Fig. 1. First, we learnt the following rule to identify book records:

$$book(X) :\!- children(X, Y), not(firstSibling(Y, Z)),$$
$$tagName(Y, V), V = \text{``img''},$$
$$containsBracketedNumber(X),$$
$$containsCurrencySymbol(X).$$

It means that a book is a node whose first child is an image node and that some of its descendants has a number in brackets and a currency symbol.

Table 3. Labelled information on the first group of datasets

CATEGORY	LABELLED INFORMATION
Books	Book (record), title, author, price, isbn, and year
Movies	Movie (record), title, director, actor, year, and runtime
Cars	Car (record), make, color, doors, engine, mileage, model, price, transmission, type, and location
Conferences	Conference (record), date, place, title, and url
Doctors	Doctor (record), name, address, phone, fax, and specialty
Jobs	Offer (record), company, location, and category
Real Estate	Property (record), address, bedrooms, bathrooms, size, and price
Sports	Player (record), name, birth, height, weight, club, country, position, age, and college

The rule to identify the titles is as follows:

$$title(X) :- not(firstSibling(X, Y)),$$
$$tagName(X, Z), Z = \text{"h1"}.$$

It means that a title is a node that does not have a left sibling and is within an H1 header.

The rule for the author label is as follows:

$$author(X) :- tagName(X, Y), Y = \text{"a"},$$
$$firstSibling(X, Z),$$
$$firstWord(Z, V), V = \text{"by"}.$$

It means that an author is a hyper-link node whose first sibling starts with the word "by".

Finally, the rule for the price label is as follows:

$$price(X) :- isPrice(X), fontWeight(X, Y),$$
$$Y = \text{"bold"}, fontSize(X, Z), Z = \text{"18.7px"}.$$

It means that a price is an element node whose text matches a price format, whose font weight is "bold", and whose font size is "18.7px".

3   Experimental Analysis

We used a computer that was equipped with an Intel Core i7-2600 that ran at 3.34 GHz, had 8 GiB of RAM, Windows 7 Pro 64-bit, Oracle's Java Development Kit 1.7.0_09, and SWI-Prolog 6.2.3. We used a collection of 25 datasets that provides a total of 657 documents. We gathered them from 20 real-world web sites and the remaining were downloaded from the RoadRunner and the RISE public repositories. For the first group of datasets, we downloaded 30 documents from each web site and handcrafted a set of annotations with the data that we wished to extract from each document, namely, data on books, cars, conferences, doctors, jobs, movies, real estates, and sport players. The information labelled from these web sites is shown in Table 3. The second group contains some of the datasets available at the RoadRunner and the RISE repository that provide

Table 4. Experimental results.

	Trinity			RoadRunner			FivaTech			SoftMealy			WIEN			OUR PROPOSAL		
Summary	P	R	F1	P	R	F1	P	R	F1	P	R	F1	P	R	F1	P	R	F1
Mean	0.968	0.968	0.966	0.575	0.567	0.571	0.829	0.882	0.848	0.867	0.691	0.724	0.694	0.617	0.627	0.985	0.986	0.984
Standard deviation	0.065	0.065	0.056	0.399	0.396	0.398	0.168	0.110	0.134	0.103	0.312	0.285	0.254	0.322	0.301	0.026	0.023	0.020
Dataset	P	R	F1	P	R	F1	P	R	F1	P	R	F1	P	R	F1	P	R	F1
Abe Books	1.000	1.000	1.000	0.648	0.585	0.615	0.917	0.991	0.953	0.873	0.578	0.696	0.516	0.160	0.245	1.000	1.000	1.000
Better World Books	0.989	1.000	0.994	-	-	-	0.989	0.958	0.973	0.979	0.985	0.982	0.428	0.350	0.385	1.000	1.000	1.000
Waterstones	0.961	1.000	0.980	0.978	0.867	0.919	1.000	0.944	0.971	1.000	1.000	1.000	0.714	0.675	0.694	0.992	1.000	0.996
IMDB	0.928	0.861	0.893	0.393	0.353	0.372	-	-	-	0.878	0.847	0.862	0.376	0.376	0.376	0.940	0.993	0.961
Disney Movies	1.000	1.000	1.000	0.667	0.667	0.667	0.712	0.671	0.691	0.970	0.970	0.970	0.718	0.718	0.718	0.984	0.992	0.988
All Movies	0.971	0.956	0.963	0.267	0.259	0.263	0.794	0.740	0.767	0.930	0.288	0.439	0.131	0.072	0.093	1.000	0.975	0.987
Auto Trader	0.992	1.000	0.996	-	-	-	-	-	-	0.889	0.870	0.880	0.889	0.000	0.000	1.000	0.981	0.990
Car Max	1.000	1.000	1.000	0.981	0.981	0.981	0.449	0.889	0.597	0.892	0.892	0.892	0.875	0.875	0.875	1.000	1.000	1.000
Classic Cars for Sale	0.859	0.904	0.881	0.493	0.493	0.493	-	-	-	0.896	0.896	0.896	0.170	0.128	0.146	0.991	0.967	0.978
All Conferences	0.984	0.992	0.988	0.722	0.722	0.722	0.840	0.900	0.869	0.993	0.253	0.404	0.800	0.400	0.533	0.939	0.992	0.961
Mbendi	1.000	1.000	1.000	0.867	0.867	0.867	0.903	1.000	0.949	0.600	0.600	0.600	0.800	0.400	0.533	1.000	1.000	1.000
Web MD	1.000	1.000	1.000	0.063	0.065	0.064	0.775	1.000	0.873	0.865	0.452	0.593	0.600	0.600	0.600	0.915	0.942	0.928
Steady Health	1.000	1.000	1.000	1.000	1.000	1.000	0.833	0.833	0.833	0.750	0.250	0.375	0.750	0.750	0.750	1.000	1.000	1.000
6 Figure Jobs	1.000	1.000	1.000	0.522	0.522	0.522	1.000	0.978	0.989	0.750	0.000	0.000	0.250	0.250	0.250	1.000	0.906	0.949
Career Builder	1.000	1.000	1.000	0.000	0.000	0.000	0.802	0.833	0.818	0.750	0.000	0.000	0.750	0.750	0.750	0.935	0.990	0.959
Homes	0.993	0.993	0.993	1.000	1.000	1.000	-	-	-	0.800	0.789	0.794	0.917	0.917	0.917	1.000	1.000	1.000
Remax	0.700	0.980	0.817	0.000	0.000	0.000	-	-	-	0.844	0.844	0.844	1.000	1.000	1.000	0.986	0.972	0.979
UEFA	1.000	1.000	1.000	1.000	1.000	1.000	-	-	-	1.000	1.000	1.000	1.000	1.000	1.000	1.000	1.000	1.000
ATP World Tour	0.971	0.994	0.982	0.967	0.967	0.967	0.990	0.878	0.930	0.943	0.943	0.943	0.714	0.714	0.714	0.949	0.970	0.958
NFL	1.000	1.000	1.000	0.978	0.978	0.978	0.530	0.813	0.642	0.714	0.714	0.714	0.857	0.857	0.857	0.994	1.000	0.997
Amazon (cars)	0.930	0.730	0.818	0.000	0.000	0.000	0.600	0.670	0.633	0.980	1.000	0.990	0.970	1.000	0.985	1.000	1.000	1.000
UEFA (teams)	0.990	0.990	0.990	0.917	0.917	0.917	0.970	0.990	0.980	0.810	0.890	0.848	0.640	0.750	0.691	1.000	1.000	1.000
Netflix	0.990	0.990	0.990	0.761	0.795	0.778	0.820	0.800	0.810	0.940	0.820	0.876	0.990	0.990	0.990	0.990	0.967	0.978
Bigbook	0.950	0.940	0.945	0.009	0.000	0.001	-	-	-	0.810	0.770	0.789	0.680	0.980	0.803	1.000	1.000	1.000
Zagat	1.000	0.860	0.925	0.000	0.000	0.000	1.000	0.980	0.990	0.810	0.630	0.709	0.810	0.720	0.762	1.000	1.000	1.000

semi-structured documents. From every web site, we selected six documents for training purposes and the remaining ones were used for validation purposes. We measured three effectiveness measures and two efficiency measures. The effectiveness measures are precision (P), which is the ratio of true positives of a rule with regard to the total number of true positives and false positives, that is, $P = tp/(tp + fp)$; recall (R), which is the ratio of true positives of a rule with regard to the total number of true positives and false negatives, that is, $R = tp/(tp + fn)$; and the $F1$ measure, which is the harmonic mean of precision and recall, that is, $F1 = 2PR/(P + R)$. The efficiency measures are the learning time (LT) and the extraction time (ET) as measured in CPU minutes; by extraction time, we mean the time required to execute a rule so that the data of interest is extracted from a web document. Since these measures are sensitive to unexpected experimentation conditions, we repeated the experiments 25 times and averaged the results after discarding some outliers using the well-known Cantelli's inequality.

We searched the Web and contacted many authors in order to have access to the implementation of as many proposals as possible. We managed to find an implementation for SoftMealy [17] and Wien [21], which are classical proposals, and RoadRunner [11], FivaTech [19], and Trinity [32], which are recent proposals. Results are presented in Table 4. A dash in a cell means that the corresponding proposal was not able to learn a rule for a given dataset. From this table, we can confirm that our proposal clearly achieves the best effectiveness. It achieved an average precision of 0.985 ± 0.026, an average recall of 0.986 ± 0.023, and an average F_1 of 0.984 ± 0.020. Only Trinity can achieve results that are comparable to ours, although we outperform it: our precision was 0.016 ± 0.091 higher, our recall 0.018 ± 0.088 higher, and our F_1 0.018 ± 0.076 higher. Furthermore, the standard deviation of these measures is smaller, which means that our approach is generally more stable than the others, that is, it does not generally produce rules whose effectiveness largely deviates from the average.

4   Related Work

The literature provides many rule-based proposals. They build on a generic algorithm that interprets rules that are specific to a web site. These rules range from regular expressions to context-free grammars, Horn clauses, tree templates, or transducers, to mention a few. They can be handcrafted [14,26], which is a tedious and error-prone approach, learnt supervisedly [5,6,8,9,12,13,15–18,20, 21,25,31,33], which requires the user to provide a training set in which he or she has annotated the positive examples, i.e., the data to extract), or unsupervisedly [2–4,10,11,19,22,23,27,28,30,34,36–38], which does not require an annotated training set, but a person to interpret the results.

In spite of the fact that learning such rules has historically been considered a text classification problem, very few proposals have explored using features of the tokens or the DOM nodes, e.g., their length, their colour, their depth, the ratio of letters, and the like. Neither have many proposals explored the idea of using relational features that allow to establish relationships amongst the examples. One of the main problems with exploring such relational features is that they cannot rely on a vector-based representation of the examples since the description of an example may involve features that are also computed from its neighbours.

Only a few proposals have explored the previous idea, namely: (a) SRV [13] starts with an overly-general rule and uses a specialisation procedure that is guided by the Information Gain function to learn a set of rules that can classify sequences of tokens; (b) Irmak and Suel's proposal [18] is an active learning approach, i.e., it learns several XPath-based rules that identify a number of records and then relies on a user to select the most appropriate ones; the procedure is repeated until the user is satisfied. (c) Bădică et al. [5] suggested to transform the input documents into a knowledge base and then use FOIL to infer rules that allow to characterise the DOM nodes that contain positive examples by means of their tags and the tags of their neighbours; (d) Fernández-Villamor et al.'s proposal [12] starts with a set of overly-specific rules and generalises them by combining pairs of rules as long as the overall F_1 measure improves.

Our proposal differs in that it relies on an extensive open catalogue of attributive and relational features whereas others hardly include a few attributive features and seldom use rendering features, which have proven to be necessary to deal with current web documents. Additionally, none of these most-related proposals allow for recursion or negated conditions; ours does, which helps learn richer and more expressive rules.

Regarding computing negative examples, SRV works on text fragments, which makes it problematic when computing negative examples since the number of negative fragments is enormous. The authors included a hard bias so as not to enumerate the whole set of negative examples. In Irmak and Suel's proposal, the user guides the search for rules because negative examples cannot be generated. In L-Wrappers [5] negative examples are computed by using the well-known Closed-World Assumption as it is implemented by the FOIL system. In practice, the authors had to reduce the number of negative examples to roughly 0.1 % for the approach to be manageable; Fernández-Villamor et al.'s and our proposal work on DOM trees and negative examples are computed as the nodes that were not annotated by the user.

Regarding the learning algorithm, SRV is more inefficient since when a token in the neighbourhood helps discern well between positive and negative examples, it is bound by a relational condition and then the system explores only one of its attributive features. Thus, a discriminatory token in the neighbourhood has to be re-bound as many times as attributive features are necessary to be explored. This worsens the efficiency of the learning process. L-Wrappers is very inefficient in practice since the authors mention that extracting records with three or more attributes is infeasible in practice; they resorted to a technique that learns to extract pairs of attributes and then combine the results.

Only L-Wrappers and our proposal include mechanisms to limit the complexity of the rules, which is a means to avoid producing very-specific rules, to prune conditions that are not promising enough, which speeds up the learning process, and to recover from bad decisions by performing backtracking, which reduces significatively the chances to produce poor-quality rules.

5   Conclusions and Future Work

In this paper, we present a very effective proposal to learn rules that allows to extract the information of interest from deep-web sources automatically, so that it can be further processed by software agents. Since the search cost was high, we devised a technique to reduce the number of negative examples that turned it into a practical approach. Our results prove that our proposal is more effective than others in the literature, but, contrarily to them it can be easily evolved since it relies on an open catalogue of features. In future, we plan on analysing a series of heuristics to speed up the learning process without loss of effectiveness. They include exploring new search heuristics and scoring functions to guide the search, exploring conditions in small chunks, sorting the features so that the most frequent in the past are explored first, and so on.

150   P. Jiménez and R. Corchuelo

Acknowledgments. Our work was funded by the Spanish and the Andalusian R&D-
&I programmes by means of grants TIN2007-64119, P07-TIC-2602, P08-TIC-4100,
TIN2008-04718-E, TIN2010-21744, TIN2010-09809-E, TIN2010-10811-E, TIN2010-099
88-E, TIN2011-15497-E, and TIN2013-40848-R, which got funds from the European
FEDER programme.

References

1. Álvarez, M., Pan, A., Raposo, J., Bellas, F., Cacheda, F.: Finding and extracting
   data records from web pages. Signal Process. Syst. **59**(1), 123–137 (2010)
2. Arasu, A., Garcia-Molina, H.: Extracting structured data from web pages. In:
   SIGMOD Conference, pp. 337–348 (2003)
3. Ashraf, F., Özyer, T., Alhajj, R.: Employing clustering techniques for automatic
   information extraction from HTML documents. IEEE Trans. Syst. Man Cybern.
   Part C **38**(5), 660–673 (2008)
4. Barbosa, J.P.D.: Adaptive record extraction from web pages. In: WWW, pp. 1335–
   1336 (2007)
5. Bădică, C., Bădică, A., Popescu, E., Abraham, A.: L-wrappers: concepts, properties
   and construction. Soft Comput. **11**(8), 753–772 (2007)
6. Califf, M.E., Mooney, R.J.: Bottom-up relational learning of pattern matching rules
   for information extraction. J. Mach. Learn. Res. **4**, 177–210 (2003)
7. Chang, C.H., Kayed, M., Girgis, M.R., Shaalan, K.F.: A survey of web information
   extraction systems. IEEE Trans. Knowl. Data Eng. **18**(10), 1411–1428 (2006)
8. Chang, C.H., Kuo, S.C.: OLERA: semisupervised web-data extraction with visual
   support. IEEE Intel. Syst. **19**(6), 56–64 (2004)
9. Cohen, W.W., Hurst, M., Jensen, L.S.: A flexible learning system for wrapping
   tables and lists in HTML documents. In: WWW, pp. 232–241 (2002)
10. Crescenzi, V., Mecca, G.: Automatic information extraction from large websites.
    J. ACM **51**(5), 731–779 (2004)
11. Crescenzi, V., Merialdo, P.: Wrapper inference for ambiguous web pages. Appl.
    Artif. Intel. **22**(1–2), 21–52 (2008)
12. Fernández-Villamor, J.I., Iglesias, C.A., Garijo, M.: First-order logic rule induction
    for information extraction in web resources. Int. J. Artif. Intel. Tools **21**(6), 20
    (2012)
13. Freitag, D.: Machine learning for information extraction in informal domains.
    Mach. Learn. **39**(2/3), 169–202 (2000)
14. Gregg, D.G., Walczak, S.: Exploiting the information web. IEEE Trans. Syst. Man
    Cybern. Part C **37**(1), 109–125 (2007)
15. Gulhane, P., Madaan, A., Mehta, R.R., Ramamirtham, J., Rastogi, R., Satpal, S.,
    Sengamedu, S.H., Tengli, A., Tiwari, C.: Web-scale information extraction with
    vertex. In: ICDE, pp. 1209–1220 (2011)
16. Hogue, A.W., Karger, D.R.: Thresher: automating the unwrapping of semantic
    content from the world wide web. In: WWW, pp. 86–95 (2005)
17. Hsu, C.N., Dung, M.T.: Generating finite-state transducers for semi-structured
    data extraction from the Web. Inf. Syst. **23**(8), 521–538 (1998)
18. Irmak, U., Suel, T.: Interactive wrapper generation with minimal user effort. In:
    WWW, pp. 553–563 (2006)
19. Kayed, M., Chang, C.H.: Fivatech: page-level web data extraction from template
    pages. IEEE Trans. Knowl. Data Eng. **22**(2), 249–263 (2010)

20. Kosala, R., Blockeel, H., Bruynooghe, M., den Bussche, J.V.: Information extraction from structured documents using k-testable tree automaton inference. Data Knowl. Eng. **58**(2), 129–158 (2006)
21. Kushmerick, N., Weld, D.S., Doorenbos, R.B.: Wrapper induction for information extraction. In: IJCAI, vol. 1, pp. 729–737 (1997)
22. Liu, B., Zhai, Y.: NET – a system for extracting web data from flat and nested data records. In: Ngu, A.H.H., Kitsuregawa, M., Neuhold, E.J., Chung, J.-Y., Sheng, Q.Z. (eds.) WISE 2005. LNCS, vol. 3806, pp. 487–495. Springer, Heidelberg (2005)
23. Liu, W., Meng, X., Meng, W.: Vide: a vision-based approach for deep web data extraction. IEEE Trans. Knowl. Data Eng. **22**(3), 447–460 (2010)
24. Meng, W., Yu, C.T.: Advanced Metasearch Engine Technology. Morgan & Claypool Publishers, USA (2010)
25. Muslea, I., Minton, S., Knoblock, C.A.: Hierarchical wrapper induction for semi-structured information sources. Auton. Agents Multi-Agent Syst. **4**(1/2), 93–114 (2001)
26. Raposo, J., Pan, A., Álvarez, M., Hidalgo, J., Viña, Á.: The wargo system: semi-automatic wrapper generation in presence of complex data access modes. In: DEXA Workshops, pp. 313–320 (2002)
27. Simon, K., Lausen, G.: ViPER: augmenting automatic information extraction with visual perceptions. In: CIKM, pp. 381–388 (2005)
28. Sleiman, H.A., Corchuelo, R.: An unsupervised technique to extract information from semi-structured web pages. In: Wang, X.S., Cruz, I., Delis, A., Huang, G. (eds.) WISE 2012. LNCS, vol. 7651, pp. 631–637. Springer, Heidelberg (2012)
29. Sleiman, H.A., Corchuelo, R.: A survey on region extractors from web documents. IEEE Trans. Knowl. Data Eng. **25**(9), 1960–1981 (2013)
30. Sleiman, H.A., Corchuelo, R.: TEX: an efficient and effective unsupervised web information extractor. Knowl.-Based Syst. **39**, 109–123 (2013)
31. Sleiman, H.A., Corchuelo, R.: A class of neural-network-based transducers for web information extraction. Neurocomputing **135**, 61–68 (2014)
32. Sleiman, H.A., Corchuelo, R.: Trinity: on using trinary trees for unsupervised web data extraction. IEEE Trans. Knowl. Data Eng. **26**(6), 1544–1556 (2014)
33. Su, W., Wang, J., Lochovsky, F.H.: ODE: ontology-assisted data extraction. ACM Trans. Database Syst. **34**(2) (2009)
34. Tao, C., Embley, D.W.: Automatic hidden-web table interpretation, conceptualization, and semantic annotation. Data Knowl. Eng. **68**(7), 683–703 (2009)
35. Turmo, J., Ageno, A., Català, N.: Adaptive information extraction. ACM Comput. Surv. **38**(2) (2006)
36. Wang, J., Lochovsky, F.H.: Data extraction and label assignment for web databases. In: WWW, pp. 187–196 (2003)
37. Zhai, Y., Liu, B.: Structured data extraction from the web based on partial tree alignment. IEEE Trans. Knowl. Data Eng. **18**(12), 1614–1628 (2006)
38. Zhu, J., Nie, Z., Wen, J.R., Zhang, B., Ma, W.Y.: Simultaneous record detection and attribute labeling in web data extraction. In: KDD, pp. 494–503 (2006)

A Novel Approach to Web Information Extraction

Antonia M. Reina Quintero$^{(\boxtimes)}$, Patricia Jiménez, and Rafael Corchuelo

ETSI Informática, Avda. Reina Mercedes, s/n., 41012 Sevilla, Spain
{reinaqu,patriciajimenez,corchu}@us.es

Abstract. Business Intelligence requires the acquisition and aggregation of key pieces of knowledge from multiple sources in order to provide valuable information to customers. The Web is the largest source of information nowadays. Unfortunately, the information it provides is available in semi-structured human-friendly formats, which makes it difficult to be processed by automated business processes. Classical propositional and ILP machine-learning techniques have been applied for this purpose. However, the former have not enough expressive power, whereas the latter are more expressive but intractable with large datasets. Propositionalisation was devised as a means to provide propositional techniques with more expressive power, enabling them to exploit structural information in a propositional way that allows them to be efficient. In this paper, we present a proposal to extract information from semi-structured web documents that uses this approach. It leverages a classical propositional machine learning technique and enhances it with the ability to learn from an unbounded context, which helps increase its precision and recall. Our experiments prove that our proposal outperforms other state-of-art techniques in the literature.

1 Introduction

Business Intelligence can be defined as the process of finding, gathering, aggregating, and analysing information for decision making [15]. The World Wide Web is an increasingly important data source for business decision making; however, extracting information from the Web remains one of the challenging issues related to Web business intelligence applications [6]. Web business intelligence is an emerging type of decision support software that "leverages the unprecedented content on the Web to extract actionable knowledge in an organizational setting" according to [20].

Web business intelligence applications should have access to structured data that could be easily extracted and incorporated into decision-making applications. However, the Web provides textual information in semi-structured formats, but those formats are intended for human consumption, which makes them difficult to be processed automatically for business purposes. Web information extractors are tools designed to extract the information of interest from such documents in a structured format.

© Springer International Publishing Switzerland 2015
W. Abramowicz (Ed.): BIS 2015, LNBIP 208, pp. 152–161, 2015.
DOI: 10.1007/978-3-319-19027-3_13

We focus on supervised rule-based information extractors, which are general-purpose algorithms that are configured by means of extraction rules that are learnt from labelled examples that must be provided by a user. Such rules are actually classifiers, since they get a piece of text or a DOM node as input and predict a user-defined class for it.

Unfortunately, many existing proposals build on specific-purpose machine-learning techniques that were specifically tailored to the problem of extracting web information. This makes it difficult to adapt them as the Web evolves because the features of the documents on which they rely and the techniques used to analyse them are built into the proposals.

Machine-learning techniques that come from the field of concept learning have proven to be the only approaches that can scale along a number of distinct sources and can infer knowledge in a general way. There exists two approaches: relational and propositional learning. The former attempts to learn first-order rules that generalise a set of examples by using rich representations that allow to exploit both attributive features of the examples and their relationships, whereas the latter only use attributive features. Therefore, relational systems are preferable since they allow to learn more expressive rules than their counterparts [9]. Unfortunately, a major drawback of relational systems is the expensive learning process, which becomes intractable with large datasets [1].

Propositionalisation [11] is a different way to conduct relational learning efficiently. It maps a relational representation onto features in propositional form, that is, in an attribute-value representation that is amenable for conventional data mining systems. Although propositionalisation has some inherent limitations, such as the inability to learn recurrent definitions, there are several reasons for using this process, the most important ones are: it deals with the combinatorial explosion of the number of potential features; it can leverage the existing knowledge regarding propositionalisation; and, finally, our results prove that our approach is very competitive in terms of effectiveness and efficiency since it outperforms the results obtained by classical and new proposals in the literature for web information extraction.

To the best of our knowledge, no propositionalisation system has been developed in the field of web information extraction. In this paper, we present the first such proposal in the literature. We have performed an extensive experimentation that proves that it outperforms other state-of-the-art proposals, which proves that it is very promising.

The intuitive idea behind our approach, named Yebra, is the following: it relies on a training set that consists of a set of nodes to extract and a subset of nodes that should not be extracted. Then, it learns the best possible rules by using only the attributive features of the nodes in the training set. If the rules thus learnt are perfect, the process finishes. Otherwise, Yebra tries to find better rules by combining attributive features of the nodes in the training set with attributive features of their neighbour nodes. Neighbour nodes are reached by applying relational features such as next sibling or parent, amongst others. At each step and whilst no perfect rules are found, the system continues combining attributive features from the previous step with attributive features of increasingly distant neighbour nodes.

Algorithm 1. YEBRA's main procedure.

```
 1: procedure YEBRA(dataSet, relations)
 2:     trainingSet ← createTrainingSet(dataSet)
 3:     testSet ← dataSet
 4:     resultRule ← learnRule(trainingSet, testSet)
 5:     resultBindings ← {(node, null, null)}
 6:     expansion ← resultRule
 7:     while ¬isPerfect(resultRule, testSet) and expansion ≠ null do
 8:         (expansion, resultBindings, trainingSet, testSet) ←
                                findExpansion(resultRule, resultBindings, trainingSet,
                                testSet, dataSet, relations)
 9:         if expansion ≠ null then
10:             resultRule ← expansion
11:         end if
12:     end while
13:     return (resultRule, resultBindings)
14: end procedure
```

The rest of the paper is organised as follows: Sect. 2 explains our proposal; then the experimental results are detailed in Sect. 3; Sect. 4 presents the related work; Sect. 5 concludes the paper.

2   Description of Our Proposal

In this section, we first describe our training and test sets, then report on Yebra's main algorithm, and then on the algorithms to learn a rule and to expand it.

2.1   The Training Set and the Test Set

The training set is a bag of vectors. Each vector refers to a node in the DOM tree, for which it holds its set of attributive features (A_{ij}) and its class (C_i). An initial training set is created by the *createTrainingSet* function (Line 2), which works as follows: first, it adds the positive examples to the training set. Positive examples are those vectors whose class is not *null*. Then, it adds some negatives examples (not all) to the training set. The negative examples are the ones whose class is *null*. The negative examples added to the training set are computed amongst the negative examples that are the nearest neighbours to the positive examples that belong to the dataset. That is, for every positive example, Yebra finds its k nearest neighbours and adds them to the training set. Note that this process helps Yebra to better discover the frontier amongst positive and negative examples. Experimentally, we have found out that setting k to three is a good trade off between efficiency and effectiveness.

2.2   The Main Procedure

Algorithm 1 presents the main procedure in Yebra. The first step is to find the best rule using attributive features only (Lines 2–4). Then, it iterates until the rule being learnt is perfect or there are not any further expansions (Line 7). An expansion is a new rule learnt by adding extra information to the training set.

The extra information is added as new components to the vectors of the training set (and also the test set). The new components are the attributive features of those nodes that are reachable by applying a relational feature or a combination of relational features. Thus, the existence of an expansion means that Yebra has found a rule that improves on the previous one. But it has to check if this new rule can also be improved (that is, if a new, better classifier can be learnt by adding more features of the neighbours to the training set). If the rule cannot be improved, then there are no expansions, which means that Yebra has found the best possible rule. Finally, note that in each iteration *resultRule* contains the best rule found so far. If it is not perfect, but can be improved, Yebra continues with the next iteration. If it is not perfect and cannot be improved, Yebra stops and returns it.

2.3   Learning a Rule

Once the initial training and test sets are created, Yebra has to learn the best possible rule using procedure *learnRule* (Algorithm 2). It works on a training set and a test set and returns a rule. The intuitive idea behind this procedure is the following: first, it learns a rule, if the rule is perfect, then the procedure stops and returns that rule; otherwise, the training set is modified by adding the misclassified examples, and the process is repeated until the rule is perfect or the rule does not improve on the previous one. That is, the training set is enriched iteratively with misclassified examples, which helps learn better rules.

The procedure starts by balancing the training set (Line 2). This step is necessary because function *learnClassifier* (Line 3) uses a PART learner [21], which does not have good results with unbalanced datasets. The *balance* function computes the number of elements of the majority class, and it then replicates as many examples of each class as needed to obtain a training set that has approximately the same number of examples of every class.

After that, it learns a PART classifier from the balanced training set (Line 3). Then, the set of misclassified examples is computed using this classifier and the test set. Note that although the classifier has been learnt from the input training set (which is a subset of the nodes in the original dataset), the misclassified examples are computed from the test set (which includes all of the nodes in the original dataset).

Then, the procedure iterates until the rule is perfect (the misclassified set is empty) or the new rule does not improve on the old rule. In each iteration:

1. A new training set is computed. It is the union of the vectors included in the previous training set and the vectors related to the nodes that have been misclassified.
2. A new classifier is learnt from the new training set.
3. A new set of misclassified examples is computed for the new classifier.

If the new classifier is perfect, the procedure ends and returns the new classifier. If the new classifier is not perfect and does not improve on the old classifier, the

Algorithm 2. Procedure LEARNRULE

1: **procedure** LEARNRULE($trainingSet, testSet$)
2:     $oldTrainingSet \leftarrow balance(trainingSet)$
3:     $oldRule \leftarrow learnClassifier(oldTrainingSet)$
4:     $oldMisclassified \leftarrow findMisclassified(oldRule, testSet)$
5:     $isPerfect \leftarrow oldMisclassified = \emptyset$
6:     $improves \leftarrow true$
7:     **while** $\neg isPerfect$ **and** $improves$ **do**
8:         $newTrainingSet \leftarrow balance(oldTrainingSet \cup oldMisclassified)$
9:         $newRule \leftarrow learnClassifier(newTrainingSet)$
10:        $newMisclassified \leftarrow findMisclassified(newRule, testSet)$
11:        $isPerfect \leftarrow oldMisclassified = \emptyset$
12:        $improves \leftarrow | newMisclassified | < | oldMisclassified |$
13:        **if** $isPerfect$ **or** $improves$ **then**
14:            $oldTrainingSet \leftarrow newTrainingSet$
15:            $oldRule \leftarrow newRule$
16:            $oldMisclassified \leftarrow newMisclassified$
17:        **end if**
18:     **end while**
19:     $result \leftarrow oldRule$
20:     **return** $result$
21: **end procedure**

procedure ends and returns the old classifier. Finally, if the new classifier is not perfect but improves on the old one, the process is repeated to see if the new classifier can be improved. (Rule R_1 improves on rule R_2 if R_1 misclassifies less examples than R_2.)

2.4   Expanding a Rule

Once Yebra has learnt the best rule using only attributive features, it tries to improve it by adding extra information to the training set. The extra information is obtained by navigating trough the DOM tree by means of relational features, in such a way that if N_i is a vector belonging to the training set with components $(A_{i1}, A_{i2},...,A_{in},C_i)$, and N_j is a vector belonging to the dataset with components $(A_{j1}, A_{j2},...,A_{jn},C_j)$, and R_p is a relational feature such that $N_j = R_p(N_i)$, then an expansion of N_i using R_p is the vector $(A_{i1}, A_{i2},...,A_{in},A_{j1}, A_{j2},...,A_{jn},C_i)$.

To deal with expansions Yebra relies on so-called bindings. A binding is a triplet $(target, relation, source)$ in which $target$ and $source$ are variable names, and $relation$ refers to a relational feature. We can think of a binding as an expression similar to $target \leftarrow relation(source)$, that is, a binding maps a source node onto a target node using a relational feature.

The procedure responsible for expanding the rule, and as a consequence, the training and test sets, is $findExpansion$ (Algorithm 3).

$findExpansion$ works on a rule, a set of bindings, the training set, the test set, the original dataset, and a set of relational features. If $findExpansion$

Algorithm 3. Procedure FINDEXPANSION

```
 1: procedure FINDEXPANSION(rule, bindings, trainingSet, testSet, dataSet, relations)
 2:     resultRule ← null
 3:     resultBindings ← null
 4:     resultTrainingSet ← null
 5:     resultTestSet ← null
 6:     for all binding b in bindings while ¬isPerfect(resultRule, testSet) do
 7:         for all relation r in relations while ¬isPerfect(resultRule, testSet) do
 8:             newBinding ← (freshVariable(), r, target(b))
 9:             newTrainingSet ← expand(trainingSet, newBinding, dataset, relations)
10:             newTestSet ← expand(testSet, newBinding, dataset, relations)
11:             newRule ← learnRule(newTrainingSet, newTestSet)
12:             if gain(rule, testSet, newRule, newTestSet) >
                       gain(rule, testSet, resultRule, resultTestSet) then
13:                 resultRule ← newRule
14:                 resultBindings ← bindings ∪ {newBinding}
15:                 resultTrainingSet ← newTrainingSet
16:                 resultTestSet ← newTestSet
17:             end if
18:         end for
19:     end for
20:     return (resultRule, resultBindings, resultTrainingSet, resultTestSet)
21: end procedure
```

cannot find an expansion that improves on the input rule, it returns a *null* expansion; otherwise, it expands the training and test sets, the bindings, and returns the new, improved, rule.

Yebra starts with the initial binding (*node, null, null*) (Algorithm 1, Line 5), that is, the algorithm starts with a variable named *node*, and no relation involved. So the first time *findExpasion* is executed, the set of bindings has only one element, the initial binding.

First, *findExpansion* initialises with *null* values the variables *resultRule*, *resultBindings*, *resultTrainingSet*, and *resultTestSet*. Note that if no expansions can be found, *resultRule* remains with a *null* value. Then it iterates on the Cartesian product of bindings and relations. For each pair:

1. It creates a new binding. The new binding is the result of applying the relational feature to that binding.
2. It expands the training set. This implies the expansion of every vector in the training set, by applying the relation specified by the new binding.
3. It expands the test set. This implies the expansion of every vector in the test set, by applying the relation defined in the new binding.
4. It learns a new rule with the expanded training and test sets. Note that we are learning the best possible rule for the expanded training set.
5. It checks whether the new rule improves on the previous one. In that case, it stores the new rule, the expanded training and test set, and the new binding.

To evaluate if the rule learnt by means of an expansion is better than the rule obtained using another different expansion, Yebra relies on the information gain function [14].

Thus, when the *findExpansion* procedure is called, the best rule obtained by expanding the training set is computed (if there is one). After that, Yebra calls *findExpansion* again to check further expansions can improve on the current rule.

3   Experimental Analysis

To prove that our proposal is worth from a practical point of view, we have performed a series of experiments in which we have collected the usual effectiveness measures, namely: precision (P), recall (R), and the $F1$ measure ($F1$). We have developed a Java 1.7 implementation of Yebra and we have run it on a four-threaded Intel Core i7 computer that ran at 2.93 GHz, had 4 GiB of RAM, Windows 7 Pro 64-bit, Oracle's Java Development Kit 1.7.0_02, and Weka 3.6.8. We used a collection of 23 datasets that provides 850 web documents regarding books, movies, conferences, cars, doctors, sports, restaurants, and so on.

The empirical comparison was performed with two classical proposals (Soft-Mealy [7] and WIEN [12]) and three recent proposals (RoadRunner [4], FiVaTech [10], and Trinity [18]). Regarding Yebra, we selected six documents from each site to learn an extraction rule, and then validated the result using the remaining documents; in average, we used 22.78 ± 9.87 % of the datasets for training purposes and the rest for validation purposes.

Table 1 summarises our results. A dash in a cell means that the corresponding proposal was not able to learn a rule for a given dataset, be it because it failed to find it, because it ran out of memory, or a bug that raised an exception. It is not surprising at all that the recent proposals outperform the classical ones regarding every effectiveness measure. Amongst the recent proposals, Trinity seems to be the one that achieves the best effectiveness. Note, however, that Yebra outperforms them all since it is able to induce rules that are more precise and have better recall in general; furthermore, the standard deviation of precision and recall is smaller, which means that our learning approach is more generally stable than the others, that is, it does not generally produce rules whose effectiveness largely deviates from the average. These results prove that Yebra is quite an effective approach to web information extraction.

4   Related Work

The majority of supervised techniques in the literature build on ad-hoc machine learning algorithms that were specifically tailored to the problem of learning extraction rules [3,16]. Most of them try to learn token or XPath patterns that are based on token lexemes, their lexical classes, or HTML tags and their attributes: [12] presented a pioneering technique to learn two patterns of tokens that characterise the left and the right context of the information to extract. Hsu and Dung [7] devised an approach that first models the structure of the information in a set of web documents using finite automata and then learns transition conditions. Soderland [19] presented a proposal that starts with an overly-general

Table 1. Experimental results

	SoftMealy			WIEN			RoadRunner			FivaTech			Trinity			Yebra		
Summary	P	R	F1	P	R	F1	P	R	F1	P	R	F1	P	R	F1	P	R	F1
Mean	0.84	0.57	0.62	0.81	0.67	0.71	0.42	0.42	0.42	0.81	0.83	0.81	0.95	0.91	0.92	0.97	0.94	0.95
Standard deviation	0.16	0.32	0.30	0.16	0.26	0.22	0.46	0.47	0.45	0.18	0.18	0.17	0.08	0.16	0.12	0.04	0.09	0.06
Dataset	P	R	F1	P	R	F1	P	R	F1	P	R	F1	P	R	F1	P	R	F1
Awesome Books	1.00	0.39	0.56	0.77	0.26	0.39	1.00	1.00	1.00	0.85	1.00	0.92	1.00	0.87	0.93	0.97	0.90	0.93
Albania Movies	0.85	0.40	0.54	0.87	0.24	0.38	0.81	1.00	0.89	0.82	0.81	0.81	0.95	0.98	0.96	0.95	0.95	0.95
All Conferences	0.99	0.25	0.40	0.80	0.40	0.53	-	-	-	0.84	0.90	0.87	0.98	0.99	0.99	0.92	0.84	0.88
Amazon (cars)	0.98	1.00	0.99	0.97	1.00	0.98	0.27	0.33	0.30	0.60	0.67	0.63	0.93	0.73	0.82	0.97	0.90	0.93
Amazon (popartist)	0.94	0.72	0.82	0.92	0.58	0.71	0.99	0.99	0.99	1.00	1.00	1.00	1.00	0.98	0.99	1.00	1.00	1.00
Ame. Medical Assoc.	0.79	0.39	0.53	0.60	0.60	0.60	-	-	-	-	-	-	0.98	1.00	0.99	0.99	0.99	0.99
ATP World Tour	0.94	0.94	0.94	0.71	0.71	0.71	0.00	0.00	0.00	0.99	0.88	0.93	0.97	0.99	0.98	0.98	0.99	0.99
Better World Books	0.98	0.99	0.98	0.43	0.35	0.39	0.00	0.00	0.00	0.99	0.96	0.97	0.99	1.00	0.99	1.00	1.00	1.00
Car Max	0.89	0.89	0.89	0.88	0.88	0.88	0.00	0.00	0.00	0.45	0.89	0.60	1.00	1.00	1.00	1.00	0.99	0.99
Car Zone	0.92	0.02	0.05	0.82	0.83	0.83	0.00	0.00	0.00	0.92	1.00	0.96	0.98	1.00	0.99	0.98	0.99	0.99
Disney Movies	0.97	0.97	0.97	0.72	0.72	0.72	0.00	0.00	0.00	0.71	0.67	0.69	1.00	1.00	1.00	0.94	0.91	0.93
Homes	0.80	0.79	0.79	0.92	0.92	0.92	0.00	0.00	0.00	-	-	-	0.99	0.99	0.99	1.00	1.00	1.00
IAF	0.37	0.43	0.40	1.00	1.00	1.00	0.00	0.00	0.00	0.53	0.69	0.60	0.84	0.38	0.52	0.92	0.70	0.79
Insight into Diversity	0.57	0.47	0.52	1.00	1.00	1.00	0.70	0.70	0.70	1.00	0.74	0.85	0.83	0.83	0.83	0.87	0.83	0.85
Job of Mine	0.75	0.03	0.06	0.50	0.50	0.50	0.86	1.00	0.93	-	-	-	0.86	1.00	0.93	0.97	0.97	0.97
LA Weekly	0.87	0.49	0.63	0.90	0.80	0.85	0.00	0.00	0.00	0.83	0.57	0.68	0.97	0.92	0.94	0.98	0.98	0.98
Major League	0.99	0.46	0.63	0.65	0.33	0.44	0.00	0.00	0.00	0.99	1.00	0.99	0.98	0.55	0.70	0.99	0.70	0.82
Mbendi	0.60	0.60	0.60	0.80	0.40	0.53	0.90	1.00	0.95	0.90	1.00	0.95	1.00	1.00	1.00	1.00	1.00	1.00
Okra	0.83	0.82	0.82	0.60	0.67	0.63	0.96	0.56	0.71	0.49	0.34	0.40	1.00	0.82	0.90	1.00	0.95	0.97
Remax	0.84	0.84	0.84	1.00	1.00	1.00	-	-	-	-	-	-	0.70	0.98	0.82	0.85	0.97	0.91
RPM Find	0.72	0.03	0.06	0.99	0.99	0.99	0.98	0.99	0.98	-	-	-	0.95	0.97	0.96	1.00	1.00	1.00
Steady Health	0.75	0.25	0.38	0.75	0.75	0.75	0.00	0.00	0.00	0.83	0.83	0.83	1.00	1.00	1.00	1.00	1.00	1.00
UEFA (players)	0.92	0.90	0.91	0.92	0.51	0.66	0.92	0.92	0.92	0.91	0.94	0.92	1.00	0.90	0.95	1.00	1.00	1.00

rule and then specialises it with patterns that match token sequences. Muslea et al. [13] presented a proposal that builds on a hierarchical schema that models the information in a web document; for every positive example, it attempts to learn an automata that can recognise the sequence of tokens from the starting of the parent example until its initial token, and an automata that recognises the sequence of tokens from the final token until the final token of the parent example. Sleiman and Corchuelo [17] presented the most recent proposal in this field; they use finite automata to represent the structure of the information to be extracted and then use neural networks to learn transition conditions.

None of the previous techniques have explored the idea of tackling information extraction using features of the tokens or the DOM trees themselves, e.g., their length, their colour, their depth, the ratio of letters, and the like; neither have they explored using features of the nodes in the neighbourhood. The only proposals that have explored this idea rely on first-order learning procedures, namely: SRV [5], which works on the textual view of the documents; it starts with the most general rule and then specialises it by adding conditions so that the rule matches as many positive examples as possible and reduces the number of negative ones matched. Irmak and Suel [8] presented a proposal that works on the DOM-tree view of the documents and their rules are sets of conditions that work on XPaths; their proposal creates several sets of extraction rules that generalise the user-annotated examples in different ways; next, the user has to select the set of records that best suits his or her interests, and the learning process is executed again to correct mistakes. Bădică et al. [2] presented a proposal that also works on the DOM-tree view of the input documents; their rules basically attempt to classify positive examples by means of their tags and the tags of their neighbouring nodes; their algorithm relies on the FOIL system [14] to learn a set of Horn clauses from a logic representation of the DOM-tree nodes and their features.

Our contribution to the state of the art is twofold: on the one hand, we use a propositional approach that relies on a extensive catalogue of features; on the other hand, our propositional approach has the ability to exploit relational features that explore an unbounded context in the neighbourhood of the DOM tree nodes to extract. None of these approaches has been explored so far in the literature regarding web information extraction, but our results prove that they are very promising.

5   Conclusions

In this paper, we have presented a new approach to information extraction; contrarily to the majority of proposals in the literature, which provide ad-hoc algorithms, our proposal maps the problem of information extraction onto the problem of learning from a set of vectors that represent the features computed on a DOM tree plus a number of relations amongst them. Our empirical results prove that our approach is very promising since it achieves better effectiveness than other state-of-the-art proposals.

Acknowledgements. Our work was funded by the Spanish and the Andalusian R&D &I programmes by means of grants TIN2007-64119, P07-TIC-2602, P08-TIC-4100, TIN2008-04718-E, TIN2010-21744, TIN2010-09809-E, TIN2010-10811-E, TIN2010-099 88-E, TIN2011-15497-E, and TIN2013-40848-R, which got funds from the European FEDER programme.

References

1. Blockeel, H., Raedt, L.D., Jacobs, N., Demoen, B.: Scaling up inductive logic programming by learning from interpretations. Data Min. Knowl. Discov. **3**(1), 59–93 (1999)
2. Bădică, C., Bădică, A., Popescu, E., Abraham, A.: L-Wrappers: concepts, properties and construction. Soft Comput. **11**(8), 753–772 (2007)
3. Chang, C.H., Kayed, M., Girgis, M.R., Shaalan, K.F.: A survey of web information extraction systems. IEEE Trans. Knowl. Data Eng. **18**(10), 1411–1428 (2006)
4. Crescenzi, V., Merialdo, P.: Wrapper inference for ambiguous web pages. Appl. Artif. Intell. **22**(1&2), 21–52 (2008)
5. Freitag, D.: Machine learning for information extraction in informal domains. Mach. Learn. **39**(2/3), 169–202 (2000)
6. Gregg, D.G., Walczak, S.: Exploiting the information web. IEEE Trans. Syst. Man Cybern. Part C **37**(1), 109–125 (2007)

7. Hsu, C.N., Dung, M.T.: Generating finite-state transducers for semi-structured data extraction from the Web. Inf. Syst. **23**(8), 521–538 (1998)
8. Irmak, U., Suel, T.: Interactive wrapper generation with minimal user effort. In: WWW, pp. 553–563 (2006)
9. Kavurucu, Y., Senkul, P., Toroslu, I.H.: A comparative study on ILP-based concept discovery systems. Expert Syst. Appl. **38**(9), 11598–11607 (2011)
10. Kayed, M., Chang, C.H.: FiVaTech: page-level web data extraction from template pages. IEEE Trans. Knowl. Data Eng. **22**(2), 249–263 (2010)
11. Kramer, S., Lavrač, N., Flach, P.: Propositionalization approaches to relational data mining. In: Džeroski, S., Lavrač, N. (eds.) Relational Data Mining, pp. 262–291. Springer, Heidelberg (2001)
12. Kushmerick, N., Weld, D.S., Doorenbos, R.B.: Wrapper induction for information extraction. In: IJCAI (1), pp. 729–737 (1997)
13. Muslea, I., Minton, S., Knoblock, C.A.: Hierarchical wrapper induction for semi-structured information sources. Auton. Agent. Multi-Agent Syst. **4**(1/2), 93–114 (2001)
14. Quinlan, J.R., Cameron-Jones, R.M.: Induction of logic programs: FOIL and related systems. New Gener. Comput. **13**(3&4), 287–312 (1995)
15. Saggion, H., Funk, A., Maynard, D., Bontcheva, K.: Ontology-based information extraction for business intelligence. In: Aberer, K., Choi, K.-S., Noy, N., Allemang, D., Lee, K.-I., Nixon, L.J.B., Golbeck, J., Mika, P., Maynard, D., Mizoguchi, R., Schreiber, G., Cudré-Mauroux, P. (eds.) ASWC 2007 and ISWC 2007. LNCS, vol. 4825, pp. 843–856. Springer, Heidelberg (2007)
16. Sleiman, H.A., Corchuelo, R.: A survey on region extractors from web documents. IEEE Trans. Knowl. Data Eng. **25**(9), 1960–1981 (2013)
17. Sleiman, H.A., Corchuelo, R.: A class of neural-network-based transducers for web information extraction. Neurocomputing **135**, 61–68 (2014)
18. Sleiman, H.A., Corchuelo, R.: Trinity: on using trinary trees for unsupervised web data extraction. IEEE Trans. Knowl. Data Eng. **26**(6), 1544–1556 (2014)
19. Soderland, S.: Learning information extraction rules for semi-structured and free text. Mach. Learn. **34**(1–3), 233–272 (1999)
20. Srivastava, J., Cooley, R.: Web business intelligence: mining the web for actionable knowledge. INFORMS J. Comput. **15**(2), 191–207 (2003)
21. Witten, I.H., Frank, E.: Weka machine learning algorithms in Java. In: Data Mining: Practical Machine Learning Tools and Techniques with Java Implementations, pp. 265–320. Morgan Kauffman Publishers (2000)

Business Process Management
and Mining

Mining Multi-variant Process Models
from Low-Level Logs

Francesco Folino, Massimo Guarascio, and Luigi Pontieri[✉]

ICAR-CNR, National Research Council, via P. Bucci 41C, 87036 Rende, CS, Italy
{ffolino,guarascio,pontieri}@icar.cnr.it

Abstract. Process discovery techniques are a precious tool for analyzing the real behavior of a business process. However, their direct application to lowly structured logs may yield unreadable and inaccurate models. Current solutions rely on event abstraction or trace clustering, and assume that log events refer to well-defined (possibly low-level) process tasks. This reduces their suitability for logs of real BPM systems (e.g. issue management) where each event just stores several data fields, none of which fully captures the semantics of performed activities. We here propose an automated method for discovering an expressive kind of process model, consisting of three parts: *(i)* a logical event clustering model, for abstracting low-level events into classes; *(ii)* a logical trace clustering model, for discriminating among process variants; and *(iii)* a set of workflow schemas, each describing one variant in terms of the discovered event clusters. Experiments on a real-life data confirmed the capability of the approach to discover readable high-quality process models.

Keywords: Business process mining · Log abstraction · Trace clustering

1 Introduction

Workflow discovery techniques [1] have gained attention in BPM applications, owing to their ability to extract (out of historical execution data) a descriptive model for the behavior of a process, which can support key process analysis/design, process improvement and strategic decision making tasks.

However, two critical issues undermine the effectiveness of traditional workflow discovery methods, when they are applied to the logs of lowly-structured processes: *(i)* the high level of details that usually characterizes log events, which makes it difficult to provide the analyst with an easily interpretable description of the process in terms of relevant business activities, and *(ii)* the presence of various execution scenarios (a.k.a. "process variants"), which exhibit different business processing logics (often determined by key context factors), and cannot be captured effectively with a single workflow model. In fact, when applied to such logs, most current workflow discovery techniques tend to yield "spaghetti-like" models, suffering from both low readability and low fitness [7].

© Springer International Publishing Switzerland 2015
W. Abramowicz (Ed.): BIS 2015, LNBIP 208, pp. 165–177, 2015.
DOI: 10.1007/978-3-319-19027-3_14

Two kinds of solution methods have been proposed in the literature to alleviate these problems: *(i)* turn raw events into high-level activities by way of automated abstraction techniques [6,10], *(ii)* partition the log into trace clusters [4,5,8,9], capturing homogenous execution groups, and then separatley model each cluster with a simpler and more fitting workflow. Since both kinds of methods assume that each log event refers to a predefined (possibly low-level) process task, they are of limited usefulness for many real-world flexible BPM applications (e.g., product management, or problem/issue tracking). In such cases, indeed, each log event takes the form of a tuple storing several data fields, none of which can be interpreted as a task label (capable to fully capture the semantics of the performed activity). Moreover, the clusters produced by current event abstraction (resp., trace partitioning) approaches are not self-descriptive, and the analyst must carry out difficult and long interpretation/validation tasks to turn them into meaningful classes of activities (resp., process instances).

Contribution. In order to overcome these limitations, a two-fold mining problem is stated in this work, for a given low-level multi-dimensional event log. On the one hand, we want to discover an event clustering function, allowing for automatically abstracting log events into (non a-priori known) event types, each of which is meant to represent a distinguished pattern for the execution of single work items. While exploiting such event abstraction, we also want to possibly detect and model different process execution variants.

Our solution approach relies, first of all, on a specialized logic-based event clustering method, which can find a clustering function encoded in terms of decision rules (over event attributes), to provide the analyst with an interpretable description of the discovered activity types (i.e. event clusters). We also find a similar conceptual clustering function for partitioning the log traces into different execution classes, by using their associated context data (e.g. cases' properties or environmental factors) as descriptive variables. To this end, an iterative trace-clustering approach is proposed, which tries to greedily maximize the (average) quality of the workflow schemas induced from the discovered trace clusters.

Expressing trace/event clusters via predictive (logical) rules is an important distinguishing feature of our approach, which makes it possibly support the implementation of advanced run-time services. This point is discussed in Sect. 6, which also provides a comparison with related work in the literature.

2   Preliminaries

Log Data. For each process case a *trace* is recorded, storing the sequence of *events* happened during its enactment. Let E and T be the universes of all possible events and traces, respectively, for the process under analysis. For each trace $\tau \in T$, $len(\tau)$ denotes the number of events stored in τ, while $\tau[i]$ is i-th event in τ, for $i \in \{1, ..., len(\tau)\}$.

For each event $e \in E$, let $prop(e)$ be a tuple data properties (over some attribute space) associated with e — each event also refers to a case ID and

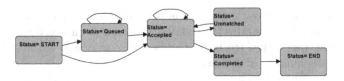

Fig. 1. Workflow schema induced from the *Problem Management* log, when using a **status**-based event abstraction.

to a timestamp, but these are useless for recognizing general activity patterns. Execution cases as well are often associated with a number of data properties. For each trace $\tau \in \mathcal{T}$, let $prop(\tau)$ be the data properties of τ. Finally, let $traces(L)$ (resp., $events(L)$) denote the set of traces (resp., events) stored in L.

Workflow Schemas. Our final aim is to describe process behavior by way of flow-oriented models, like those commonly used to express the control-flow logics of a business process. For the sake of concreteness, we here focus on the language of *heuristics nets* [14], where a *workflow schema* is essentially a directed graph, where each node represent a process activity, and each edge (x, y) encodes a dependency of y on x. In addition, one can express cardinality-based fork (resp., join) constraints for nodes with multiple outgoing (resp., incoming) edges.

For any workflow schema W, let $\mathcal{A}(W)$ be the set of all activities featuring in W. The definition of $\mathcal{A}(W)$ is a crucial point in our setting, where it may well happen that no predefined business activities exist for the process. In such a case, our approach consists in partitioning log events into event classes, representing distinguished activity patterns, which can be eventually used to label the nodes of a workflow schema.

Example 1. Let us consider a real-life Problem Management application (used as a case study in Sect. 5), where each log trace registers the history of a ticket. Each log event e is a tuple featuring 8 data attributes: the **status** (*accepted, queued, completed,* or *closed*) and **substatus** (*unmatched, awaiting_assignment, assigned, in_progress, wait, cancelled,* or *closed*) of the ticket when e happened; the **resource** who generated e, and her nationality (**res_country**); the **support team**, **functional division** and **organization line** which the resource was affiliated to; the country where the line was located (**org_country**). Since such representation carries no information on what process task was performed in each event, we need to infer activity types from event tuples, to grasp some suitable abstraction level on process behavior. A common approach consists in abstracting each event tuple into just one of its attributes, and interpret it as an activity label. Figure 1 shows the model discovered (by algorithm Heuristics Miner [14]) after replacing each event with its respective **status**'s value.   ◁

Figure 1 evidences that a one-attribute event abstraction is likely to produce a lowly informative (or even trivial) process model for a log storing fine-grain execution traces. On the other hand, using a combination of multiple attributes to the same purpose may well yield a cumbersome and overfitting workflow schema, as confirmed by our tests. The latter solution is, in fact, also unviable in many real application scenarios for scalability reasons, seeing as the computation time of typical approaches (to both workflow discovery and log abstraction) is quadratic in the number of activities.

Behavioural Profiles. Behavioural profiles [13] are basic ordering relationships over activity pairs, which can summarize a workflow schema's behavior, and can be computed efficiently for many classes of workflow specification languages.

Let W be a workflow schema, and $\mathcal{A}(W)$ be its associated activities. Let $\succ^W$ be a "weak order" relation inferred from W, such that, for any $x, y \in \mathcal{A}(W)$, $y \succ^W x$ iff there is at least a trace admitted by W where y occurs after x. Then, the *behavioral profile matrix* of W, denoted by $\mathcal{B}(W)$, is a function mapping each pair $(x, y) \in \mathcal{A}(W) \times \mathcal{A}(W)$ to an ordering relation in $\{\leadsto, +, \|\}$, as follows: (i) $\mathcal{B}(W)[x, y] = \leadsto$, iff $y \succ^W x$ and $x \not\succ^W y$ (*strict order*); (ii) $\mathcal{B}(W)[x, y] = +$, iff $x \not\succ^W y$ and $y \not\succ^W x$ (*exclusiveness*); (iii) $\mathcal{B}(W)[x, y] = \|$, iff $x \succ^W y$ and $y \succ^W x$ (*observation concurrency*).

Let τ be a trace in $\mathcal{T}$ (with event universe E), x and y be two event classes (i.e. disjoint subsets of E), and $\mathcal{B}$ be a behavioral profile matrix. Then we say that τ *violates* (resp., *satisfies*) $\mathcal{B}[x, y]$, denoted by $\tau \not\vdash \mathcal{B}[x, y]$ (resp., $\tau \vdash \mathcal{B}[x, y]$), if the occurrences of x and y in τ infringe (resp., fullfill) the ordering constraints in $\mathcal{B}[x, y]$. Precisely, it is $\tau \not\vdash \mathcal{B}[x, y]$ iff there exist $i, j \in \{1, ..., len(\tau)\}$ such that $\tau[i] = y$, $\tau[j] = x$, and: either *(i)* $\mathcal{B}[x, y] = +$, or *(ii)* $\mathcal{B}[x, y] = \leadsto$ and $i < j$.

Behavioral profiles help us measure workflow conformance, as in what follows.

Definition 1. *Let L be an event log, W be a workflow schema, and $\sigma \in \mathbb{N}$ be a (lower) noise threshold. Then the compliance of W w.r.t. L, denoted by $compl(W, L)$, is defined as: $compl(W, L) = \frac{1}{|\mathcal{A}(W)|^2} \times |\{(x, y) \in \mathcal{A}(W) \times \mathcal{A}(W) | \neg (\exists^{\geq \sigma} \tau \in traces(L) \text{ s.t. } \tau \not\vdash \mathcal{B}(W)[x, y])\}|$. Moreover, the precision of W w.r.t. L, denoted by $prec(W, L)$, is defined as: $prec(W, L) = \frac{1}{|\mathcal{A}(W)|^2} \times |\{(x, y) \in \mathcal{A}(W) \times \mathcal{A}(W) | \mathcal{B}(W)[x, y] = \| \text{ and } \neg(\exists^{\geq \sigma} \tau \in L \text{ s.t. } |\{x, y\} \cap tasks(\tau)| = 1)\}|$ where $\exists^{\geq \sigma}$ is a counting quantifier, asserting the existence of at least σ elements in a given set.[1]* □

3   Problem Statement

As discussed above, we want to discover two interrelated clustering models: one for log events, and another for log traces (intended to recognize process variants).

For the sake of interpretability, in both cases we seek a clustering model that can be encoded by decision rules. Let us assume that each rule is a conjunctive boolean formula of the form $(A_1 \in V_1) \wedge (A_2 \in V_2) \wedge \ldots \wedge (A_k \in V_k)$, where, for each $i \in \{1, \ldots, k\}$, A_i is a descriptive attribute defined on some given set Z of data instances, and V_i is a subset of A_i's domain. For any $I \subseteq Z$ and for any such a rule r, let $cov(r, I)$ denote the set of all I's instances that satisfy r.

---

[1] In the tests described in Sect. 5 we always set $\sigma = 0.01 \cdot |traces(L)|$.

A *conceptual clustering model* for Z is a list $\mathcal{C} = \langle r_1, ..., r_n \rangle$ of conceptual clustering rules (for some positive integer number n), which defines a partitioning of Z into n parts $P_1, \ldots, P_n$, where $P_i = cov(r_i, Z) / \bigcup_{j=1}^{i-1} cov(r_j, Z)$.

Our ultimate goal is to find a multi-variant process model leveraging two such clustering functions (for events and traces, resp.), as it is formally defined next.

Definition 2. *Let L be a log, and $\mathcal{T}$ (resp., E) the associated trace (resp., event) universe. Then, a High-Level Process Model (HLPM) for L is a triple $\langle \mathcal{C}^E, \mathcal{C}^T, WS \rangle$ such that: (i) $\mathcal{C}^E = \langle r_1^E, \ldots, r_p^E \rangle$ is a conceptual clustering model for E, where $p \in \mathbb{N}$ is the number of event clusters, and r_i^E (for $i = 1, .., p$) is the clustering rule of the i-th event cluster; (ii) $\mathcal{C}^T = \langle r_1^T, \ldots, r_q^T \rangle$ is a conceptual clustering model for $\mathcal{T}$, where $q \in \mathbb{N}$ is the number of trace clusters, and r_j^T (for $j = 1, .., q$) is the clustering rule of the j-th trace cluster; $WS = \langle W_1, \ldots, W_q \rangle$ is a list of workflow schemas, where W_k (for $k = 1, .., q$) models the k-th trace cluster, using the classes yielded by $\mathcal{C}^E$ as activities.* □

Clearly enough, sub-model $\mathcal{C}^E$ plays as an event abstraction function, which maps each event $e \in E$ to an event class, based on logical clustering rules (expressed on e's properties). Analogously, model $\mathcal{C}^T$ partitions historical execution traces into behaviorally homogenous clusters (via logical clustering rules over traces' properties). Each discovered trace cluster, regarded as a distinct process variant, is also equipped with a workflow schema, where some of the clusters of $\mathcal{C}^E$ feature as (high-level) activity nodes.

Our discovery problem can be stated conceptually as the search for an optimal HLPM that maximizes some associated quality measure, such as those defined in the previous section.

4 Solution Approach

Technically, we rephrase the discovery of conceptual clustering models (over either events or traces) as a predictive clustering problem. Basically, predictive clustering approaches [3] assume that two kinds of data attributes characterize each element z in a given space $Z = X \times Y$ of instances: *descriptive* attributes and *target* attributes, denoted by $descr(z) \in X$ and $targ(z) \in Y$, respectively. The goal of these approaches is to find a logical partitioning function (of the same nature as our conceptual clustering models) that minimizes $\sum_{C_i} |C_i| \times Var(\{targ(z) \mid z \in C_i\})$, where C_i ranges over current clusters, and $Var(S)$ is the variance of set S.

Different predictive clustering models exists in the literature. Owing to scalability and readability reasons, we preferred *Predictive Clustering Trees* (PCTs), where the clustering function is a (propositional) decision tree.

We next introduce ad-hoc propositional encodings for events and traces, allowing for inducing a clustering model by reusing a PCT learner.

Definition 3. *Let L be a log, over event (resp., trace) universe E (resp., $\mathcal{T}$). Then, the e-view of L, denoted by $\mathcal{V}_E(L)$, is a relation containing a tuple $z_{i,j}$ for each $\tau_i \in traces(L)$ and for each $j \in \{1,\ldots,len(\tau_i)\}$ (i.e. for each event $\tau[j] \in events(L))$, such that: (i) $descr(z_{i,j}) = prop(\tau_i[j])$, and (ii) $targ(z_{i,j}) = \langle \frac{j}{len(\tau_i)}, \frac{j}{maxL}, \frac{len(\tau_i)-j}{maxL} \rangle$, where $maxL = \max(\{ \; len(\tau) \mid \tau \in traces(L)\})$.* □

In such a log view (acting as training set for predictive clustering), the descriptive (input) attributes each instance $z_{i,j}$ are the data fields in $prop(\tau_i[j])$ (see Sect. 2), while the target ones are three indicators (derived from the relative/absolute position j of $\tau_i[j]$ in its surrounding trace). Intuitively, we want events with similar intra-trace positions (and similar properties) to be put together.

For the clustering of traces, we introduce another log view, named *t-view*, where the context data and abstract activities of each trace play as descriptive features, while the target variables express local activity relationships.

Specifically, let $\mathcal{B}$ be a behavioral profile matrix (derived from some workflow schema), and $\alpha : E \to A$ be an event clustering function. Then, for each trace τ and any activities a_i and a_j in A we define a target variable $v_\mathcal{B}(\tau, a_i, a_j)$ as follows: (i) $v_\mathcal{B}(\tau, a_i, a_j) = 1$ if $\tau \not\vdash \mathcal{B}[a_i, a_j]$; (ii) $v_\mathcal{B}(\tau, a_i, a_j) = \frac{f(\tau, a_i, a_j)}{2 \times c(\tau, a_i, a_j)}$ if both a_i and a_j occur in τ, where $c(\tau, a_i, a_j) = |\{(i', j') \mid i', j' \in \{1,\ldots,len(\tau)\} \wedge i' \neq j' \wedge \alpha(\tau[i']) = a_i \wedge \alpha(\tau[j']) = a_j\}|$, and $f(\tau, a_i, a_j) = \mathbf{sum}(\{\mathbf{sgn}(j' - i') \mid i', j' \in \{1,\ldots,len(\tau)\} \wedge \alpha(\tau[i']) = a_i \wedge \alpha(\tau[j']) = a_j\})$, where $\mathbf{sgn}$ denotes function *signum*; and (iii) $v_\mathcal{B}(\tau, a_i, a_j) = \mathbf{null}$ if τ does not contain both a_i and a_j. This way, $v_\mathcal{B}(\tau, a_i, a_j)$ can capture a violation to a behavioral profile (case i), or keep information on the mutual positions of a_i and a_j, if both occur in τ (case ii).

Definition 4. *Let L be a log over event and trace universes E and $\mathcal{T}$, $A = \{a_1,\ldots,a_k\}$ be a set of event clusters (regarded as abstract activities), and $\alpha : E \to A$ be an event clustering function. Let also $\mathcal{B} : A \times A \to \{\rightsquigarrow, +, \|\}$ be a behavioral profile matrix. Then, the t-view of L w.r.t. α and $\mathcal{B}$, denoted by $\mathcal{V}_\mathcal{T}(L, \alpha, \mathcal{B})$, is a relation containing, for each $\tau \in traces(L)$, a tuple z_τ such that: (i) $descr(z_\tau) = prop(\tau) \oplus act_\alpha(\tau)$, where $\oplus$ denotes tuple concatenation, and $act_\alpha(\tau)$ is a vector in $\{0,1\}^k$ such that, for $i = 1,..,k$, $act_\alpha(\tau)[i] = 1$ iff activity a_i occurs in τ (i.e. iff $\exists j \in \{1,\ldots,len(\tau)\}$ s.t. $\alpha(\tau[j]) = a_i$); and (ii) $targ(z_\tau) = \langle v_\mathcal{B}(\tau, a_1, a_1), \ldots, v_\mathcal{B}(\tau, a_1, a_k), v_\mathcal{B}(\tau, a_2, a_2), ..., v_\mathcal{B}(\tau, a_2, a_k), \ldots, v_\mathcal{B}(\tau, a_i, a_i), \ldots, v_\mathcal{B}(\tau, a_i, a_k), \ldots, v_\mathcal{B}(\tau, a_k, a_k) \rangle$.* □

Algorithm `HLP-mine`*.* Our approach to the discovery of a HLPM is illustrated as an algorithm, named HLPM-mine and reported in Fig. 2. The algorithm follows a two-phase strategy: it first finds a conceptual clustering model (Steps 1–3) for the events, and then computes a collection of trace clusters, along with their associated clustering rules and workflow schemas (Steps 4–24).

Conceptual clustering models are found through function `minePCT`, which leverages a PCT-learning method in [3]. In the algorithm, this function is applied to both an *e-view* (Step 2), and a *t-view* (Step 12). Two further parameters allow to constrain a PCT's growth: the minimal coverage a node must have to be possibly split, and the maximal number of leaves, respectively.

Input: Log L, maximal number $m \in \mathbb{N}$ of trace clusters, minimal clusters'
        coverage $minCov \in [0,1]$, minimal quality-gain $\gamma \in [0,1]$
Output: An HLPM for $\mathcal{T}$

---

1. $V := \mathcal{V}_E(L)$; // compute L's e-view (cf. Def. 3)
2. $T := \mathtt{minePCT}(V, minCov, \infty)$;
3. $C_E := \mathbf{extractRules}(T, minCov, \gamma)$; // C_E is a clustering model for E
4. Create an initial cluster c_0 gathering all L traces;
5. $w_0 := \mathtt{mineWF}(L, C_E)$; compute $\mathcal{B}(w_0)$;
6. $TClust := \{c_0\}$; $W(c_0) := w_0$; $rule(c_0) := [\mathbf{true}]$;
7. $Q := \{c_0\}$; // Q is a queue of candidate trace clusters
8. **while** $(0 < |TClust| < m)$ and $Q \neq \emptyset$ **do**
9. $c^* = \arg\min_{c \in Q}(|c| \times \mathtt{compl}(W(c), c))$;
10.   **if** $|c^*| \geq minCov \times |traces(L)|$ **then**
11. $V := \mathcal{V}_T(c^*, C_E, \mathcal{B}(W(c^*)))$; // build a t-view (cf. Def. 4)
12. $T := \mathtt{minePCT}(V, minCov, m - |TClust| + 1)$;
13.   Let ΔCl be the clusters associated with T's leaves
14.   **for each** cluster c in ΔCl **do**
15.   Let $traces(c)$ be the traces assigned to c;
16. $W(c) := \mathtt{mineWF}(traces(c), C_E)$; compute $\mathcal{B}(W(c))$;
17. $rule(c) := \mathtt{mergeRules}(rule(c^*), rule(c))$;
18.   **end for**
19.   **if** $\{W(c) \mid c \in \Delta Cl\} \prec^\gamma W(c^*)$ **then** //cf. Def. 5
20. $TClust := TClust \cup \Delta Cl - \{c^*\}$; $Q := Q \cup \Delta Cl$;
21.   **end if**
22. $Q := Q - \{c^*\}$;
23.   **end if**
24. **end while**
25. $C_\mathcal{T} := \langle\rangle$; $WS := \langle\rangle$; // inititalize the trace-clustering model and workflow list
26. **for each** cluster $c \in TClust$ **do**
27. $C_\mathcal{T}.\mathrm{append}(rule(c))$; $WS.\mathrm{append}(W(c))$;
28. **end for**
29. **return** $\langle C_E, C_\mathcal{T}, WS \rangle$

Fig. 2. Algorithm $\mathtt{HLPM\text{-}mine}$

The PCT found for log events is turned into a conceptual clustering model through an iterative bottom-up procedure, named $\mathtt{extractRules}$, omitted for lack of space. Basically, starting with the rules of the tree leaves, this procedure replaces each rule r with that of the parent, if $|cov(r, \mathcal{V}_E(L))| < minCov \times |\mathcal{V}_E(L)|$, and the variance of $cov(r, \mathcal{V}_E(L))$ is higher than, or nearly equal to, the parent's variance (precisely, the former is lower than the latter of a fraction γ at most). Notice that, in current implementation this test is performed on a separate pruning set. Whenever a rule r is removed, all the instances in $cov(r, \mathcal{V}_E(L))$ are assigned to the parent rule (i.e. to the rule of the parent of r's node), which is appended to the clustering model, if it does not appear in it yet.

Trace clusters are computed through an iterative partitioning scheme (Step 4–24), where $TClust$ is the set of current trace clusters, and Q just contains the ones that may be further split. For any trace cluster c, $W(c)$ and $rule(c)$ store the associated workflow schema and clustering rule, respectively.

Before the loop, $TClust$ just consists of one cluster, gathering all L's traces. At each iteration, we try to split the cluster c^* in Q that has been given the lowest compliance score by the (profile-based) conformance measure $compl$ (see Definition 1), among all those overcoming the coverage threshold $minCov$. To this end, a PCT model T is induced from the propositional view of cluster c^*, according to the discovered event clustering $\mathcal{C}_E$, and the behavioral profiles of the schema associated with c^* (see Definition 4).

For each leaf cluster c in T, we extract a workflow schema and the associated behavioral profiles, and merge the rule of c with that of the cluster (namely, c^*) T was induced from (Steps 16–17).

At this point, the algorithm verifies if the workflows discovered after splitting cluster c^* really allowed to model more effectively the behavior of c^*'s traces. This check relies on an ad-hoc quality relationship, named γ-improve, which is defined next, based on the conformance metrics $compl$ and $prec$ (see Definition 1).

Definition 5. *Let c be a set of traces, $\{c_1, .., c_k\}$ be a partition of c, W be a workflow schema for c, and $WS = \{W_1, .., W_k\}$ be a set of workflow schemas s.t. W_i models c_i and $\mathcal{A}(W_i) \subseteq \mathcal{A}(W)$, for $i = 1, .., k$. Then, for $\gamma \in [0, 1]$, we say that WS γ-improves W, denoted by $WS \prec^\gamma W$, iff (i) $\sum_{i=1}^{k} \frac{|traces(c_i)| \cdot compl(W_i, c_i)}{|traces(c)|} >$ $(1 + \gamma) \cdot compl(W, c)$, or (ii) $compl(W, c) \leq \sum_{i=1}^{k} \frac{|traces(L_i)| \cdot compl(W_i, c_i)}{|traces(c)|} \leq$ $(1 + \gamma) \cdot compl(W, c)$ and $\sum_{i=1}^{k} \frac{|traces(L_i)| \cdot prec(W_i, c_i)}{|traces(c)|} > (1 + \gamma) \cdot prec(W, c)$.* $\square$

We hence assume that the workflows of c^*'s children clusters model the behavior of $traces(c^*)$ better than $W(c^*)$ alone, if they get, in the average, higher compliance (on their respective sub-clusters) than W (on c^* as a whole). When the compliance score keeps unchanged, we still prefer the new schemas if they are more precise than $W(c^*)$.

The split of c^* is eventually kept only if the above improvement relationship holds. In any case, c^* is removed from the list of candidates for further refinement.

Steps 25–28 just build the list of trace clustering rules, and the list of workflow schemas, based on the set of trace clusters eventually left in $TClust$.

5   Experiments

The approach described so far was implemented into a Java prototype system, and tested on the log[2] of a real problem management system, encompassing 1487 traces recorded from January 2006 to May 2012. As explained in Example 1, each log event stores eight data attributes (i.e., status, substatus, resource, res_country, functional_division, org_line, support_team,

---

[2] Available at http://www.win.tue.nl/bpi/2013/challenge.

Table 1. Results (avg±stdDev) obtained with the event clustering method (no trace clustering) and manual abstraction criteria. The best value of each column is in bold.

Event clustering method	Fitness	BehPrec	#nodes	#edges	#edgesPerNode
status	0.676±0.010	0.543±0.004	**6.0±0.0**	**9.0±0.0**	3.0±0.0
substatus	0.715±0.057	0.546±0.003	9.0±0.0	18.0±0.0	4.0±0.0
all	0.723±0.032	**0.921±0.005**	1464.2±18.5	2224.6±23.9	3.0±0.0
HLPM-mine(m=1)	**0.815±0.024**	0.736±0.017	17.0±0.0	32.2±0.8	3.8±0.1

Table 2. Results (avg±stdDev) yielded by different trace clustering methods, after abstracting events with the clusters found by HLPM-mine(m=1). Best scores are in bold.

Trace clustering method	Fitness	BehPrec	#nodes	#edges	#edgesPerNode
ACTITRAC[8]	0.690±0.027	**0.796±0.012**	12.6±0.1	20.0±0.4	3.1±0.0
TRMR[4]	0.571±0.075	0.751±0.021	10.3±0.2	17.0±0.4	3.3±0.0
A-TRMR[4]	0.706±0.015	0.764±0.007	11.6±0.2	18.1±0.4	3.1±0.0
KGRAM[5]	0.663±0.023	0.725±0.008	10.7±0.2	16.7±0.4	3.1±0.0
HLPM-mine(m=∞)	**0.851±0.013**	0.742±0.007	**10.0±0.2**	**13.8±0.3**	**2.6±0.0**

and org_country). For each problem case p, two attributes are associated with p's trace: p's impact (*medium, low,* or *high*), and the product affected by p.

We enriched each trace τ with further context-oriented attributes: (*i*) firstOrg, indicating the team associated with τ's first event; (*ii*) workload, quantifying the number of problems open on the time, say t_τ, when τ started; (*iii*) several time dimensions (namely, week-day, month and year) derived from t_τ.

In all the tests discussed next, we ran HLPM-mine with $minCov = 0.01$, and $\gamma = 0.1$, while using plugin FHM [14] to instantiate function mineWF.

Evaluation Metrics. Different kinds of metrics (fitness, precision, generalization) exist for evaluating the quality of a workflow schema. In particular, fitness metrics quantify the capability to replay a given log, and represent the main evaluation criterion [7], while the other metrics serve finer-grain comparisons.

In our tests, we measured the fitness of each discovered heuristics-net according to the *Improved Continuous Semantics Fitness* (named *Fitness* hereinafter) defined in [12]. Basically, the fitness of a schema W w.r.t. a log L (denoted by *Fitness(W,L)*) is the fraction of L's events that W can parse exactly, with a special punishment factor for benefitting the schemas that yield fewer replay errors in fewer traces.

Moreover, the behavioral precision of schema W w.r.t. log L, denoted by *BehPrec(W, L)*, was simply measured as the average fraction of activities that are not enabled by a replay of L in W.

As to schema's complexity, we considered: the numbers of nodes (*#nodes*) and of edges (*#edges*), and the average number of edges per node (*#edgePerNode*).

Test Bed. As no current approach mixes automated event abstraction and trace clustering, we tested these two facets incrementally, for the sake of comparison.

cluster	rule
e_1	substatus=assigned $\wedge$ org_line=G199 3rd $\wedge$ org_country=us $\wedge$ resource_country=usa
e_3	substatus=in_progress $\wedge$ org_line=G199 3rd $\wedge$ org_country=us $\wedge$ resource_country=usa
e_4	substatus=in_progress $\wedge$ org_line=G199 3rd $\wedge$ org_country=us $\wedge$ resource_country=poland
e_5	substatus=closed $\wedge$ org_line=G199 3rd $\wedge$ resource_country=poland
e_7	substatus=awaiting_assignment $\wedge$ org_line=G199 3rd $\wedge$ org_country=us

cluster	rule
t_1	$\neg e_4 \wedge \neg e_9 \wedge \neg e_{13}$
t_2	$\neg e_4 \wedge e_9$
t_3	$\neg e_4 \wedge \neg e_9 \wedge e_{13}$
t_4	$e_4 \wedge e_2 \wedge$ month ≤ 5
t_5	$e_4 \wedge e_2 \wedge$ month $> 5 \wedge$ workload ≤ 355
t_6	$e_4 \wedge e_2 \wedge$ month $> 5 \wedge$ workload > 355
t_7	$e_4 \wedge \neg e_2 \wedge$ product $\in \{PROD98, PROD96\}$
t_8	$e_4 \wedge \neg e_2 \wedge$ product $\in \{PROD428, PROD97\}$

Fig. 3. Some event clusters (left) and trace clusters (right) found by HLPM-mine.

First, we assessed the ability of our event abstraction method to help discover higher-quality workflow schemas. To this end, we evaluated the workflow schemas extracted by algorithm FHM to an abstract version of the log, obtained by replacing the original events with the event clusters found by HLPM-mine, ran without any trace clustering (i.e. by setting $m = 1$). As a term of comparison, we considered three "manual" event abstraction criteria: *(i)* replacing each event with the associated value of status, or *(ii)* of substatus, and *(iii)* using each distinct event 8-ple as an activity type (all).

As to trace clustering, we considered four competitors: algorithm Actitrac [8]; the sequence-based and alphabet-based versions of the approach in [4] (TRMR and A-TRMR, resp.); and the approach in [5] (KGRAM), exploiting a k-gram representation. Since all of these competitors lack any mechanisms for abstracting log events and for selecting the number of clusters, we provided each of them with the abstraction function and the number of trace clusters found by our approach.

Quantitative Results. Table 1 reports the quality scores obtained by our event clustering approach (without trace clustering), here viewed as an enhanced data-driven event-abstraction criterion, compared with the "manual" basic ones described before. To this end, we first partitioned all log events via the Steps 1–3 of algorithm HLPM-mine, and then replaced each event with the label of its respective cluster. All measures were computed by averaging the results of 10 trials, performed each in 10-fold cross-validation.

Interestingly, our event abstraction method gets the best fitness outcome (0.815), and a satisfactory precision score, at the cost of little increase in structural complexity (8 more nodes and 14 more edges) w.r.t. the substatus abstraction, which is the most effective among all 1-attribute abstractions. By the way, only the dummy abstraction all gets higher precision, but it returns overly complex and overfitting workflows.

The benefit of clustering log traces is made clear by Table 2, which reports the quality results obtained by our approach and by the competitors. More precisely, in each clustering test, we computed an overall *Fitness* (resp., *BehPrec*) measure for each method, as the weighted average (with cluster sizes as weights) of the *Fitness* (resp., *BehPrec*) scores received by the workflow schemas induced (with FHM) from all the trace clusters discovered in the test. As all cross-validation

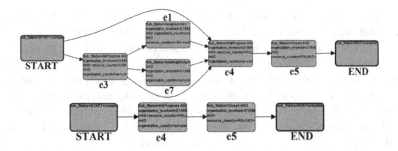

Fig. 4. Workflow schemas induced from trace clusters t_7 (top), and t_8 (bottom).

trials HLPM-mine (ran with $m=\infty$) yielded 8 trace clusters, the same number of clusters was given as input to the competitors.

Notably, HLPM-mine managed to neatly improve the average fitness (0.851) w.r.t. the case where no trace clustering was performed (0.815), and surprisingly outperformed all competitors on this fundamental metrics, despite it can only split the log by way of boolean formulas over trace properties. HLPM-mine also managed to achieve good precision (0.742), compared to its competitors, and to find quite readable workflow schemas — exhibiting, indeed, the lowest average numbers of nodes (10), edges (14) and edges per node (2.6).

Qualitative Results. In order to give a more concrete idea of the models that our approach can extract, we ran algorithm HLPM-mine on the entire log (without cross validation), while still setting $m=\infty$. As a result, 8 trace clusters were found, and equipped with separate workflow schemas, each describing a distinct problem-management scenario. For instance, Fig. 4 shows the models of two trace clusters — notice that each node is labelled with the identifier of an event cluster, whose clustering rule can be found in Fig. 3 (left).

Besides being more compact than the models discovered without trace clustering (see the last row in Table 1), these schemas help us reckon that problem cases follow different execution scenarios. Specifically, the schema of cluster t_8 captures a particular scenario, where two activities (i.e. event clusters e_4 and e_5) are executed in sequence. Moreover, the trace clustering rules on the right of Fig. 3 let us reckon that these scenarios differ both for the value of context factors, and for the occurrence of certain activity types (i.e. event clusters).

6   Discussion and Future Work

We have presented a clustering-based process discovery method for low-level multi-dimensional logs, where event and trace clusters are used to capture different activity types and process variants, respectively. Tests on a real-life log showed that the approach is able to find high-quality readable workflow models.

Novelty Points. Several features distinguish our proposal from current process mining solutions, in addition to the very idea of combining activity abstraction and trace clustering — which has only be explored in [4] so far. Fist of all, each event/trace clustering function discovered by our approach is natively encoded in terms of logical rules, which the analyst can easily interpret, to distillate a semantical view of process behavior and of its dependence on relevant properties of process events/cases (and on other context-related factors). Moreover, we removed the common assumption that each log event explicitly refers (or can be easily mapped) to some predefined process activity. In fact, even most current activity abstraction methods [6,10] rely on the presence of activity labels (possibly defined at a high level of granularity) associated with log events, when trying to aggregate them into higher-level activities (or sub-processes). We pinpoint that no high-level activity types are assumed to be known in advance, differently from [2], where a method was proposed to map log events to a-priori given activity types. The problem faced here is, in fact, more challenging, in that event types are to be learnt inductively from scratch.

Owing to the predictive nature of our models, they can help implement advanced run-time services, beside serving descriptive analyses. For example, one can dynamically assign any novel process case, say c, to one of the discovered process variants (i.e., trace clusters). The workflow schema associated with that cluster can be then presented as a customized (context-adaptive) process map, showing how c could proceed. Moreover, by continuously evaluating the degree of compliance between c and its reference schema deviating can be detected.

Future Work. Implementing the advanced run-time services above (useful in flexible BPM settings) is left for future work. Inspired by the proposal in [11], we will also try to extend our approach to deal with interleaved logs, which mix traces of different processes without keeping information on which process generated each of them. This would let us release another common assumption — orthogonal to that considered in our work, on the existence of a-priori knowledge on the mapping of log events to process activities — of most process mining approaches. Finally, we plan to test our approach on a wider range of logs.

References

1. van der Aalst, W.M.P., van Dongen, B.F., et al.: Workflow mining: a survey of issues and approaches. Data Knowl. Eng. **47**(2), 237–267 (2003)
2. Baier, T., Mendling, J., Weske, M.: Bridging abstraction layers in process mining. Inf. Syst. **46**, 123–139 (2014)
3. Blockeel, H., De Raedt, L.: Top-down induction of first-order logical decision trees. Artif. Intell. **101**(1–2), 285–297 (1998)
4. Bose, R.P.J.C., van der Aalst, W.M.P.: Trace clustering based on conserved patterns: towards achieving better process models. In: Rinderle-Ma, S., Sadiq, S., Leymann, F. (eds.) BPM 2009. LNBIP, vol. 43, pp. 170–181. Springer, Heidelberg (2010)
5. Bose, R.P.J.C., van der Aalst, W.M.P.: Context aware trace clustering: towards improving process mining results. In: SDM 2009, pp. 401–412 (2009)

6. Jagadeesh Chandra Bose, R.P., van der Aalst, W.M.P.: Abstractions in process mining: a taxonomy of patterns. In: Dayal, U., Eder, J., Koehler, J., Reijers, H.A. (eds.) BPM 2009. LNCS, vol. 5701, pp. 159–175. Springer, Heidelberg (2009)
7. Buijs, J.C.A.M., van Dongen, B.F., van der Aalst, W.M.P.: On the role of fitness, precision, generalization and simplicity in process discovery. In: Meersman, R., et al. (eds.) OTM 2012, Part I. LNCS, vol. 7565, pp. 305–322. Springer, Heidelberg (2012)
8. De Weerdt, J., van den Broucke, S., et al.: Active trace clustering for improved process discovery. IEEE Trans. Knowl. Data Eng. **25**(12), 2708–2720 (2013)
9. Greco, G., Guzzo, A., et al.: Discovering expressive process models by clustering log traces. IEEE Trans. Knowl. Data Eng. **18**(8), 1010–1027 (2006)
10. Günther, C.W., Rozinat, A., van der Aalst, W.M.P.: Activity mining by global trace segmentation. In: Rinderle-Ma, S., Sadiq, S., Leymann, F. (eds.) BPM 2009. LNBIP, vol. 43, pp. 128–139. Springer, Heidelberg (2010)
11. Liu, X.: Unraveling and learning workflow models from interleaved event logs. In: ICWS 2014, pp. 193–200 (2014)
12. de Medeiros, A.A.: Genetic Process Mining. Ph.D. thesis, TU/e (2006)
13. Weidlich, M., Polyvyanyy, A., et al.: Process compliance analysis based on behavioural profiles. Inf. Syst. **36**(7), 1009–1025 (2011)
14. Weijters, A.J.M.M., Ribeiro, J.T.S.: Flexible heuristics miner (FHM). In: CIDM 2011, pp. 310–317 (2011)

On Improving the Maintainability of Compliance Rules for Business Processes

Sven Niemand[1]([⊠]), Sven Feja[2], Sören Witt[2], and Andreas Speck[2]

[1] Provinzial Nord Brandkasse AG, 24097 Kiel, Germany
sven.niemand@provinzial.de
[2] Institut für Informatik, Christian-Albrechts-Universität zu Kiel,
24098 Kiel, Germany
{svfe,swi,aspe}@informatik.uni-kiel.de

Abstract. Business process regulatory compliance management (RCM) is ensuring that the business processes of an organization are in accordance with laws and other domain-specific regulations. In order to achieve compliance, various approaches advocate checking process models using formal compliance rules that are derived from regulations. However, this shifts the problem of ensuring compliance to the rules - for example, the derived rules have to be updated in the case that regulations are changed. In this paper we show how existing RCM solutions can be extended with traceability between compliance rules and regulations. Traceability supports the alignment of regulations and rules and thus helps improving the overall maintainability of compliance rules.

Keywords: Compliance management · Business process management · Compliance checking · Regulations

1 Introduction

Organizations have to comply with several regulations that are subject to change, such as laws (e.g. Sarbanes-Oxley Act[1], Basel III[2]) and standards (e.g. ISO 9000[3]). These *regulations* are documents written in natural language, containing a set of guidelines specifying constraints and preferences pertaining to the desired structure and behavior of an enterprise. Non-compliance with regulations can lead to serious consequences, including loss of business reputation and penalties levied against the organization. *Regulatory compliance management* (RCM) is the problem of ensuring that enterprises are in accordance with laws and domain-specific regulations [1]. In the field of RCM, regulations are usually decomposed in *compliance requirements*, which are pieces of text extracted from a regulation, specifying expected behavior as well as tolerated and non-tolerated

---

[1] http://www.soxlaw.com/ [Accessed 13 March 2015].
[2] http://www.bis.org/bcbs/basel3.htm [Accessed 13 March 2015].
[3] http://www.iso.org/iso/iso_9000 [Accessed 13 March 2015].

© Springer International Publishing Switzerland 2015
W. Abramowicz (Ed.): BIS 2015, LNBIP 208, pp. 178–190, 2015.
DOI: 10.1007/978-3-319-19027-3_15

deviations. The extraction of compliance requirements requires the intervention of regulatory (e.g. juridical) and enterprise experts (e.g. business analysts).

All value-adding activities inside organizations are realized and supported by business processes. Hence, compliance to regulations must be ensured at the level of business processes in particular [2]. In order to analyze, simulate, execute, monitor, and optimize their business processes, more and more organizations use business process management (BPM) [3], where the concept of a process model is fundamental. Process models may be used to configure information systems and serve as a means to analyze, understand, and improve the processes they describe [3].

RCM of process models has been identified as one of the core challenges in the discipline of BPM [4], and a variety of proposals have been made to facilitate it. Existing approaches to ensure compliance at design time, i.e. before the processes are executed, suggest automating compliance validation of process models in order to avoid manual auditing procedures (e.g. [5–9]). The models are checked using predefined *compliance rules*, which are semi-formal representations of compliance requirements. However, the specification of compliance rules shifts the problem of ensuring compliance from the process models to the rules. It remains unclear how regulatory compliance of derived rules is guaranteed, e.g. in the case that regulations are changed. It can be a complex and error-prone task to identify manually all rules that need to be updated.

In this paper, we present an approach to extend existing RCM solutions with the concept of traceability between regulations and compliance rules. Based on the principle of establishing links between rules and the corresponding text of regulations, we describe the architecture and adaptations of the RCM life-cycle in order to realize traceability. Among other advantages, this enables us to integrate tool support for regulation change detection which helps providing the assurance of up-to-date and better maintainable compliance rules. The remainder of this paper is structured as follows: Sect. 2 outlines a motivating scenario and contains a short introduction to the state-of-the-art of business process RCM solutions. In Sect. 3, we present our approach to improving the maintainability of compliance rules in business process RCM solutions. Some details of a proof-of-concept prototype and evaluation results are provided in Sect. 4. We finish with a conclusion and outlook in Sect. 5.

2   Motivation and the State-of-the-Art

It is the task of RCM to ensure compliance of business processes and regulations which, in the context of BPM, can be realized in three main ways. The first way is to ensure compliance of the process models before the execution of the processes (static verification at design time). At run time, compliance can be checked monitoring the running process instances. Finally there is a way of backward compliance checking by analyzing the traces of completed process executions [2]. In this paper, we focus only on the design time aspects of RCM, which we prefer due to their preventative focus. In this section, we will introduce the state-of-the-art of business process RCM using an example which is taken from a case study we conducted in cooperation with an insurance company (cf. Sect. 4.2).

2.1 Motivating Scenario

In Germany, a series of typical contractual rights and obligations of insurer and policyholder is regulated by different laws - particularly by the German Insurance Contract Act (IC Act)[4]. This is done in order to ensure a fair balance between the interests of the insurer and the policyholder [10]. An excerpt from the IC Act is shown in the following snippet. The text regulates documentation requirements of the insurer and its duty to advise the policyholder before the time of the conclusion of the contract. In the excerpt, activities are highlighted in **bold**, conditions are *italicized*, and temporal indications are underlined.

Excerpt from the IC Act 2008[4]: Sect. 6 Advising the policyholder

(1) *If the difficulty in assessing the insurance being offered or the policyholder himself and his situation gives occasion thereto*, the insurer must **ask him about his wishes and needs** and, also bearing in mind an appropriate relation between the time and effort spent in providing this advice and the insurance premiums to be paid by the policyholder, the insurer shall **advise the policyholder and state reasons for each of the pieces of advice** in respect of a particular insurance. He shall **document this**, taking into account the complexity of the contract of insurance being offered.

(2) Before the contract is concluded, the insurer shall **provide the policyholder with the advice in writing**, clearly and comprehensibly stating reasons. This **information may be provided verbally** *if the policyholder so wishes or if and insofar as the insurer guarantees provisional cover*. In such cases the **information shall be provided in writing** to the policyholder without undue delay as soon as the contract has been made [...]

Regulations such as the IC Act impact on the business processes of an enterprise. Besides their use in BPM, process models help share the understanding of a process with other people. Thus, in order to show business experts how a compliant business process for advising policyholders would look like, juridical experts could develop a process model based on the IC Act excerpt - a proposal is shown in Fig. 1. BPMN 2.0 [11] was chosen as modeling language because it is an important industry standard which is supported by most BPM solutions. From a business standpoint, however, this legally sound model is not practicable. For example, it lacks the case that no contract is concluded after the advising of the policyholder. Business experts could, however, use this model as a reference process model and adapt it according to the specific requirements of the insurer. Modifications in turn can lead to breaches of compliance and call for checking compliance of process models each time they are changed. This could be done

---

[4] http://www.gesetze-im-internet.de/englisch_vvg/ [Accessed 13 March 2015].

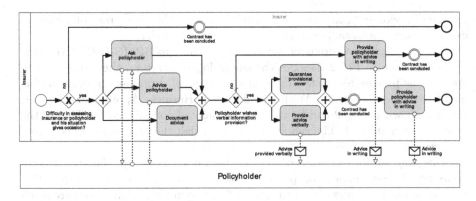

Fig. 1. Business process model derived from the German Insurance Contract Act

manually by juridical experts, but manual checking activities are time expensive and error-prone - especially when there are lots of compliance requirements that should be applied to many process models [9]. This issue is addressed by current business process RCM solutions as shown in Sect. 2.2.

2.2 Compliance Rules for Checking Process Models

Compliance requirements often originate in legal texts and have informal and abstract character. For example, the following (simplified) compliance requirement in natural language is derived from subsection (2) in the IC Act snippet: *Whenever the advice had been provided verbally and the contract has been concluded, the information shall be provided in writing to the policyholder without undue delay.* This requirement applies to the business process and its model shown in Fig. 1. In order to avoid manual auditing procedures of process models with respect to informal requirements, current RCM solutions suggest to represent compliance requirements in a formal and structured notation. These formal *compliance rules* enable tool support for automatic compliance checking of process models. E.g., the Business Application Modeler [5] allows for the specification of graphical rules in Computational Tree Logic, called G-CTL. It provides the ability to define rules on basis of process elements using a rule model editor and to verify process models against these rules through model checking.

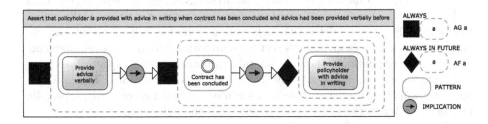

Fig. 2. BPMN-G-CTL rule as a formal representation of a compliance requirement

In case of a violated rule the model checker delivers the error trace of a counter example, which is visualized directly in the process model.

The compliance requirement example represented as BPMN-G-CTL [12] rule model is shown in Fig. 2. Besides the use of logic operators, the figure demonstrates how process elements can be placed inside of containers in order to formulate atomic expressions. With the aid of the Business Application Modeler, the rule shown in Fig. 2 can be used for automated compliance checking of the process model in Fig. 1. The validation will show that the process model is correct with respect to the rule.

This way of formalizing compliance requirements is an effective way of verifying the compliance of process models according to compliance rules. Furthermore, it enables the separation of specifying compliance requirements and business processes as advised in [13]. However, the feasibility of maintaining large sets of compliance rules is still limited regarding updates of rules and regulations. Imagine the case that a regulation is changed: manually identifying all rules that are related to the updated parts of the regulation can be challenging and time expensive. And the fact that regulations regularly change means that compliance rules have to be updated frequently. Empirical research proves that updates of existing compliance frameworks to comply with new obligations are seen as a key problem by industry compliance experts [14]. Thus, we believe that existing RCM solutions have to be extended in order to improve the overall maintainability of compliance rules.

2.3   Related Work

An overview on the state-of-the-art of business process compliance approaches based on a literature review is provided in [15]. As in [5], where a graphical notation for temporal logic (CTL) used for modeling compliance requirements is introduced (cf. Sect. 2.2), a large part of these approaches focuses on the design time aspects of RCM and on verification and validation techniques. In [7], the authors propose using predefined rule patterns (BPMN-Q queries) which are translated into temporal logic (PLTL) for compliance checking. The approach in [6] advocates using a visual pattern-based language grounded on temporal logic (LTL) that enables using compliance patterns, which are high-level abstractions of frequently used compliance requirements. Another approach that enables to model graphical compliance requirements introduces compliance rule graphs (CRG) used for compliance verification as presented in [9]. Deontic logic is used in [4,8], where process models are enriched by control tags to visually annotate and analyse the models. Although these approaches help to ensure compliance of business processes with respect to (semi-formal) compliance requirements derived from regulations, they do not take into account the aspect of assuring that compliance requirements stay in line with the regulations. Thus, in this paper we will show how such approaches can be extended accordingly, especially in order to address the issue of the regulation's changing nature.

On the other hand, research exists on establishing a connection between process models and regulations. The authors in [16] propose to bookmark structural parts in regulation documents and to annotate process models with references to these bookmarks. Work in [17] proposes to translate laws into process models using the Semantic Process Language (SPL) and to establish traceability through the documentation of compliance requirements within the process model itself. In [18] an approach is presented to increase the traceability between law and processes using the Visual Law and Process Modeller (VLPM). The tool supports translating laws into process models and linking laws and models. These approaches help to understand the impact of regulation or process changes to their counterparts. In comparison with previously mentioned RCM solutions, however, they lack the possibility for the (semi-formal) specification of compliance requirements and the (automated) verification of process models, thus requiring manual compliance checking activities inducing the aforementioned drawbacks.

3   Improving the Maintainability of Compliance Rules

Traceability between business processes and regulations and a change detection mechanism for regulations have been identified as important means to facilitate up-to-date business process compliance to regulations [16]. However, in the design of rule based RCM solutions these aspects have been neglected so far. In order to improve the maintainability of compliance rules, we advocate to integrate these aspects into existing RCM solutions by expanding them with tool support that helps to perform related activities.

3.1   Conceptualization

With *traceability* in the context of RCM solutions we mean the bi-directional ability to interrelate regulations and compliance rules: It should be possible to easily trace back compliance rules to the parts of regulations where the rules are derived from. On the other hand, it should be easily possible to spot all compliance rules that are derived from a particular part of a regulation. As a mechanism to realize traceability, we propose the establishment of *links* between compliance rules and regulations, i.e. human and machine interpretable relations where rules or even elements in these rules refer to the respective parts of regulations. For example, the task "Provide advice verbally" is an element of the compliance rule shown in Fig. 2. It is derived from the text "information may be provided verbally" in the IC Act snippet in Sect. 2.1. Thus, we would link the task in the rule with the part "Sect. 6, subsection (2)" of the IC Act. We will build on this simple principle when we realize further tool support.

3.2   Architecture

We propose the integration of four specific components into existing RCM solutions. These components are (1) a version control system for storing regulation documents and modeled compliance rules, (2) a tool for modeling compliance rules with the ability to establish links to structural parts in the regulation documents, (3) a component for identifying compliance rules linked with selected structural parts in regulation documents, which we will call *link tracer*, and (4) a change detection tool that identifies changes in regulation documents as well as the compliance rules affected by these changes. Within existing RCM solutions, components specific for our approach will be integrated with the other components of the RCM solution, such as a process model editor, a compliance checking infrastructure utilizing generic techniques like model checking for ensuring that process models are in accordance with compliance rules, a business process management system for coordinating the enactment and execution of the process models, etc. A high-level view on these components integrated in a RCM solution and the relations between the components are shown in Fig. 3. In the following, approach-specific components are described in further detail.

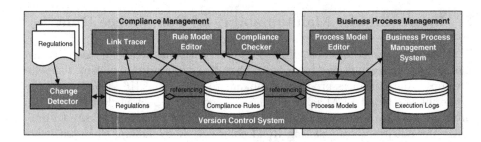

Fig. 3. High-level view on the components used in the approach

Version Control System. Multiple versions of various files can be stored in a version control system. This allows for tracing the changes between different versions of files. We will use a version control system for storing regulation documents, compliance rules and process models.

Rule Model Editor with Regulation Linking Extension. Compliance requirements are represented as formal compliance rules that can be used for checking process models, e.g. using BPMN-G-CTL [12] as shown in Sect. 2.2. As a specific functionality in the presented approach, the compliance rule editor should allow establishing links between elements of the rule models and the text of regulation documents. This requires for the provision of the documents in a way that the regulations can be referenced by the rule model editor. The link should be stored together with the rule model in the version control system.

Link Tracer. A regulation link tracer tool helps identifying all rule models that contain links referencing to the particular part of a regulation. In order to identify these rules, all rule models stored in the version control system can systematically be parsed for links that point to the respective part of the regulation document (or more fine-grained parts that are comprised by the respective part, e.g. subsections of a regulation when a particular section is of interest). We do not postulate a separate tool for tracing links from a rule model to the linked regulation part, as the links are already integrated in the models and can be examined in the rule model editor.

Change Detector. The change detection component is used to detect changes in the text of regulations. This can be done by a comparison of the actual regulation documents stored in the version control system and newer versions of these documents, e.g. new versions of law that have been published on the web page of a governmental organization. For each change that has been detected, affected rule models are identified in the version control system based on the established links pointing to that regulation document.

3.3 Life Cycle Integration

Both RCM and BPM can be supposed to follow a life cycle of phases [13]. Each life cycle consists of the phases elicitation, modeling (design and formalization), enactment and controlling (optimization) of compliance rules and business processes, respectively. The design phases of both life cycles are required to be synchronized to create consistent process models and compliance rules [13] - remember, rules can be designed on the basis of process model elements. Compliance checking of process models using compliance rules is the way back from RCM to BPM, it initiates the process elicitation phase in case of a process model violating a rule. In our approach, we extend the RCM life cycle with a *change detection* and *synchronization* phase. This is the phase where new regulation documents are provided and existing documents are updated in the version control system, which will subsequently initiate the rule elicitation phase. The integrated and extended life cycles are shown in Fig. 4. In the following, we outline the specific steps covered by our approach that will be enabled through the components described in Sect. 3.2, and allocate the steps to the phases of the extended RCM life cycle as shown in Fig. 4.

Linking Regulations and Process Constraints. The first step is to select relevant regulation documents. They are stored and provided in the version

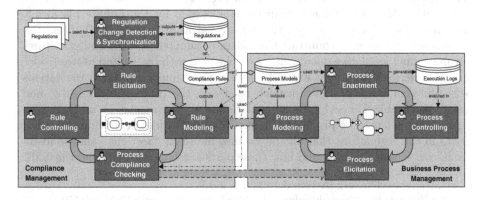

Fig. 4. Managing maintainable compliance rules in a BPM environment

control system. Then, compliance rules are derived from the text of the regulation. The rules are modeled using the rule model editor and linked with the corresponding regulation text using the editor's linking functionality. The rules are stored in the version control system. These steps can be allocated to the phases *Regulation Synchronization, Rule Elicitation* and *Rule Modeling.*

Change Detection. The change detection component is used to detect external changes in the regulations that have been stored in the version control system. If changes were detected, these changes and affected compliance rules are shown to the modeler. The modeler has to decide on necessary adjustments and should implement them if necessary. The final step is to transfer the new versions of compliance rules into the version control system and update the regulation documents in the version control system. These steps can be allocated to the phases *Regulation Change Detection, Regulation Synchronization, Rule Elicitation* and *Rule Modeling.*

Link Tracing Between Regulations and Process Constraints. The compliance rule model editor is used for the optimization of rules or their adaptation to changes in process models. Due to the existing links, the text representation of the regulation is recognizable directly in the editor and consequently can be taken into account when applying changes to the rules. In order to analyze regulations on the other hand, e.g. performing completeness checks with regard to the existing rules, the appropriate regulatory document in the version control system is opened with the link tracer tool. Any granular text part in the document can be selected to serve as a basis for performing link tracing where all referencing rules are identified. These steps can be allocated to the phases *Rule Controlling* and *Rule Elicitation.*

4   Evaluation

4.1   Proof-of-Concept Prototype

The described approach has been implemented as an extension of the Business Application Modeler, which is realized as a plug-in for the Eclipse platform[5]. The regulation documents need to be provided as structured XML files. The governments of several countries have already adopted XML-based structures for publishing legislative documents, as is the case with the United States Code[6] and the German laws[7], for example. In order to support different regulations with different structures, we decided to design the prototype extensible and user-configurable. For each type of regulation (e.g. "German law"), an XML configuration file has to be provided. These configuration files determine structural parts in the regulations that should be linkable (e.g., "Sections" and "Subsections" in "German laws"), and how these parts can be extracted from the XML

---

[5] http://www.eclipse.org [Accessed 13 March 2015].

[6] http://uscode.house.gov/download/download.shtml [Accessed 13 March 2015].

[7] http://www.gesetze-im-internet.de [Accessed 13 March 2015].

regulation documents. The extraction of parts is configured using *XML Path Language* (XPath) which is a query language for selecting nodes from an XML document. Regulation documents and configuration XML files are matched via the regulation's *document type definition* (DTD).

The rule editor of the Business Application Modeler has been extended with a corresponding linking functionality. A regulation link can be established for each element in the compliance rules. The user interface of the linking functionality is shown in Fig. 5. The figure shows a link between the rule model element selected in the rule editor (not shown) and Sect. 6, subsection 2 of the IC Act.

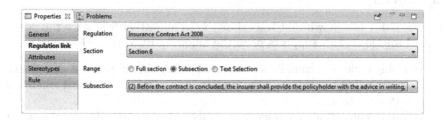

Fig. 5. Regulation linking extension of the compliance rule editor

The link tracing and change detection component have been realized as one tool on the basis of the Eclipse platform's compare editor. The tool is configured with the same XML configuration files as the rule editor. It supports exploring regulation documents, identifying referencing compliance rules and synchronizing external regulation documents with those stored in the version control system. The task of adapting compliance rules according to changes in regulations is supported by populating the Eclipse platform's task list with detected changes. Changes are categorized into new, changed, and deleted parts. The associated tasks can be scheduled, enriched with notes, and finally be marked as completed in the task list. Besides, the change detection component can be used to perform some activities in an automated way, e.g. deleting rules when the corresponding regulation parts have been deleted.

4.2 Case Study

In order to evaluate the approach, we conducted a case study in cooperation with a German insurance company. The prototype implementation has been used to model the complete process of insurance contract conclusion on the basis of respective one year old regulation documents. Figure 1 shows an excerpt from this process. The whole process covers additional aspects like declaration of intent, contractual agreement, periods for acceptance, revocation, etc. It is divided in 5 subprocesses and comprises 72 tasks in total. We then derived 43 compliance requirements pertaining to the process model from mandatory law (ius cogens). We modeled them as compliance rules and linked them with the regulations. Figure 2 shows one of these rules. After this, we adopted the process to the specific requirements and context of the insurer. The compliance rules were used to verify that the process remains compliant despite the adjustments. Finally, we took the current versions of the regulation documents and used the change detection tool to discover relevant changes in the linked laws. Several changes were detected and shown as tasks to the modeler. Most of the changes in law did not affect the compliance rules or the process, anyway, two changes required to change rules, and one change finally required to adapt the process model in order to be compliant with the rule and thus the regulation again.

4.3 Results

The case study has shown that traceability is an important feature in the design of RCM solution which supports us in the following RCM-related activities:

- Performing reviews of compliance rules and justify the existence of a particular compliance rule by providing a reference to the parts of regulation where the rule is derived from.
- Performing compliance completeness analysis for a particular regulation by identifying all compliance rules that are derived from this particular regulation.
- Analyzing the validity of changes in a compliance rule. As the parts of regulation can be traced back, it is easier to check if the rule is still in compliance with the regulation after changes are applied to the rule.
- Analyzing the impact of a change in the regulation. As all derived compliance rules for that regulation can be identified, it is possible to decide which rules have to be adapted according to the changes in regulation.

The overall maintainability of compliance rules is increased significantly by these aspects. However, the evaluation has also revealed some limitations. A set of restrictions is inherited from existing compliance management solutions. Issues such as the model construction problem when creating formal models or the state explosion problem connected with model checking [12] apply equally to the presented approach. Another substantial aspect is the level of automation. Change detection helps in detecting changes in regulations, but the decision if these changes require the adaptation of compliance rules and the adaptations themselves still remain mainly manual tasks. A higher level of automation is still desirable.

5 Conclusion

Traceability and change management are important requirements on RCM solutions with regard to minimizing the cost of maintenance of compliance requirements. In this paper, we presented an approach to integrate these aspects into existing RCM solutions. The components and steps comprised in the approach were presented, and the feasibility of the approach has been verified in a case study in the context of insurance companies.

All in all, the approach is a feasible way to ensure up-to-date compliance rules. Traceability supports the alignment of regulations and compliance rules, which improves the overall maintainability of compliance requirements. Actually, we are working on increasing the level of automation of the approach.

References

1. El Kharbili, M.: Business process regulatory compliance management solution frameworks: a comparative evaluation. In: APCCM 2012, CRPIT, vol. 130, pp. 23–32. ACS (2012)
2. El Kharbili, M., Stein, S., Markovic, I., Pulvermüller, E.: Towards a framework for semantic business process compliance management. In: GRCIS 2008, pp. 1–15 (2008)
3. van der Aalst, W.M.P.: Business process management: a comprehensive survey. ISRN Softw. Eng. **2013**, 1–37 (2013)
4. Sadiq, S., Governatori, G., Namiri, K.: Modeling control objectives for business process compliance. In: Alonso, G., Dadam, P., Rosemann, M. (eds.) BPM 2007. LNCS, vol. 4714, pp. 149–164. Springer, Heidelberg (2007)
5. Feja, S., Witt, S., Speck, A.: BAM: A requirements validation and verification framework for business process models. In: QSIC 2011, pp. 186–191. IEEE (2011)
6. Elgammal, A., Turetken, O., van den Heuvel, W., Papazoglou, M.: Formalizing and Appling [SIC] Compliance Patterns For Business Process Compliance. Software & Systems Modeling. Springer, Berlin (2014)
7. Awad, A., Decker, G., Weske, M.: Efficient compliance checking using BPMN-Q and temporal logic. In: Dumas, M., Reichert, M., Shan, M.-C. (eds.) BPM 2008. LNCS, vol. 5240, pp. 326–341. Springer, Heidelberg (2008)
8. Sadiq, S., Governatori, G.: Managing regulatory compliance in business processes. In: vom Brocke, J., Rosemann, M. (eds.) Handbook on Business Process Management 2, pp. 265–288. Springer, Berlin (2015)
9. Ly, L.T., Knuplesch, D., Rinderle-Ma, S., Göser, K., Pfeifer, H., Reichert, M., Dadam, P.: SeaFlows toolset – compliance verification made easy for process-aware information systems. In: Soffer, P., Proper, E. (eds.) CAiSE Forum 2010. LNBIP, vol. 72, pp. 76–91. Springer, Heidelberg (2011)
10. Schimikowski, P.: Versicherungsvertragsrecht. C.H. Beck, München (2014)
11. Chinosi, M., Trombetta, A.: BPMN: an introduction to the standard. In: Computer Standards & Interfaces, pp. 124–134 (2012)
12. Feja, S., Fötsch, D.: Model checking with graphical validation rules. In: ECBS 2008, pp. 117–125. IEEE Computer Society Press, Washington (2008)
13. Ramezani, E., Fahland, D., van der Werf, J.M., Mattheis, P.: Separating compliance management and business process management. In: Daniel, F., Barkaoui, K., Dustdar, S. (eds.) BPM Workshops 2011, Part II. LNBIP, vol. 100, pp. 459–464. Springer, Heidelberg (2012)
14. Syed Abdullah, N., Sadiq, S., Indulska, M.: Emerging challenges in information systems research for regulatory compliance management. In: Pernici, B. (ed.) CAiSE 2010. LNCS, vol. 6051, pp. 251–265. Springer, Heidelberg (2010)
15. Fellmann, M., Zasada, A.: State-of-the-art of business process compliance approaches: a survey. In: ECIS 2014 (2014)
16. Rudzajs, P., Buksa, I.: Business process and regulations: approach to linkage and change management. In: Grabis, J., Kirikova, M. (eds.) BIR 2011. LNBIP, vol. 90, pp. 96–109. Springer, Heidelberg (2011)

17. Olbrich, S., Simon, C.: Process modelling towards e-Government - visualisation and semantic modelling of legal regulations as executable process sets. Electron. J. e-Gov. **6**, 43–54 (2008)
18. Ciaghi, A., Mattioli, A., Villafiorita, A.: A tool supported methodology for BPR in public administrations. Int. J. Electron. Gov. **3–2**, 148–169 (2010)

Evolutionary Computation Based Discovery of Hierarchical Business Process Models

Thomas Molka[1,2]([✉]), David Redlich[1], Wasif Gilani[1], Xiao-Jun Zeng[2], and Marc Drobek[1]

[1] HANA Cloud Computing, Systems Engineering, SAP (UK) Ltd, Belfast, UK
thomas.molka@sap.com
[2] School of Computer Science, The University of Manchester, Manchester, UK

Abstract. Business process models that describe how the execution of work in a business is structured are an important asset of modern enterprises. They serve as documentation, and, if easily understandable, allow process stakeholders to make better decisions on the business process. Traditionally, these models have been created manually after analyzing the process, which can lead to outdated information when changes are introduced into the process. Today, information systems connected to the business processes log event data reflecting the real execution of the processes, and process discovery techniques have been developed to automatically extract models from these event logs. Most of these techniques discover well formalized models such as Petri nets, which can be hard to understand in case of larger process models. The evolutionary computation based approach presented in this paper discovers process models complying to the specification of BPMN, one of the most used but not well formalized notations for documenting business processes. Our approach limits the set of possible process models to hierarchically structured models, and therefore facilitates well structured and simple results. An evaluation with eight event logs shows that, despite the limitation to well structured and simple models, the approach delivers competitive results when compared with other process discovery techniques.

Keywords: Business process management · Process mining · Evolutionary algorithms

1 Introduction

Today's enterprises face an increasingly competitive market characterized by changing customer demands and newly released products or special offers placed by competitors, forcing enterprises to constantly adapt their businesses. This requirement for adaptation extends to the enterprises' business processes (BPs), i.e. the way work is executed. Due to non-documented behaviour occurring during execution, e.g. through the introduction of a new BP or structural changes in existing BPs, a deviation from the originally documented models is often observed. These changes lead to documentation (e.g. in the shape of BP

© Springer International Publishing Switzerland 2015
W. Abramowicz (Ed.): BIS 2015, LNBIP 208, pp. 191–204, 2015.
DOI: 10.1007/978-3-319-19027-3_16

models) becoming outdated, thus hindering insightful decision making, process performance analyses, and optimizations. At the same time, computerization in enterprises is growing and information systems support the execution of BPs. These systems are logging events describing the actual execution of the processes.

The lack of documentation and the increasing amount of logged information motivates the need for process mining [1,2], a field which tries to address the research question of which process models best represent a given event log. Techniques in this area focus on mining well formalized models such as causal nets or Petri nets [3–5]. When representing BPs, these models tend to be difficult to understand for business analysts and process stakeholders who prefer to make decisions based on the models they understand such as BPMN [6]. Furthermore, the process models mined by causal net or Petri net based techniques get large and complex for processes with many process steps and high variability (so called Spaghetti processes [7]). We present an approach to mine models specific to the business process domain along with a hierarchical structure. The approach allows for dealing with noise in event logs and the set of business process constructs permitted in the resulting model is configurable.

In the remainder of this paper we define the model used for representing BPs (Sect. 2), and explain the evolutionary computation based approach to mine them (Sect. 3). We conclude by setting our approach into the context of related work (Sect. 5), evaluating it (Sect. 4), and summarizing our results (Sect. 6).

2   Model

The process models we mine consist of basic business process control-flow elements. Additionally we take a hierarchical view on the process that limits the mining results to well-structured processes.

The set of basic business process control-flow elements we use are: a *start* event and an *end* event, a set of *activities*, and gateways. A gateway can be a *decision* which splits a branch into exclusive choice branches, a *merge* which reconnects exclusive branches, a *fork* which splits a branch into parallel branches, or a *join* which reconnects parallel branches. Moreover, a connection relation stores how these control-flow elements are interconnected.

Furthermore, we use a hierarchical view on process models. Rather than being a process model itself, this is a perspective on a control-flow model which allows for viewing the control-flow model at different levels of hierarchy (e.g. zooming in to see every detail of the model, or showing only the constructs at top level). Moreover, this perspective enables the definition of simple yet powerful operations on the process model as will be shown in Sect. 3.1. This hierarchical perspective consists of two types of elements: *atomic graphs* and *sequences of atomic graphs*. Both atomic graphs and sequences of atomic graphs are views on parts of a control-flow model which do not contain a start and end element, i.e. whose set of business process elements is limited to activities, forks, joins, decisions, and merges. Furthermore, both are required to be *self-contained*, i.e. capable of constituting a valid BP model if they had an end element appended and a start element prepended. Thus, all parts in our hierarchical view of a process are valid process models and may only be made up of a combination of parts which are valid process models themselves.

Atomic graphs are views on parts of the process model that are *atomic*, i.e. from which one cannot remove any sequence of control-flow elements at the front or back without violating the property of the respective atomic graph being self-contained. More specifically, being self-contained means that every atomic graph has a exactly one defined first and one last control flow element. The first element has at most one source control-flow element outside the atomic graph and the last element has at most one target control-flow element outside the atomic graph. Replacing this source and target by a start and end element would make the graph a valid business process model. We distinguish four types of graphs: activities, parallel graphs, exclusive graphs, and loops. Activities represent just a simple activity from the set control-flow elements, whereas the latter three consist of two branch sequences of atomic graphs. Parallel graphs have fork as first element and join as last element, exclusive graphs have a decision as first and a merge as last element, and loops have a merge as first and a decision as last element.

A sequence of atomic graphs is an ordered list of atomic graphs, where for any two contiguous atomic graphs the last control-flow element of the first graph is connected to the first control-flow element of the second graph. A sequence of atomic graphs can either be the process or a branch sequence. In case of the process sequence, the first control-flow element of the first graph in the sequence is connected to the start element, and the last control-flow element of the last graph is connected to the end element. For branch sequences, the first and last elements are connected to the corresponding gateways elements of the atomic graph which contains the sequence as a branch.

This way, a hierarchically structured view on the control flow model is achieved, i.e. the atomic graphs in a branch sequence are one level deeper in the hierarchy than the atomic graph containing the branch sequence. Figure 1 shows an example of the hierarchical view on a business process model. The two

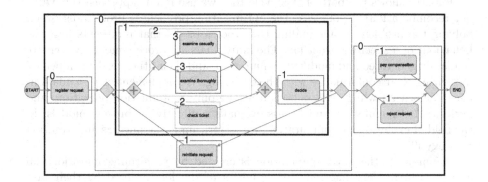

Fig. 1. Example of a process model and its hierarchical view.

parts framed by black rectangles (thicker strokes) are sequences containing more than one atomic graph. The other rectangles (thinner strokes, coloured) show atomic graphs (graphs on the same hierarchy level are framed with the same colour) and a number indicating the hierarchy level they are in. The process sequence consists of three atomic graphs: an atomic activity, a loop, and an exclusive structure (all with hierarchy level 0 as they are not part of any other atomic graph). Inside the loop, the hierarchy level stretches down to 3 (the two activities framed by purple rectangles are part of an exclusive graph which is part of a parallel graph which is part of a loop).

3   Mining Hierarchical Process Models

For discovering process models we use evolutionary computation, which aims at solving an optimization problem by imitating biological evolution and concepts such as natural selection/survival of the fittest, mutation and recombination. Solution candidates for the optimization problem are generated in a randomized fashion and evaluated by how well they solve the optimization problem (i.e. how *fit* they are). Generally, evolutionary algorithms keep a population of solution candidates (or: *individuals*), from which some are selected to reproduce (*parent selection*), i.e. to generate new individuals by *mutation* (creation of a modified version of one parent individual) or *recombination* (creation of a new individual by crossing two parent individuals). After this reproduction step, a *survivor selection* (also: environmental selection) is applied to the (now grown) population in order to shrink it down to the original size. Typically, this survivor selection chooses either strictly the fittest individuals, or uses a random distribution where individuals with a higher fitness have a higher chance of being selected. This cycle of selecting parents of one *generation* (a population at a specific point of the evolution), mutation/recombination, and selecting the best individuals into the next generation is carried out until an individual of sufficient fitness is found, or some other termination condition is reached (e.g. after a predefined number of generations).

Figure 2 shows the part of this cycle that we use in our approach (i.e. without recombination). In our case the optimization problem is process discovery. Solving the problem means finding the process model that represents best the behaviour of a specified event log. The individuals are process models which are randomly generated and modified by mutation, and their fitness is determined by measuring their conformance with the event log. We implemented a population-based algorithm which uses mutation as the only reproduction operation, and performs both mutation and fitness evaluation directly on process models. In the field of evolutionary computation, the algorithm is classified as an Evolution Strategy [9].

In our case, the initial population is created by performing mutations on a sequential model containing all activities of the log, ordered by their first appearance in the log. Parent selection is performed randomly (uniform random selection). In the environmental selection, the best individuals are taken into

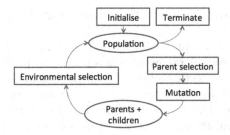

Fig. 2. Evolutionary algorithm cycle as used in this paper

the next generation (best selection). This offers an adequate balance of *selection pressure*: the goal of the evolutionary algorithm is to pressure the population towards improving the fitness from generation to generation, but to do this in a way that does not converge to a local optimum (i.e. a solution that is much better than surrounding or similar solutions but far from the global optimum) too fast. The reason for this is that mutations generally do not turn the process model into a completely different one, but make smaller modifications. By using a larger population of models and allowing also bad models to reproduce (uniform parent selection), the optimization has the chance to explore into many different directions even if they do not seem promising from the beginning. The selection of the best, on the other hand, adds selection pressure and drives the improvement to converge to an optimal solution.

3.1   Mutation

The mutation of a parent individual creates a modified clone of the parent individual. In our case, a process model is cloned to create a child model which is then modified by the application of one or more model operators. We have defined the following types of model operators:

- *move (⇝):* parts of the process are moved to another place in the process
- *complexify (⊕, ⊗, ↺):* parts of the process are set to run in parallel or alternative (exclusive) branches, or in a loop
- *simplify (◯):* parts of the process running in parallel or alternative branches, or in a loop, are set to run in a sequential manner

Note that operators for inserting and deleting activities are not required, assuming all activities are present in the model. As every hierarchical model is backed by a control-flow model, operations on the hierarchical model also take effect in the control-flow model. In the following we describe these operators exemplary.

The first operator moves atomic graphs from one place in the model to another. This includes removing a subsequence of atomic graphs from one sequence of atomic graphs, inserting it in another, and updating the control-flow element connection relation according to the new structure. To avoid creating invalid models, we introduce a preprocessing step to find only valid destinations for each move (e.g. avoiding a situation where an atomic graph is moved into itself).

The second type of operators aims at adding complexity to the control-flow model by inserting a new hierarchy level at a particular point in the hierarchical model. This is done by selecting a list of atomic graphs in a sequence, splitting it into two parts and defining these two parts to be running in parallel or exclusive branches, or in a loop. This involves creating the corresponding gateways in the control-flow model and their connections to the elements in the two branches.

Lastly, the simplify operator is used to reduce complexity by removing a hierarchy level. This is done by deleting a parallel graph, an exclusive graph, or a loop, deleting the two branches of the graph and inserting the contents of the two branches concatenated into the graph's parent sequence at the position of the graph before the operation. The gateways of the graph being "simplified" are deleted from the set of control-flow elements, and new connections are created between the contents of the two branches and the elements which are one level higher in the hierarchical structure.

An example illustrating the effects of all aforementioned operators can be seen in Fig. 3. When applying the four complexify operations and one move operation

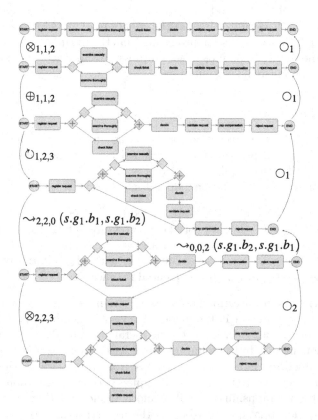

Fig. 3. Examples of operators on a hierarchical model

as shown on the left side of the picture, the model evolves from the sequential process on the top into the more complex form at the bottom of the picture. On the right side, five inverse operations (four simplify and one move) show a possible way of devolving the model back into the sequential model from the bottom of the picture to the top. For visual purposes, the operations are applied on the main process sequence, unless otherwise specified by parentheses. The three indices at the complexify operators specify the indices of the atomic graphs used to create two branches, e.g. $\otimes_{1,1,2}$ means that the graphs from index 1 (inclusive) to index 1 (inclusive) in the process sequence constitute the first branch and the graphs from index 1 (exclusive) to index 2 (inclusive) in the process sequence constitute the second branch of the exclusive graph to be created. The index of the simplify operator represents the index of the atomic graph in the sequence to be simplified. For the move-operator, in $\leadsto_{2,2,0} (s.g_1.b_1, s.g_1.b_2)$ the first two indices in $2, 2, 0$ describe the range of atomic graphs in the sequence from which they are moved, and the last index describes the position in the sequence into which they are inserted. In this example, the last atomic graph in the first branch of the loop is moved to the beginning of the second branch of the loop. The loop is the second atomic graph in the process sequence and thus denoted as $s.g_1$ ($s.g_0$ is the first graph).

Note that with a certain combination of the operators all possible BP models can be constructed which conform to our definition of the hierarchical perspective. This means the search space can be completely covered by the evolution strategy, enabling it to (potentially) always find the best conforming process model within the restrictions imposed by the hierarchy definition.

When a mutation is performed, one of the five operators ($\leadsto$, $\oplus$, $\otimes$, $\circlearrowleft$, and $\bigcirc$) is selected using a uniformly distributed random variable (this default distribution can be modified by the user, e.g. to disallow loop constructs in the resulting model the user may deactivate the $\circlearrowleft$ operator). Thereafter, the operator parameters, which define the parts of the hierarchical model to be modified, are randomly chosen.

In case of the $\leadsto$ operator, a sequence of atomic graphs is randomly selected from the hierarchical model. From this sequence, a part is randomly selected, forming the list of atomic graphs to be moved. Then, the hierarchical model is searched for potential destination sequences (suitable sequences to which the atomic graphs can be moved without violating model validity). From these sequences, one is randomly selected as destination sequence, as well as a position in this sequence where the graphs are inserted. If the mutation applies one of the complexify operators ($\oplus$, $\otimes$, or $\circlearrowleft$), then a sequence of atomic graphs is randomly selected from the model. A part of this sequence is randomly selected, forming the list of atomic graphs to which the complexify operator is applied. In this list a random position is chosen at which the list is split. The two resulting parts form the two branches of the created parallel, exclusive, or loop construct. In case of simplify, an atomic parallel or exclusive graph or a loop is randomly selected from the model and the $\bigcirc$ operator is applied.

3.2 Fitness

We define the fitness of a process model as a function of its conformance with a reference event log (i.e. the event log of the process it aims to reflect). Models which conform well to the event log are considered fit. An event log is the result of the execution of a BP. It consists of *traces* which represent a single instance of executing the process from start to end. Each trace is a sequence of events, where each of those events reflects the execution of a single activity of the BP. Hence, if an event pertaining to activity a is directly followed by an event pertaining to activity b in a trace, a good process model should also allow for activity b being executed directly after activity a. In order to measure the conformance of a log and process model, we first extract what is called the *directly-follows footprint* from the event log. This is a relation which stores for every activity represented by an event in the log, the activities it is directly followed by, thus describing the behaviour of the event log in an abstract manner. From the BP model we extract the same kind of footprint, describing what kind of behaviour the model allows. This enables us to measure the conformance between model and log in two directions:

- *recall*: how much of the log behaviour is supported by the model[1]
- *precision*: how much of the model behaviour is evident in the log

We further refer to the level of conformance measured in this way as *footprint conformance*. Since this is describing model-to-log conformance on a very "local" level, i.e. how activities directly follow one another, we further use a measure for determining how well whole process instances in the form of traces in a log are reproducible by a model. This "global" level of conformance is measured by replaying traces of the log on the process model. This measure only expresses the recall direction, and is further referred to as *replay conformance*. In [8] a detailed description of the algorithms used for measuring footprint and replay conformance between event logs and BP models is provided.

In addition to the conformance between model and log, the complexity of the process model is taken into account when evaluating its fitness. We favour process models if they are simple and conform well to their reference log. For a given process model m and its reference log l the fitness is defined as:

$$\mathtt{fit}(m, l) = \mathtt{conf}(m, l) * 0.99^{\#\mathtt{cs}(m)}$$

The conformance of a model m and its log l is defined as the average of replay and footprint conformance:

$$\mathtt{conf}(m, l) = \frac{\mathtt{conf}_{rp}(m, l) + \mathtt{conf}_{fp}(m, l)}{2}$$

---

[1] in the process mining domain, this is often called (trace) fitness, whereas in our approach the term *fitness* refers to the overall quality of a model.

where replay conformance considers the proportion of perfectly replayable traces and the proportion of replayable events:

$$\mathtt{conf}_{rp}(m,l) = \left(\frac{\#\mathtt{rp}_{traces}(m,l)}{\#\mathtt{traces}(l)} + \frac{\#\mathtt{rp}_{events}(m,l)}{\#\mathtt{events}(l)} \right)/2$$

and footprint conformance considers recall and precision of the directly-follows footprint, as well as the occurrence footprint conformance:

$$\mathtt{conf}_{fp}(m,l) = \frac{\mathtt{fp}_{rec}(m,l) + \mathtt{fp}_{pre}(m,l) + \mathtt{occ}(m,l)}{3}$$

This ensures $\mathtt{conf}(m,l) \in [0,1]$ for any model m and log l, where 0 indicates that model and log behaviour are disjunct, and 1 means model and log fit perfectly.

The second term of the function $\mathtt{fit}$ reduces the fitness of a model according to the complexity of the model. We measure model complexity by the number of constructs (such as loop, parallel, or exclusive) in a model, denoted as $\#\mathtt{cs}(m)$. The term $0.99^{\#\mathtt{cs}(m)}$ penalizes the fitness of a model by about 1 % per construct for less complex models while ensuring a value $\in [0,1]$ even if there are many constructs (63 % penalty for models with 100 constructs).

Size-Dependent Complexity Penalty. A drawback of this term is that model complexity is penalized regardless of the number of activities in the model, which can lead to the introduction of local optima in case of models with a high number of activities. Consider the upper model in Fig. 4 and its directly-followed footprint matrix. The model is a sequence of 26 activities, thus its footprint matrix consists of 729 entries. An application of the exclusify operator on the part of the model which contains only activity b results in a model with one additional construct. Also, one of the 729 cells in the model's footprint matrix changes, since activity c may now follow after activity a. Assuming this one cell has now been changed to reflect the respective cell in the reference log, this increases the model's conformance by $\frac{1}{729} * \frac{1}{6}$ (since the footprint recall $\mathtt{fp}_{rec}$ contributes to footprint conformance with a weight of $\frac{1}{3}$, and to the overall model conformance with a weight of $\frac{1}{6}$). A constant number for penalizing model complexity (such as 0.99, reducing the fitness by about 1 %), however, will likely cause this

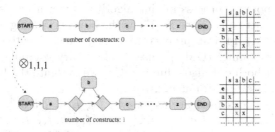

Fig. 4. Effect of making activity "b" optional

conformance improvement to be ignored in the optimization. A fitness function which takes into account the number of activities can be defined as follows:

$$\texttt{fit}(m, l) = \texttt{conf}(m, l) - \frac{1}{6} * \frac{\#\texttt{cs}(m)}{(\#\texttt{act}(l) + 1)^2 + 1}$$

Here, the complexity penalty consists of a number of constructs multiplied with the weight of footprint precision/recall in the overall conformance, and a term expressing less than the proportion of a single cell in the directly-follows footprint matrix ($\#\texttt{act}(l)$ is the number of activities occurring in the log l, and $(\#\texttt{act}(l) + 1)^2$ the number of cells in the footprint matrix). Our experiments show that this way of penalizing complexity makes the algorithm more robust to local optima.

4   Evaluation

In order to evaluate our approach, we mine models from eight different event logs (simple and complex processes, artificial and real-life logs) and convert the results into a format readable by the process mining toolkit ProM [13]. We use ProM to determine the quality of the resulting process models by transforming them into Petri nets and measuring their conformance with the corresponding event log. Note that these conformance measures are based on Petri nets and are different from the measures used to determine model fitness in our evolution strategy.

The four quality metrics we use are *trace fitness*, *precision*, *generalization*, and *simplicity*, which are described as standard quality metrics for process models in [15]. Trace fitness is a purely recall-based measure, determining how well the event log aligns with the model when trying to replay the traces on the model. Precision checks if the model allows for other behaviour than present in the traces while replaying the event log on the model. Generalization punishes overfitting models which fit everything in the log, even if it is exceptional behaviour/noise. For these three measures we use plugins of the *PNetReplayer* package in ProM (described in [15]). To measure simplicity, we count the number of model elements in the Petri net (places, transitions, and arcs). The lower the number, the simpler (i.e. better readable) is the model.

In order to put the results into context, we perform the same evaluation with three other approaches related to our work. The heuristic miner (HM) [4] is one of the most popular process mining algorithms which mines models in the shape of heuristic nets (described in [17] as the best performing algorithm). The inductive miner (IM) [14] uses process trees, a notion similar to our hierarchical model. The genetic miner (GM) [5] uses evolutionary computation and mines heuristic nets (for both our algorithm and GM we evaluate the best model they can produce within a time frame of 2h).

The eight event logs used are as follows: log $L_1 = \{(b, a)^4, (a, b, d, e)^5, (b, a, e, d)^4, (b, a, c, a, b, c, b, a, d, e, e, d)^6, (g, g)^2, (f, h)^3, (f, f, h, f, g, h, g, f, h)^8, (g, h, f)^2\}$ consists of 8 different activities. The logs EX5 and REP are the two examples "exercise5" and "repairExample" taken from the ProM website (consisting of 14 and 8 activities, respectively). BE1 - BE3 are event logs created by simulating complex models consisting of 20 activities. DF is a portion of real-life event log from an eHealth process [16] with 18 activities, which does not contain all of the process behaviour. FL_A is a real-life log representing the subprocess "A" (consisting of 10 activities) of the a loan application process available on the Business Process Intelligence Challenge 2012 website.

Table 1. Conformance results of the different discovery algorithms

Log	Trace Fitness				Precision				Generalization				Simplicity			
	HM	IM	EM	GM	HM	IM	EM	GM	HM	IM	EM	GM	HM	IM	EM	GM
L_1	0.679	0.863	**1.000**	0.868	**0.532**	0.529	0.442	0.497	0.638	0.422	**0.755**	0.681	86	91	**69**	94
EX5	0.985	0.935	**1.000**	**1.000**	0.495	0.560	**0.681**	0.480	0.931	0.996	**0.999**	0.928	155	102	**81**	262
REP	**1.000**	**1.000**	**1.000**	**1.000**	0.905	**0.955**	**0.955**	0.943	0.998	**0.999**	**0.999**	0.998	72	**46**	49	70
BE1	0.991	**1.000**	**1.000**	0.721	**0.838**	0.814	0.821	0.707	0.998	**1.000**	**1.000**	**1.000**	192	132	**130**	167
BE2	0.924	0.981	**1.000**	0.882	**0.737**	0.594	0.684	0.635	**1.000**	**1.000**	**1.000**	0.992	196	156	**115**	203
BE3	0.822	0.983	**1.000**	0.981	**0.891**	0.443	0.635	0.732	**1.000**	**1.000**	**1.000**	**1.000**	178	149	**125**	222
DF	**1.000**	0.911	**1.000**	0.906	0.563	0.559	**0.588**	0.276	**0.914**	0.906	0.829	0.336	177	136	**133**	176
FL_A	0.974	**1.000**	**1.000**	**1.000**	0.920	0.695	0.953	**0.965**	**0.925**	0.825	0.811	0.804	98	62	77	193
AVG	0.922	0.959	**1.000**	0.920	**0.735**	0.644	0.720	0.654	**0.925**	0.893	0.924	0.842	144.3	109.3	**97.4**	173.4

Table 1 shows the results of our evaluation (values rounded half up to three digits after the comma, one digit for simplicity average). The bottom row shows for each mining algorithm and quality measure the average over all event logs. The best results are marked bold. Our approach (labelled EM) scores consistently well in trace fitness and simplicity (best or second best in for all event logs), and often well in precision and generalization. Due to the limitation to well-structured models in the inductive miner and our approach, they find far simpler models than the heuristic and genetic miner. The genetic miner finds the model with the best recall and precision for FL_A. This, however, is achieved at the cost of simplicity, which is poor for most event logs. The heuristic miner performs well in generalization and precision and poor in simplicity.

5 Related Work

Most existing approaches in process discovery focus on the extraction of Petri nets [10]. The genetic miner [5] is an approach which, similar to ours, uses an evolutionary computation based algorithm to mine process models. The models are created in the shape of heuristic nets which can be transformed into Petri nets. In this approach, the models are not limited to a hierarchical structure. The conformance checking is done on heuristic nets by counting how many events the net is able to replay. It tries to avoid overfitting by penalizing a high number of enabled activities during replay. It uses a mutation which randomly changes activity bindings in the heuristic net, and a recombination of two parent nets which produces a copy of one parent and then replaces some activity bindings with bindings found in the other parent. A drawback is that this approach may lead to invalid models which need to be repaired after the recombination.

In [11] process structure trees are defined. This type of model stores the processes in a tree representation similar to the hierarchical model used in this paper. Process structure trees are stand-alone models whereas the hierarchical model is this paper is a view on a BPMN based control-flow model.

The inductive miner [14] is a deterministic algorithm which mines process structure trees from event logs by using a divide-and-conquer approach. It builds a directly-follows graph of activities and then continues to apply a set of rules, splitting the graph into two parts every time a rule is applied. The rules for the creation of parallel constructs in this approach are very strict as they require every possible combination of activities in parallel branches to occur in the event log. This requirement is rarely fulfilled in real-life event logs which leads to an under-representation of parallel behaviour in the results of the inductive miner.

The evolutionary tree miner [12] also uses process structure trees, and is, as our approach, based on evolutionary computation. It uses four quality dimensions for determining model fitness as a weighted average, giving the user some configuration ability to allow for guided process discovery. The precision measure used in [12] is limited in the detection of underfitting (one of the examples allows for loop behaviour which is not present in the event log, even though the model has a precision of 100 %). The mutation used by the evolutionary tree miner can add, remove, and swap the type of nodes in a process tree. In order to express the ↝ operator defined in Sect. 3.1, the evolutionary tree miner would require a sequence of add and remove operations. It defines a recombination which crosses two parent trees by using some parts of each parent to create a new tree. The approach allows for duplicate activities which makes it easier to express certain behaviour in some cases, but can also lead to unrealistic overfitting models (e.g. an exclusive choice construct with one branch for each trace in the event log).

6   Conclusion

This paper describes a process mining algorithm that mines BPMN models based on evolutionary computation. The models are structured in a hierarchical manner which allows for improved readability and supports viewing the process models at different levels of abstraction. Our evaluation shows that, when quantifying readability in terms of the number of model elements needed to express the behaviour in a event log, the algorithm is often able to mine models which are better readable than the results of related approaches, and still competitive in terms of accuracy. In the default configuration, the algorithm tends to find models with a very high score in recall, and consistently good precision. The hierarchical representation of the models and the operations defined on it restrict the search space in the optimization to well-structured models, avoiding the need for examining and repairing invalid models (e.g. as done in [5]), and therefore enabling the algorithm to perform well even in cases of large event logs and complex processes.

Future work includes an evaluation of how certain sequences of operators lead to better fitting process models, as well as an improvement of the operators towards a goal-oriented setting (i.e. concentrating evolutionary operators on specific parts of the process which are identified as the source of poor model-to-log conformance).

References

1. van der Aalst, W.M.P.: Process Mining: Discovery, Conformance and Enhancement of Business Processes. Springer, Heidelberg (2011)
2. van der Aalst, W.M.P., et al.: Process mining manifesto. In: BPM 2011 International Workshops (2011)
3. van der Aalst, W.M.P., Weijters, A., Maruster, L.: Workflow mining: discovering process models from event logs. IEEE Trans. Knowl. Data Eng. **16**(9), 1128–1142 (2004)
4. Weijters, A.J.M.M., Ribeiro, J.T.S.: Flexible heuristics miner (FHM). In: IEEE Symposium on Computational Intelligence and Data Mining (CIDM) (2011)
5. Alves De Medeiros, A.K., Weijters, A.J.M.M., van der Aalst, W.M.P.: Genetic process mining: an experimental evaluation. Data Min. Knowl. Disc. **14**(2), 245–304 (2007)
6. Allweyer, T.: BPMN 2.0 - Introduction to the standard for business process modeling. In: BoD (2010)
7. van der Aalst, W.M.P.: Process mining discovering and improving spaghetti and lasagna processes. In: Chawla, N., King, I., Sperduti, A. (eds.) Proceedings of the IEEE Symposium on Computational Intelligence and Data Mining, pp. 13–20 (2011)
8. Molka, T., Redlich, D., Drobek, M., Caetano, A., Zeng, X.-J., Gilani, W.: Conformance checking for BPMN-based process models. In: Proceedings of the 29th Annual ACM Symposium on Applied Computing (2014)
9. Rechenberg, I.: Evolutionsstrategie. Optimierung technischer Systeme nach den Prinzipien der biologischen Evolution, Frommann-Holzboog (1973)
10. van Dongen, B.F., Alves de Medeiros, A.K., Wen, L.: Process mining: overview and outlook of petri net discovery algorithms. In: Jensen, K., van der Aalst, W.M.P. (eds.) Transactions on Petri Nets and Other Models of Concurrency II. LNCS, vol. 5460, pp. 225–242. Springer, Heidelberg (2009)
11. Vanhatalo, J., Völzer, H., Koehler, J.: The refined process structure tree. In: Dumas, M., Reichert, M., Shan, M.-C. (eds.) BPM 2008. LNCS, vol. 5240, pp. 100–115. Springer, Heidelberg (2008)
12. Buijs, J.C.A.M., Van Dongen, B.F., van der Aalst, W.M.P.: A genetic algorithm for discovering process trees. In: 2012 IEEE Congress on Evolutionary Computation (2012)
13. van der Aalst, W.M.P., van Dongen, B.F.: ProM: the process mining toolkit. Ind. Eng. **489**, 1–4 (2009)
14. Leemans, S.J.J., Fahland, D., van der Aalst, W.M.P.: Discovering block-structured process models from event logs containing infrequent behaviour. In: Business Process Management Workshops (2013)
15. van der Aalst, W.M.P., Adriansyah, A., van Dongen, B.F.: Replaying history on process models for conformance checking and performance analysis. WIREs data mining and knowledge discovery **2**(2), 182–192 (2012)

16. Galushka, M., Gilani, W.: DrugFusion - retrieval knowledge management for prediction of adverse drug events. In: Abramowicz, W., Kokkinaki, A. (eds.) BIS 2014. LNBIP, vol. 176, pp. 13–24. Springer, Heidelberg (2014)
17. De Weerdt, J., De Backer, M., Vanthienen, J., Baesens, B.: A multi-dimensional quality assessment of state-of-the-art process discovery algorithms using real-life event logs. Inf. Syst. **37**(7), 654–676 (2012)

Collaboration

Empowering End-Users to Collaboratively Structure Processes for Knowledge Work

Matheus Hauder[1](✉), Rick Kazman[2], and Florian Matthes[1]

[1] Software Engineering for Business Information Systems,
Technical University Munich, Munich, Germany
{matheus.hauder,matthes}@tum.de
[2] Software Engineering Institute, Carnegie Mellon University,
Pittsburgh, USA
kazman@sei.cmu.edu

Abstract. Knowledge work is becoming the predominant type of work in many countries and is involved in the most important processes in organizations. Despite its increasing importance business information systems still lack appropriate support for knowledge-intensive processes, since existing workflow management solutions are too rigid and provide no means to deal with unpredictable situations. Future business information systems that attempt to improve this support need to solve the problem of facilitating non-expert users to structure their processes. The recently published Case Management Model and Notation (CMMN) might overwhelm non-expert users. Our research hypotheses is that end-users can be empowered to structure processes for knowledge work. In an evaluation with two student teams working on a software development project our solution improved the information structure, reproducibility, progress visualization, work allocation and guidance without losing the flexibility compared to a leading agile project management tool.

Keywords: Knowledge-intensive processes · CMMN · Adaptive processes · Adaptive case management · Business information systems

1 Introduction

Globalization, digital transformation as well as an ever-increasing use of information technology leads to an automation and replacement of workplaces from less skilled workers in organizations [1]. At the same time the ability to develop innovations and constantly adapt to changing market requirements is crucial for the sustainable success of an organization. As a result today's work environments require highly trained experts that can perform many complex tasks autonomously. These experts are referred to as "knowledge workers" and their processes have a tremendous impact on the success and add the most value in organizations. Davenport describes this development as follows: *"I've come to the conclusion that the most important processes for organizations today involve knowledge work. In the past, these haven't really been the focus of most organizations – improving administrative and operational processes has been easier – but they must be in future"* [2].

© Springer International Publishing Switzerland 2015
W. Abramowicz (Ed.): BIS 2015, LNBIP 208, pp. 207–219, 2015.
DOI: 10.1007/978-3-319-19027-3_17

Workflow management solutions are not suitable to support these knowledge-intensive processes, since they are too rigid and provide no means to deal with unpredictable situations [3]. Due to a large amount of exceptions resulting from the unpredictable nature of knowledge-intensive processes, traditional workflow management models would be too complex to manage and maintain [4]. Support for knowledge-intensive processes requires a balance between structured processes for repetitive steps as well as unstructured processes to facilitate creative aspects that are necessary to solve complex problems [5]. Case management has been initially promoted in 2001 by Van Der Aalst and Berens as a new paradigm to support this flexibility for knowledge workers during a process [6]. In contrast to workflow management which focuses on what *should* be done, case management focuses on what *can* be done to achieve a business goal [7].

The Object Management Group (OMG) issued a request for proposal (RFP) in 2009 to create a standard modeling notation for case management[1]. The main goal of this request is the development of a complement for the Business Process Model and Notation (BPMN) that supports a data-centric approach which is based on business artifacts [6, 7, 9]. In 2014 the result for this request was published as the Case Management Model and Notation (CMMN) [8]. The main difference of CMMN compared to BPMN is that it builds on a declarative instead of an imperative process model [9]. These declarative process models specify what should be done without specifying how it should be done to facilitate the required flexibility for more dynamic processes [10]. After the initial release of BPMN several subsets of the BPMN language were proposed by Zur Muehlen et al. [11]. Similarly, Marin et al. [12] recently proposed that future research needs to identify subsets of the CMMN language that are less complex for end-users who may not have a computer science background, and who do not need to understand the complete CMMN specification.

Our goal is to facilitate non-expert users to structure knowledge-intensive processes. Recent literature frequently emphasizes this challenge as *knowledge worker empowerment*, which is indispensable since these processes cannot be completely predefined by process designers but require on the fly adaptation by end-users while they are being executed [5, 13, 14]. Organizations are increasingly using wikis as shared knowledge repositories that can be used for collaborative gathering of information. We view these wikis as promising tools for the collaborative and self-organizing structuring of knowledge-intensive processes by end-users. Although wikis can be dynamically adapted to new needs, there are limitations of existing solutions and general research challenges for this problem:

- the notations of existing process modeling languages (e.g. CMMN) are not familiar to end-users with limited computer science background
- (lightweight) declarative process constructs like constraints for logical dependencies between tasks with their linkage to data are not supported [5, 13]
- late data modeling that allows end-users to add new data to the information model during the process enactment is not supported [5]
- mechanisms for the collaborative documentation, adaptation and instantiation of work templates for recurring knowledge-intensive tasks are not supported [14].

---

[1] http://www.omg.org/cgi-bin/doc?bmi/09-09-23, last accessed on: 2015-01-23.

2 Related Work

Templates facilitate the reuse of content on wiki pages that can be used for the explicit organizational knowledge creation (and sharing) in an organization [16, 17]. In the Darwin Wiki a template consists of types, attributes, and tasks that are collaboratively added by end-users. These templates are used to structure and reuse knowledge-intensive processes and our goal is to make adaptations of these templates as easy as possible for end-users without computer science background. The Semantic Media-Wiki is a prominent project that enables users to structure page content on top of an existing wiki with templates [15]. This tool also includes mechanisms for searching, browsing, and aggregating content through queries. Our approach differs from the Semantic MediaWiki since it supports the creation of *tasks* that can have logical dependencies and assigned attributes to guide knowledge-intensives processes. Furthermore, our solution provides a template mechanism with *roles* that can be used to restrict access rights for attributes and tasks, which is often inevitable in an organizational context.

The Organic Data Science Wiki (ODSW) extends the Semantic MediaWiki with new user interface features to support open science processes [18]. The ODSW focuses on scientific collaborations which revolve around complex science questions that require significant coordination, enticing contributors to remain engaged for extended periods of time, and continuous growth to accommodate new contributors. All user interface features in the ODSW are based on social design principles and observed best practice patterns from successful online communities. These features include tasks that are attached to wiki pages and can be decomposed into smaller subtasks. Every user in the ODSW has his own profile page with all allocated tasks that are structured along the time dimension, e.g., currently active tasks, future tasks and completed tasks. Semantic properties of the Semantic MediaWiki can be used to describe data elements that are attached to tasks. Although there are some commonalities with the usage of social design principles and patterns for successful online communities, the solution presented in this paper provides important functionalities for the structuring of knowledge-intensive processes that are missing in the ODSW. Our approach provides more fine-grained access rights, declarative process model constructs for tasks dependencies and reusable work templates.

Voigt et al. [19] detail application oriented use cases for structured wikis and take the ICKEwiki as an example. In their work the authors mentioned that *"the presentation of (role-based) tasks within the structured wiki seems to be an important feature to adequately display processes"*. The solution proposed with the ICKEwiki contains simple procedural steps. We extend this approach by taking into account dependencies between tasks and more sophisticated roles, e.g., skip and delegate roles. In addition to these aspects, the solution presented in this paper also proposes and implements a subset of the CMMN standard. In previous work [20] we proposed the Hybrid Wiki to allow users to collaboratively structure information. This paper builds on the experiences of the Hybrid Wiki and extends it with new structuring concepts for tasks, complex information structure hierarchies, new user interface concepts, more detailed role concepts for content as well as a declarative process model integrated with a subset of CMMN to support knowledge-intensive processes.

Case management can be distinguished into Production Case Management (PCM) and Adaptive Case Management (ACM) [21]. While cases in PCM are predefined at design-time, ACM proposes an adaptation of cases at run-time through end-users [21]. The main goal of ACM is to support unpredictable processes in which knowledge workers are not controlled but responsible to perform decisions autonomously. Processes with high control flow complexity and high variation are easier to implement in ACM compared to PCM and traditional workflow management. Despite the increasing interest in ACM, as measured by the number of publications on this topic since 2012 (cf. literature review in [13]), there are only few solutions available that address the challenges presented by Hauder et al. [13]. To the best of our knowledge this is the first paper that addresses the challenges of knowledge worker empowerment, data integration as well as knowledge storage and extraction for processes in knowledge work using a wiki approach.

Schönig et al. recently investigated [22] how process models can be learned from cases in knowledge-intensive processes that are described with CMMN. In their work the authors proposed to simplify the formalization of cases by using CMMN process skeletons. These skeletons are further specified by automatically learning from process execution history. As a result process models evolve and become more and more complete with every iteration. This approach could be applied in the Darwin Wiki, since the history of wiki pages could be analyzed to improve the templates. Furthermore, the quality of the learned models could be improved through our solution since our goal is to empower end-users to create more structure that can be learned. In Marin et al. [12], the authors recently investigated the meta-model based method complexity of CMMN. According to the authors future research is needed to identify subsets of the specification that reduce the complexity for end-users.

3   The Darwin Wiki

In this section, our solution to the problem of facilitating non-expert users to structure knowledge-intensive processes is presented. Our solution is illustrated with: (1) the design rationale, (2) the structuring concepts, and (3) the implementation, including the meta-model of our solution.

3.1   Design Rationale

Our primary research goal is to make the structuring of knowledge-intensive processes easier, so that non-experts without knowledge about a dedicated process notation can contribute to the modeling process. This is indispensable since one of the most challenging requirements in case management is the empowerment of knowledge workers to facilitate a self-organizing working mode [5, 13, 14]. We achieve this goal by accepting limited modeling capabilities of end-users and use only simple structuring elements as metaphors for processes. Thereby knowledge-intensive processes emerge bottom-up and modeling experts can use these emerging structuring elements (e.g. through mechanisms proposed in [22]) to maintain reusable work templates that become more and more detailed with every iteration of the process.

3.2 Structuring Concepts

In the Darwin approach wiki pages consist of rich-text for unstructured data as well as additional concepts that add more structure. These additional concepts enable the collaborative structuring and instantiation of knowledge-intensive processes for end-users which are referred to as work plans. The main structuring concepts for wiki pages are shown in Fig. 1 and described subsequently:

Tasks. We use tasks attached to wiki pages as an organizational mechanism to coordinate work. Tasks can be decomposed to exhibit goal-oriented hierarchical structures. Tasks for a page are shown in a side window for every wiki page (cf. ⓐ in Fig. 1), which allows users to browse tasks and scroll through the wiki page at the same time. Attributes ⓒ that document results can be assigned to tasks with drag and drop. After a task is selected mandatory attributes for this task are shown on the page.

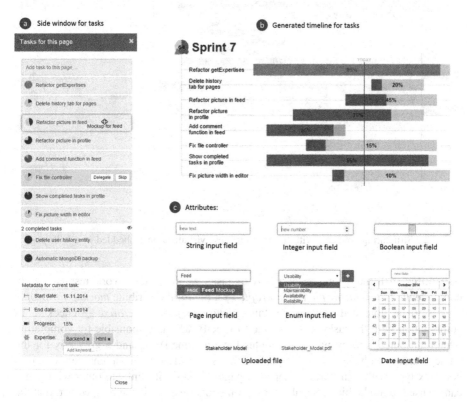

Fig. 1. Concepts for end-users in the Darwin Wiki to structure knowledge-intensive processes

Furthermore, tasks help to track the progress of activities through meta-data for start and end date, progress, and the required expertise for every task. Based on this meta-data a timeline similar to a Gantt chart is automatically generated on every page showing the current progress of all enabled tasks (and subtasks) for this page ⓑ Tasks can have attribute values assigned that reference wiki pages. These referenced wiki pages can again have tasks assigned to enable hierarchical task structures. In case a task containing subtasks is selected in the side window, the corresponding timeline with the subtasks is shown on the wiki page. The progress of tasks is either automatically computed based on the progress of the subtasks and the completion of mandatory attributes or can be manually edited by the users. Subtasks with start and end dates that exceed the dates of the parent task are marked as yellow. Overdue tasks that are not completed before the end date are marked as red. End-users can easily create new tasks on the fly at run-time in the side window by defining a name for the task and (optionally) adding attributes using the drag and drop functionality (cf. ❶ in Fig. 2). Similar to case handling, tasks can also be skipped or delegated to another responsible person if necessary [3].

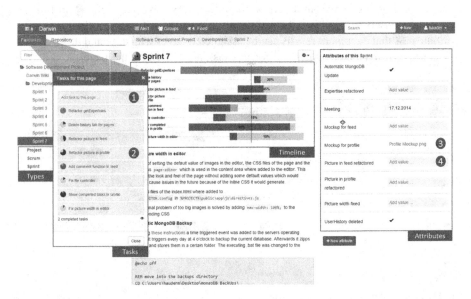

Fig. 2. Structuring concepts with the task progress visualization embedded in a wiki page

Attributes. Attributes can be assigned to tasks as mandatory work results, i.e. to complete the task *refactor picture in profile* ❷ a file for the attribute *mockup for profile* ❸ has to be uploaded and the attribute *picture in feed refactored* ❹ checked. In combination with the tasks it is possible to specify who is responsible for completing which attributes, and by when. Attributes have a name string and a grey box for the value. Values can have a simple type, e.g. file, string, date, boolean, enumeration, or a complex type with a link to another wiki page. Values with links to other wiki pages can be used to model hierarchical information structures, e.g. it can be specified that the linked wiki pages need to have certain types. Similar to our previous work, attributes and tasks from other users can be suggested to increase consistent usage of terms [20]. During the creation of an attribute end-users specify access rights to define who is allowed to edit and read attribute values. This flexible information structuring mechanism is used to integrate data in the knowledge-intensive processes which is indispensable for knowledge work (cf. data integration challenge and late data modeling requirement identified in the literature reviews in [5, 13]).

Types. End-users can assign a type for every wiki page to describe what kind of content is presented on the page. An example of a type is shown at the top of the attribute box in Fig. 2 as *sprint*. The main purpose of the type is to dynamically determine what attributes and tasks are predefined or have been added by other users. Predefined attributes and tasks of a type are immediately added when a new wiki page with this specific type is created. In addition, end-users can collaboratively extend these pages with new tasks and attributes at run-time. Predefined tasks can be created based on the collaboratively defined extensions by end-users or in advance by modeling experts in an interactive and web-based CMMN editor. This editor is only visible for modeling experts and can be accessed by clicking on the name of a particular type, e.g. *sprint* in the file explorer or above the attribute list in Fig. 2.

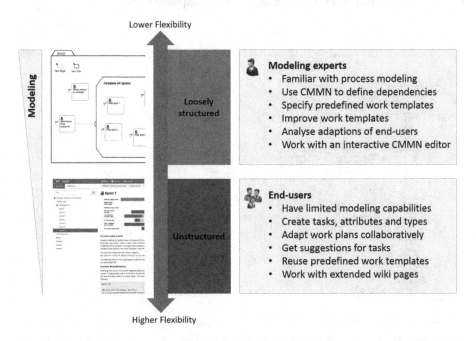

Fig. 3. Modeling experts use an interactive CMMN editor to define dependencies for tasks

Within this CMMN editor dependencies between tasks as well as completely new tasks and stages can be created by a modeling expert (cf. Fig. 3). Every task in this CMMN editor corresponds to a task that is shown in the side window for tasks on wiki pages. CMMN allows an experienced modeling expert to define the behavior of a type with stages and define logical dependencies, i.e. the task *define features for the development* has to be completed before all tasks in the stage *complete all sprints* are enabled. Tasks created by end-users are shown with simple usage statistics and the modeling experts can make them predefined for future work plans of this type. The main advantage of our approach is that end-users are able to structure and adapt a knowledge-intensive process without having to learn or understand CMMN. The Darwin Wiki uses only Stages, HumanTasks, CaseTasks, and Sentries from the CMMN 1.0 specification. The relationships between these types with the previously explained tasks and attributes are presented with our meta-model next.

3.3   Implementation

The Darwin Wiki builds on the experiences made with Hybrid Wikis that were developed and used as productive system in our group since 2009 [20]. Darwin extends Hybrid Wikis with structuring concepts for tasks, complex information structure hierarchies, user interface concepts to visualize the progress of tasks, more detailed role concepts for attributes as well as a declarative process model. The underlying meta-model of the Darwin Wiki is shown in Fig. 4 and explained next.

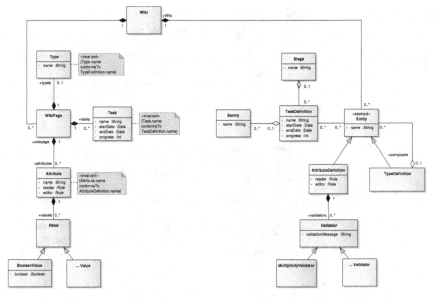

Fig. 4. Simplified conceptual meta-model of the Darwin Wiki showing the core elements in the prototype to support the collaborative structuring of processes for knowledge work

Wiki pages are extended with Tasks, Attributes, and a Type to structure knowledge-intensive processes in work plans. Additional concepts for Task-Definition, TypeDefinition, and AttributeDefinition are used to specify integrity constraints for predefined work templates. Work plans are structured by end-users, while work templates are defined by modeling experts. A type definition is loosely coupled to a type to enable the creation of new types by end-users on the fly at run-time. The same applies to attribute definition and attribute as well as task definition and task. Tasks are mapped either to HumanTasks or CaseTasks in CMMN depending on the assigned `Entities`. `Stages` are concepts that serve as containers for TaskDefinitions and track the behavior (or "episodes" [8, p. 28]) of a wiki page at run-time. The `TypeDefinition` is a composite that facilitates the modeling of complex hierarchical information and task structures. Tasks that reference other pages are automatically finished after all subtasks on these referenced pages are completed. `Sentries` specify logical entry and exit criteria between tasks and stages, e.g. a task is only enabled after certain other tasks are completed.

4 Findings

Figure 5 shows the main result of our evaluation with the work template that emerged for the development process within a project with four students in a practical course at the Technical University Munich over a time span of three months. This practical course is for master's level students in the computer science faculty and the goal for these students is the development of a software application. The project started on 22nd October 2014 and will continue until the end of February 2015. The students are following a simple Scrum process and have regular sprint meetings every week in which tasks are assigned. One team is using the Darwin Wiki to support the development process, while another team with four students is working with the project management tool Trello[2]. The students started with an empty wiki and they collaboratively structured this knowledge-intensive process gradually with a work template in just 14 days. The advisor of the course served as modeling expert. None of these students had previous experience in business process modeling.

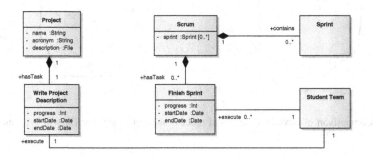

Fig. 5. Emerged work template for the software development project with four students

---

[2] https://trello.com/, last accessed on: 2014-12-04.

Each `project` is described on a wiki page and contains a predefined task `write project description`. The `sprints` are structured in a composite type definition called `scrum`, which contains all sprints with a predefined task `finish sprint` for every sprint that is assigned to the `student team`. The finish sprint tasks are automatically completed as soon as all subtasks on the assigned sprint are completed. Within the wiki pages for a sprint the required deliverables are documented as attributes and associated with tasks, e.g. mockups, development documents, checklist items consisting of boolean attribute values. Students resolve and create new tasks on their own if necessary to enable a self-organizing working mode during the sprints, which would be very difficult to achieve with a rigid and predefined workflow. Within seven weeks the students completed seven sprints, in which they created 7.3 tasks on average for every sprint. There were 0.8 attributes associated with each task on average. Examples for tasks that have been created are 'Refactor getExpertises', 'Add comment function in feed', and 'Show completed tasks in profile' (cf. Fig. 2). Attributes that were created by the students are e.g. 'Mapping of features to design principles' (as file), 'Mockup for feed' (as file), and '(Project) acronym' (as string).

Although statistical conclusions and hypothesis testing is not possible due to the limited number of participants, we have observed that non-expert users are able to easily structure a knowledge-intensive process. In addition, we conducted group interviews using open-ended and closed-ended questions to compare the Trello and the Darwin Wiki teams. We used group interviews to leverage the interactions of the group members to stimulate their experiences and filter out extreme answers. The results of the closed-ended questions are illustrated in Fig. 6 as spider diagrams and we used the open-ended questions to further elaborate these results next. The results indicate that the proposed solution is able to improve process support without losing the flexibility which is required for this knowledge-intensive process.

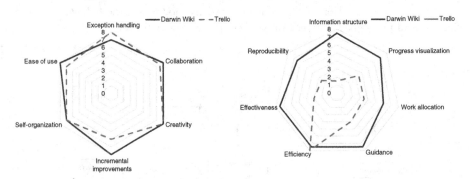

Fig. 6. Comparison of flexibility (left) and process support (right) for Darwin Wiki and Trello

We measured the flexibility of the tools with six variables. In general both teams confirmed that Trello as well as the Darwin Wiki provide very good means to adapt their processes. Trello has its strength in particular in the handling of exceptions with simple cards that are organized in subsequent columns e.g. to do, doing, and done. New cards can be easily added by end-users and moved between the columns when the project progresses. The Darwin Wiki is slightly more suitable for incremental improvements through the reuse of previously created structures e.g. sprints. Large differences between both solutions could be observed for process support, which was measured with seven variables. The Darwin Wiki allows the structuring of information to be organized e.g. presentations, mockups or sprints. Due to limitations of the Trello tool the team was using a separate web storage solution to share files and only copied the links to the files in the tool. Visualization of task progress is only possible on a coarse-grained level in Trello. In the Darwin Wiki expertise tags and the progress of the tasks were used to allocate work to students that have certain skills, while Trello only supports colors that can be added on cards to indicate priorities. Guidance during the process is worse in Trello since it is not possible to define complex hierarchies and dependencies between cards, whereas this was achieved through subtasks and different task start and end dates in Darwin. The efficiency was rated similarly for both tools and the effectiveness was rated as better in Darwin due to the attributes that were used to specify mandatory results for tasks. Reproducibility was rated in favor of Darwin because of the reuse of sprints every week.

5 Conclusion

We presented a solution that empowers non-expert users to structure knowledge-intensive processes. These process structures emerge bottom-up through contributions of end-users at run-time. Our implementation Darwin Wiki achieves this by employing lightweight structuring concepts such as attributes, tasks and types as metaphors that require no understanding of process modeling notations for end-users.

Evaluation: Based on a practical application with students conducting a software development project with Scrum, it was shown that such non-expert users were readily able to structure knowledge-intensive processes. A qualitative comparison to a widely used agile project management solution that can be applied for these kinds of processes (Trello) was conducted. As compared with Trello, the Darwin Wiki dramatically improves the organization of information, reproducibility, and guidance within the software development process without limiting users' flexibility.

Limitations: Due to the small number of students in this practical setting the results for this evaluation cannot be generalized. Knowledge-intensive processes with substantially different dimensions might be more difficult to structure for end-users or have other requirements that we didn't experience in our evaluation. Furthermore, end-users with more experience and different backgrounds have to be investigated.

Future Work: Guidance and recommendation techniques for tasks, constraints and attributes that are automatically suggested on wiki pages have to be improved. These recommendations can be based on activities performed by other users in related contexts. Automated learning of process models could be investigated to improve the work templates with every iteration.

Acknowledgement. This research has been sponsored in part by the German Federal Ministry of Education and Research (BMBF) with grant number TUM: 01IS12057.

References

1. Brynjolfsson, E., McAfee, A.: Race against the Machine: How the Digital Revolution is Accelerating Innovation, Driving Productivity, and Irreversibly Transforming Employment and the Economy. Digital Frontier Press, Lexington (2011)
2. Davenport, T.H.: Thinking for a living: how to get better performances and results from knowledge workers. Harvard Business Press, Boston (2005)
3. Van der Aalst, W., Stoffele, M., Wamelink, J.: Case handling in construction. Autom. Constr. **12**(3), 303–320 (2003)
4. Strong, D.M., Miller, S.M.: Exceptions and exception handling in computerized information processes. ACM Trans. Inf. Syst. (TOIS) **13**(2), 206–233 (1995)
5. Di Ciccio, C., Marrella, A., Russo, A.: Knowledge-intensive processes: characteristics, requirements and analysis of contemporary approaches. J. Data Semant. 1–29 (2013)
6. Van Der Aalst, W.M.P., Berens, P.J.S.: Beyond workflow management: product-driven case handling. In: Proceedings of the 2001 International ACM SIGGROUP, pp. 42–51. ACM Press, New York (2001)
7. Reijers, H.A., Rigter, J., Van Der Aalst, W.M.P.: The case handling case. Int. J. Coop. Inf. Syst. **12**(3), 365–391 (2003)
8. OMG. Case Management Model and Notation (CMMN), version 1.0, May 2014. Document formal/2014-05-05
9. Marin, M., Hull, R., Vaculín, R.: Data centric BPM and the emerging case management standard: a short survey. In: La Rosa, M., Soffer, P. (eds.) BPM Workshops 2012. LNBIP, vol. 132, pp. 24–30. Springer, Heidelberg (2013)
10. Pesic, M., van der Aalst, W.M.: A declarative approach for flexible business processes management. In: Eder, J., Dustdar, S. (eds.) BPM Workshops 2006. LNCS, vol. 4103, pp. 169–180. Springer, Heidelberg (2006)
11. Muehlen, Mz, Recker, J.: How much language is enough? theoretical and practical use of the business process modeling notation. In: Bellahsène, Z., Léonard, M. (eds.) CAiSE 2008. LNCS, vol. 5074, pp. 465–479. Springer, Heidelberg (2008)
12. Marin, M.A., Lotriet, H., Van Der Poll, J.A.: Measuring method complexity of the case management modeling and notation (CMMN). In: Proceedings of the Southern African Institute for Computer Scientist and Information Technologists Annual Conference 2014 on SAICSIT 2014 Empowered by Technology. ACM (2014)
13. Hauder, M., Pigat, S., Matthes, F.: Research challenges in adaptive case management: a literature review. In: 3rd International Workshop on Adaptive Case Management and other non-workflow approaches to BPM (AdaptiveCM), Ulm, Germany (2014)
14. Mundbrod, N., Reichert, M.: Process-aware task management support for knowledge-intensive business processes: findings, challenges, requirements. In: 3rd International Workshop on Adaptive Case Management and other non-workflow approaches to BPM (AdaptiveCM), Ulm, Germany (2014)
15. Bry, F., Schaffert, S., Vrandečić, D., Weiand, K.: Semantic Wikis: approaches, applications, and perspectives. In: Eiter, T., Krennwallner, T. (eds.) Reasoning Web 2012. LNCS, vol. 7487, pp. 329–369. Springer, Heidelberg (2012)
16. Haake, A., Lukosch, S., Schümmer, T.: Wiki-templates: adding structure support to wikis on demand. In WikiSym 2005, pp. 41–51. ACM Press, New York (2005)
17. Nonaka, I., Takeuchi, H.: The Knowledge-Creating Company: How Japanese Companies Create the Dynamics of Innovation. Oxford University Press, New York (1995)

18. Michel, F., Gil, Y., Ratnakar, V., Hauder, M.: A task-centered interface for on-line collaboration in science. In: Proceedings of the 20th International ACM Conference on Intelligent User Interfaces (IUI), Atlanta, USA (2015)
19. Voigt, S., Fuchs-Kittowski, F., Gohr, A.: Structured Wikis: application oriented use cases. In: Proceedings of the International Symposium on Open Collaboration. ACM (2014)
20. Matthes, F., Neubert, C., Steinhoff, A.: Hybrid Wikis: empowering users to collaboratively structure information. In: ICSOFT (1) (2011)
21. Swenson, K.D.: Designing for an innovative learning organization. In: 17th IEEE International Enterprise Distributed Object Computing Conference (EDOC). IEEE (2013)
22. Schönig, S., Zeising, M., Jablonski, S.: Supporting collaborative work by learning process models and patterns from cases. In: 2013 9th International Conference on Collaborative Computing: Networking, Applications and Worksharing (CollaborateCom). IEEE (2013)

Social Collaboration Solutions as a Catalyst of Consumer Trust in Service Management Platforms - A Research Study

Łukasz Łysik, Robert Kutera, and Piotr Machura[(✉)]

Wrocław University of Economics, ul. Komandorska 118/120,
53-345 Wrocław, Poland
{lukasz.lysik, robert.kutera, piotr.machura}@ue.wroc.pl

Abstract. We can observe a relatively low popularity of *service management platforms* (SMPs). Therefore is a need for searching causes of this situation and remedies. In the course of the conducted research, the authors identified a significant problem of lack of social trust in SMPs and service providers on the Internet. One of the ways to reduce its scale is to use social collaboration solutions (SCSs). Relying on transparency of activities and on social and professional relations, they can bring the consumer closer to the service provider and its customer community by receiving various stimuli resulting from social interactions.

The aim of this paper is to determine consumer attitudes towards SMPs and to identify and analyse the impact of determinants of the usefulness of SCSs in building trust in service providers and SMPs. The questionnaire survey conducted by means of CSAQ, the scope of which embraced the Polish market, identified consumer experiences with SMPs to date and their expectations of such platforms.

This paper is part of a project funded by the National Science Centre awarded on the basis of the decision number DEC-2011/03/HS4/04291.

Keywords: *Service management platform* (SMP) · e-marketplace · Brick-and-mortar services · e-services · Consumer behaviour · Social collaboration solution (SCS) · Trust

1 Introduction

The development of information society has resulted in a greater interest in using services ordered or even provided online. For the purpose of streamlining trade, electronic marketplaces have emerged, ensuring access to a multitude of service providers' offers, on the one hand, and a large group of prospective customers, on the other hand. A special case of that marketplace is *service management platforms* (SMPs). The transaction and communication mechanisms introduced as part thereof serve the purpose of streamlining the service ordering process (including service personalisation and lead time specification). However, an important problem of social trust in service providers and SMPs themselves has been identified, which is a challenge for the popularisation of such solutions. According to the authors, extremely important factors

© Springer International Publishing Switzerland 2015
W. Abramowicz (Ed.): BIS 2015, LNBIP 208, pp. 220–232, 2015.
DOI: 10.1007/978-3-319-19027-3_18

which reduce consumer anxiety might be a wider opening to communities and the provision of tools through which consumers would be able to support each other in making purchase decisions by exchanging their experiences with using services providers' offers and the platform itself. Such tools, owing to their features (e.g. personal character) might help building credibility, especially when the persons who recommend/criticise speak under their names (e.g. in the case of authenticated social login, e.g. through Facebook Connect, or Google Account). It is also worth emphasising here the possibility to gain information and knowledge from contents shared by other members of the community (crowdsourcing).

Hence, it seems essential to find answers to the following questions:

- What are the concerns of consumers in the context of using SMPs?
- What mechanisms and features of SMPs determine trust in service providers and the platform itself?
- Can the use of social solutions contribute to increased trust in service providers and the platform itself and, if so, to what extent?

The aim of this paper is to determine consumer attitudes towards SMPs and to identify and analyse the impact of the determinants of the usefulness of *social collaboration solutions* in building trust in service providers and SMPs.

2 Related Work

It becomes a necessity for R&D to be involved in development of service information systems, while there are convincing arguments identified by Boehmann (e.g. service systems cannot be meaningfully validated in laboratory experiments, novel service systems can only emerge with actors acceptance and involvement) [1]. Most of the work concerning SMPs have been mainly concentrated in the area of electronic marketplace as a sales channel for bricks-and-mortar services [2]. There are also conducted studies on the relevance of an e-marketplace of social and healthcare services for the inhabitants in general, and in particular for people with special needs [3]. Related work concerning social solutions focuses in general on social networks, social media and their application in modern companies marketing communication [4]. Furthermore, trust-related work incorporates trust in social networks, however its lack of correlation with SMPs opens area of interest to the new approaches and researches on this subject [5]. Our work aims to incorporate those factors in one concise concept.

Issues described above became also an important case in academic studies, e.g. in the Competence Center Services for Society (CC S4S), which is an initiative of the Institute of Business Informatics at Wrocław University of Economics. Its mission is to conduct research for enablement and empowerment of IT/society alignment through development of Lifestyle Engineering & Health Management (LE&HM) platform offering concierge-mediated, requestor/supplier-negotiated, and cyber/physical-based delivery of LE&HM services to requestors (consumers, patients) [6].

3   Evolution of the Service Market – Electronic Marketplaces

The technological progress and, most of all, the IT revolution have significantly affected the operation of the service market and the method of their provision, as well as on the satisfaction of needs by consumers and their households. Many services have gained their electronic form, while others – which are difficult or impossible to provide remotely – have been given a broad range of visual and interactive solutions (including communication and transaction ones) so that the consumer can become familiar with the description of a given service, engage in a discussion with the service provider's representative or other customers of the company, place an order and pay online. Service providers make efforts to enable the user to get more, as quickly as possible, and using an increasingly more intuitive interface. A modern service provision process must be oriented also towards originality and individualism and be open to society. New opportunities to locate and identify image and sound permit creation of new services offering new channels of contact to their users. The offered web solutions eliminate problems related to interoperability of different systems, additionally providing access to data, regardless of the place and channel of communication [7].

The market have evolved from the initial state, being simple news websites and corporate catalogues or classifieds websites, to interactive solutions supporting all transaction stages, oriented towards cost reduction, flexibility (e.g. price negotiations) and building mutual trust of the parties to the transaction [3]. The sales model is similar here to the one applied in the case of the online sales market, from placing a service description, to enabling selecting and ordering a service in the transaction module, online payment, to online after-sales service. However, attention should be given to the significant difference between the product and the service in the context of inside sales. The product is strictly defined in a specification and reproducible, while the service is always unique and cannot be tested before use [8]. Service provision involves consumer's trust in the service provider as regards quality, reliability, punctuality and commitment.

SMP can be defined as a specific form of a highly interactive electronic marketplace, where brick-and-mortar services and e-services are made available to customers, offering access to contact details and service portfolio, interaction with the service provider and lead time negotiations (including postponement of the original deadline), multi-channel notification system, or the possibility to rate the service provision [9]. Examples of such platforms are – in Poland: wizyta.pl, zamowwizyte.pl, sirlocal.pl, and worldwide: helpouts.google.com, myhammer.de, redbeacon.com, or Amiona – a platform developed as part of the Independent Living research project [10]. From the viewpoint of the service provider, it enables expanding the group of prospective customers, coordinate service provision (timetable management, work time optimisation), and it ensures abundance in forms of contact with customers and aid in online transaction management.

The requirements for an SMP are derived by analyzing customer needs, based on the principles of customer orientation as described in [11]. What is more, according to customer orientation should be present at every phase of the value creation and is the key success factor. These set of principles should be taken into account [12]:

- Everything – SMP should support different kinds of services,
- Everywhere & Anyhow – should be accessible at every device a customer is used to despite the location,
- Anytime – access to services should be possible at all time so that customer would feel convenient,
- One-Stop – customers can get all they need in just one location,
- Segment-of-One – all services should be personalized,
- One-face-to-the-customer – complexity of SMP should be customized and have alike front-end.

SMP use is not widespread yet although increasingly more frequent local and national implementation initiatives concerning such solutions are being observed. One of the causes of this state of affairs could be various types of barriers, whose specificity is presented in the next chapter of this paper.

4 Identification of Barriers to Using Services in the Internet Environment

Both in the case of brick-and mortar services (searched and/or bought online) and in the case of e-services various types of barriers that could cause reluctance to use ICT in order to support the service provider selection process might be encountered. The barriers can be discussed in the technological, educational or combined behavioural and psychological dimension [13].

The technological barrier is primarily the lack of access of some consumers to new, emerging technologies, that facilitate access to SMPs. Devices such as a tablets, laptops or smartphones allows for easy access to SMPs at the convenient for the user time and place. It is also important to provide a good quality Internet access (broadband cable infrastructure, as well as mobile radio telephony and other wireless broadband technologies) which can ensure communication with a proper capacity. Although this percentage decreases year after year, 25.2 % of households in Poland still did not have Internet access in 2014 [13]. From the viewpoint of affluence, the lack of a proper device (PC, laptop, tablet or smartphone) might also prove an impediment. As reported by the Central Statistical Office in Poland (GUS), 77.1 % of Polish households are equipped with computers (including as many as 94.8 % of households with children have computers) [13]. Availability of such devices is growing (the quick technology development rate results in developing successive generations of products), and high competition among producers results in price reductions, and thus broader consumer access to new technologies. The diversity of the devices from which consumers can view and order services online causes also the emergence of a barrier for service providers or SMPs – such websites might not be adjusted to individual types of devices, e.g. the service provider's website which is not developed in the RWD (Responsive Web Design) technology could be displayed improperly in the consumer's device.

Educational barriers include the level of awareness and skills of consumers being Internet users. This is because the skill of navigating the Internet does not guarantee the familiarity with SMP solutions themselves or even awareness of their existence. Some Internet users might not realise that there are any SMP solutions until they find it out from advertisements, friends or organic search. Other consumers might not have any knowledge about correct search and selection of keywords. It is also possible that the internet user finds an SMP solution, yet he or she will not be able to fully use the capabilities provided by a given platform due to the limited knowledge of web applications. The service provider's website or SMP could be poorly designed in terms of user experience (UX) or graphic user interface (GUI), which in turn might result in the use of the website or platform requiring a certain level of skill in browsing the Internet or using new technologies.

Behavioural and psychological barriers also exert a significant impact on the use of SMP solutions. They can be connected with individual constraints related to personal characteristics, the nearest social and work environment, or some cultural factors, that can influence on readiness to use such a technology or business model. Tendency to participate or stay invisible may be different in emergent communities than in online groups [14]. Barriers to participation in the SMPs may correlate with fears that accompany consumers during the decision-making processes in the real world. With reference to the barriers discussed above, unfamiliarity with the SMP environment might cause caution in using them or even reluctance to order any services online. As a result of our analysis, the following behavioural and psychological barriers are distinguished:

- fear of fraud – both in terms of loss of money, theft of personal data, undiligent service provision, and failure to provide the service whatsoever,
- lack of possibility to review the quality of the service in person before making the decision – the consumer can rely only on the information presented by the provider and other users (purchasers of services within SMPs), lack of trust in electronic payment systems (fear of transaction security, loss of data used for payment),
- lack of required competences for managing the service ordering system,
- reluctance to share personal data (the necessity to authenticate both service providers and buyers could result in the requirement to submit sensitive personal data, which – if not properly secured – might fall into the wrong hands),
- unintuitive solutions (platforms might not be designed with the comfort of the end user, including the service buyer, in mind, which could contribute to an increased rejection rate among users taking their first steps in SMP solutions),
- insufficient popularity among friends – lack of interest in SMP solutions in the consumer's environment might result in the lack of opinions and valuable information coming directly from a given person's friends, which is extremely important from the viewpoint of sales; this is because other people's opinions about using e-services exert great impact [15].

These all barriers, especially the last group of behavioural and psychological ones, are in general connected with lack of trust to many referents—salesperson, market organizer, service, service provider, technology etc. In this paper a trust is defined as *a consumer's subjective belief that the selling entity will fulfill its transactional obligations as the consumer understands them* [16].

Given the range of barriers that could hamper the use of services via the Internet, such solutions need to be sought that would eliminate them by employing tools, popular in the social media environment, which streamline the consumer–service provider interaction on the electronic marketplace.

5 A Model of User Interaction on the Web – Social Collaboration Solutions

Social media and other digital communication technologies integrated with them (e.g. mobile technologies) have recently become an integral part of pop culture. Users have begun to use social online platforms for communication, searching and sharing information, entertainment, education, etc. Such a significant social change was possible due to the employment of appropriate web tools called *social collaboration solutions* (SCS). In this paper, We define them as tools and mechanisms which enable relation building, social interaction and cooperation in the digital environment. They constitute a technological platform enabling the establishment and development of virtual communities whose members can freely create the directions and extent of interaction, thus determining its nature. Hence, the basic difference with regard to the solutions applied in traditional media is that the crucial objective of SCSs is to provide readers with the opportunity of interaction in the form of posting comments, debating, sharing with other prospective readers, evaluating, and at times even co-creating, rather than to merely publish information [17].

Due to a considerable diversity of the offered solutions, numerous attempts to classify such solutions are made. One of them is Kaplan and Haalein classification, which divides these solutions into collaboration tools (ensuring joint creation of content by multiple users), content-sharing services (focused on enabling content sharing with other users), blogs (virtual diaries), social networking tools (enabling establishment and maintenance of various types of relations), and virtual worlds (moving the user to an alternative virtual reality, which can be a game, e.g. World of Warcraft, or social, e.g. Secondlife) [18]. The solutions and mechanisms mentioned are aimed at fulfilling the following functions: networking (building social relations, reflecting the manner and planes in which users "meet" on the web), collaborating (cooperating in pursuit of a specific goal), co-creating (using various resources shared by other users in order to achieve better results), crowdsourcing (obtaining the needed content or ideas by encouraging numerous users to make their contribution), debating (exchanging opinions about a given topic in an open and informal talk), engaging (encouraging users to connect, provide and share content) [19].

Such solutions can assume the form of independent solutions operating as separate websites, yet they are increasingly more frequently integrated as components, modules or plugins for content management systems. They enrich the value of a given website by ensuring it interactivity and building a community around the subject matter of the website.

The social media environment, built on the basis of SCSs, has become an important component in the process of satisfying needs of consumers, who gain new possibilities to obtain and share information and opinions, which contribute to the improvement of consumer general satisfaction and positive experiences with the purchased product or service, brand or producer/distributor/service provider. It affects consumer behaviours and their purchase decisions. Consumers are willing to seek information in that environment because they can learn about other consumers' experiences in the context of the actual quality of and subjective impressions about a given product/service, opinions about the reliability of the product itself and customer service procedures. They can interact and exchange views, supporting each other when making purchase decisions. SCSs have also provided new possibilities of multi-channel both pre- and after-sales customer service. They have contributed to developing new channels of customer interaction with companies and transformed the marketing models applicable to date by moving the burden of consumer contact with the product/service or brand from the (physical or virtual) sales location to the stage of an extensive search of information and evaluation of variants – the Zero Moment of Truth (ZMOT) concept (cf. [20]). SCSs enable also the use of its social identity for logging in numerous independent websites, e.g. online shops. From the viewpoint of the consumer, such a solution permits avoiding the multiplication of passwords, ensures the fundamental validation requirement, and provides substantial possibilities of offer personalisation, at the same time bringing, however, a certain risk with regard to loss of privacy and anonymity, as well as security hazards [21]. It is essential for companies to collect a valuable database of those who already are their customers and ones who are interested in their offers [22]. Social media might considerably enrich such databases both in terms of quantity and quality. They do so by social validation of identity and sharing unique data, which cannot be gathered in the ordinary process of registration in a website, such as lifestyle, manner of expressing oneself, group of friends, online activity, actual interests, leisure activities, etc.

As has been mentioned above, one of the causes of the fast SCS adaptation in consumer applications is the existence of the possibility to obtain additional information, apart from that provided by the supplier, or reviews of some unknown experts, from consumers themselves, people who have already gained experience in using a given product/service or have a broader general knowledge in a given area. This results from relying more on opinions of ordinary users, the value of which is even greater if they are expressed by members of one's social network. Therefore, it is crucial to examine the power of that factor in the context of offering services within SMPs. This will be the object of the next part of this paper.

6   Analysis of SCS Impact on Building Consumer Trust in Services Offered Online

In the foregoing parts of this paper, We identified the most important problems involved in extensive use of SMPs and indicated SCSs as ones extensive use of which might contribute to the popularity of SMP solutions and the increased trust in service providers. The goal of the empirical part of this study is to describe consumer attitudes towards SMPs as well as to identify and analyse the impact of determinants of the usefulness of social solutions in building trust in service providers and SMPs.

For the problem formulated above, there has beed developed an appropriate survey questionnaire constructed on the basis of the CSAQ (computerised self-administrated questionnaire) method [23]. In the survey, selected respondents received an invitation to complete the questionnaire online at research project's website (socialmediabusiness.pl/ ankieta). It was composed of three major parts: "Demographics" (4 questions), "Social media" (5 questions), "SMP online platform" (11 questions). Owing to this composition, the questionnaire is clear and enables obtaining answers to the most important questions asked in this paper. Part one identifies the consumer, part two analyses consumer attitude towards social media and method of using them. Part three is supposed to identify user attitude towards ordering services online and practices employed in this regard. Additionally, in this part, We make an attempt to determine the SMP components owing to which buyers of such solutions are more willing to use them.

The questionnaire survey was addressed to a selected group of people concentrated on the issues of social media and ones using the Internet in the everyday life actively and obtaining most information through available online channels. The survey was conducted at the turn of 2014 and 2015, and covered 110 respondents. The survey was promoted via direct contacts, personal interviews, and distributed through third party websites. In the survey, We tried not to discriminate users on the grounds of sex, place of residence, or level of income so that the obtained results are organic. Only the distribution of the survey questionnaire enabled it to be addressed to people who actively use social media and online tools to seek service providers – such an approach will allow not only examination of the current situation on the SMP solution market but also obtaining suggestions from users as to what should be improved in the present platforms. The demographic structure of the analysed population was women (45 %) and men (55 %), with as many as 95 % of the respondents being 18–40 years old, while the remaining 5 % - over 40. The vast majority of the respondents (86 %) were persons with a higher education degree and 23 % of them graduated from secondary schools. Most answers were obtained from large Polish cities – 85 %, whereas 11 % of the respondents were from medium cities and towns, and 7 % of them were from both small towns and rural areas. As regards net monthly income, the most numerous group were persons with earnings falling into the range of PLN 2,001–3,500 net (32 %) and below PLN 2,000 net (31 %) - 25 % of the respondents declared that they earned PLN 3,501–6,000 net, while 13 % - more than PLN 6,000 net. Based on the distribution of earnings, it can be noticed that 39 % of respondents earned more than PLN 3,500, who can be considered affluent.

In the part dedicated to the social media environment, the most frequently used media were identified: community portals (98 %), multimedia sharing websites (85 %), blogs (59 %), opinion and review sharing websites (55 %). The obtained answers indicate that Polish users are very willing to use social media; it can be assumed that at least two of them are the daily routine. It is more and more frequent and popular to share opinions and reviews as well as to seek them actively. Through multimedia sharing websites, users can easily seek and share reviews of service providers and the services delivered by them.

Another issue is to determine user experience based on the time of using social media. As follows from the carried out survey, 81 % of the respondents had used such media for over 5 years, and 26 % declared that they had used them for at least 2 years. Such a distribution is an effect of disseminating the survey questionnaire; the authors wanted to examine the current situation regarding SMPs and to obtain users' suggestions about how the operation of such platforms could be improved.

The survey made an attempt to identify the purpose for which people use social media and it was found that 94 % of respondents used them for communication and gaining information, 79 % for maintaining family relations, and over half of them, that is 51 %, for making purchase decisions. Such a distribution suggests that family is an essential environment for everyday communication for us, which – coupled with the fact that we take advantage of other people's opinions when making purchase decisions – seems to be extremely significant.

Consumers were also asked in the survey to specify what kind of information they sought when choosing a service provider via the Internet. The answer prevailing among the respondents was other people's opinions (87 %), followed by contact details (82 %), a specific description of a service (73 %), portfolio (64 %). As can be easily observed, it is confirmed that we seek other people's/customers' opinions when choosing the service provider. Placing customers' opinions about service providers in the place where offers are presented can become a cure-all for the lack of users' trust in service providers' offers published online.

In the next question, the users were asked to assess selected factors affecting the decision about choosing a service provider online. The survey indicates that recommendations by friends are very important to 67 % of the respondents, while other customers' opinions – for 38 % of them (with 55 % considering them important). Apart from that, a professional description of services, attractive price of the offer, short lead time, and reputation of the service provider are also appreciated by the respondents – more than half of them regard these as significant criteria when selecting the service provider.

The respondents also shared their opinions about the important barriers in ordering services online, pointing primarily to fear of fraud (68 %), impossibility to review the quality of the service in person (65 %). 35 % of the respondents are reluctant to share personal data. Such a distribution might suggest that putting a greater emphasis on obtaining other users' opinions and the willingness to share the effects of a given service provider's works, for instance on opinion sharing websites, could exert positive influence on the process of ordering services online.

The study also made an attempt to identify the factors that could convince the respondent to apply SMP solutions, and the obtained structure of answers is presented in Fig. 1.

Fig. 1. Factors encouraging the consumer to use SMPs

The obtained results indicate that it is recommendations of friends (as many as 81 %) that could convince the respondents to use an SMP, and thus to order services via the Internet. Collaboration with popular brands, which reflects certain standards represented by the brand (46 %), and recommendations of strangers who use the service provider's offers (34 %) might also encourage using SMPs.

Another aspect of the conducted survey is the identification of the determinants of the selection of a given offer from among others within SMPs. The results are presented in Fig. 2.

According to the respondents, the greatest (75 %) impact is exerted by customers' opinions and reviews; 60 % pointed to the number of services provided, while 57 % - to the service provider's rating on a given website, and 49 % are interested also in other customers' comments or recommendations.

Fig. 2. Determinants of the selection of a service provider within SMPs

To sum up the results of the survey, it is worth quoting the results of the last of the analysed issues, namely the question asked to the respondents: can building an active community of consumers and service providers around SMPs help building greater trust in the services offered? This question was answered "definitely yes" by 62 % of the respondents, and "rather yes" by 33 % of them.

7   Conclusion

As demonstrated by the conducted analysis and survey, the issues of SMPs are already present in the consciousness of Polish users. Offering services with the use of Internet tools is welcomed with increasingly greater enthusiasm and has focused on communities of consumers, families and close friends thus far. This description of the market justifies the employment of mechanisms which are widespread in all types of social tools, which is confirmed also by the results of the conducted questionnaire survey. SCSs enable users to freely exchange thoughts and opinions about e-service providers and thus considerably eliminate the lack of users' trust in both SMPs and, as a consequence, in service providers. Transparency of information and activities of both users and service providers brings them closer to each other, shortens the distance and makes their relationship mutually beneficial.

Finally, it is worth adding that, despite technological and psychological barriers, SMP solutions are gaining more and more users. The technological progress enables increasing the capabilities of modern communication tools on the one hand – this results in reducing their prices and makes the technology more available, and permits streamlining the interaction between the human being and the computer by applying more and more efficient algorithms. Employment of SCSs, in turn, allows elimination of all consumer fears related i.e. with the loss of privacy or the impossibility to verify the service provider.

The presented work is the basis for in-depth study related to sales driven activities in the field of SMPs and technical feasibility study. Within further research identified factors can be analyzed empirically in the real SMP environment. The observations enumerated in this paper presents the current state of Polish SMPs, their issues and opportunities. Polish SMPs are usually focused on appointment coordination and services itself, leaving SCS as an underestimated fulfilment to the main subject. The results of analysis could then serve as guidelines for making SMP solutions better in various dimensions – for all parties involved: consumers, service providers, and SMP management units.

References

1. Boehmann, T., Leimeister, J.M., Moeslein, K.: Service systems engineering: a field for future information systems research. Bus. Inf. Syst. Eng. **2**, 73–79 (2014)
2. Schenkel, P., Osl, P., Österle, H.: Towards an electronic marketplace for bricks-and-mortar services. In ACIS 2013 Proceedings (2013)

3. Cruz-Cunha, M.M., Varajão, J., Miranda, I., Lopes, N., Simoes, R.: An e-marketplace of healthcare and social care services: the perceived interest. Procedia Technol. **5**, 959–966 (2012)
4. North, N.S.: Social media's role in branding: a study of social media use and the cultivation of brand affect, trust, and loyalty (2011)
5. Adali, S., Escriva, R., Goldberg, M.K., Williams, G.: Measuring behavioral trust in social networks. In: 2010 IEEE International Conference on Intelligence and Security Informatics (ISI), pp. 150–152. IEEE (2010)
6. Maciaszek, L.A.: Competence Center Services for Society (CC S4S) Research Agenda (2015). http://s4s.ue.wroc.pl/index.php/research-agenda
7. Polska Agencja Rozwoju Przedsiębiorczości: Rozwój sektora e-usług na świecie – II edycja (2012). www.web.gov.pl/e-booki/272_1789.html
8. Blythe, J.: Consumer Behaviour. SAGE Publications, London (2013)
9. Schenkel, P., Osl, P., Oesterle, H.: Towards an electronic marketplace for bricks-and-mortar services. In: Deng, H., Standing, C. (eds.) ACIS 2013: Information Systems: Transforming the Future: Proceedings of the 24th Australasian Conference on Information Systems, Melbourne, Australia, 4–6 December 2013, pp. 1–11 (2013)
10. Amiona, Business Engineering Institute St. Gallen AG. http://www.amiona.com/
11. Kagermann, H., Österle, H., Jordan, J.M.: IT-Driven Business Models, p. 220. Wiley, New Jersey (2010)
12. Flis, R., Szut, J., Mazurek-Kucharska, B., Kuciński, J.: Bariery hamujące rozwój i globalizację e-usług. Badanie zapotrzebowania na działania wspierające rozwój usług świadczonych elektronicznie (e-usług) przez przedsiębiorstwa mikro i małe, PARP, Warszawa, Poland (2009)
13. Główny Urząd Statystyczny: Społeczeństwo informacyjne w Polsce w 2014 r (2014)
14. Ridings, C., Gefen, D., Arinze, B.: Psychological barriers: Lurker and Poster motivation and behavior in online communities. Commun. Assoc. Inf. Syst. **18**, 329–354 (2006)
15. Wolny, R.: Innowacje w świadczeniu usług – potrzeba czy konieczność? Handel Wewnętrzny, tom 2. Marketingowe sposoby kreowania wartości dla klienta, pp. 302–303 (2013)
16. Kim, D.J., Ferrin, D.L., Raghav Rao, H.: A trust-based consumer decision-making model in electronic commerce: the role of trust, perceived risk, and their antecedents. Decis. Support Syst. **44**(2), 544–564 (2008)
17. Knight, M., Cook, C.: Social Media for Journalists – Principles and Practice, p. 4. Sage Publications Ltd., London (2013)
18. Kaplan, A.M., Haenlein, M.: Users of the world, unite! the challenges and opportunities of social media. Bus. Horiz. **53**, 59–68 (2010)
19. Jespersen, L.M., Hansen, J.P., Brunori, G., Jensen, A.L., Holst, K., Mathiesen, C., Rasmussen, I.A.: ICT and social media as drivers of multi-actor innovation in agriculture–barriers, recommendations and potentials. European Commission, Directorate-General for Research (2014)
20. Kutera, R., Łysik, Ł., Machura P.: Zero moment of truth: a new marketing challenge in mobile consumer communities. In: Proceedings of European Conference on Social Media (ECSM 2014), The Journal of Information, Communication and Ethics in Society (JICES), Emerald, Brighton, UK (2014)
21. Gafni, R., Nissim, D.: To social login or not login? - exploring factors affecting the decision. Issues Inf. Sci. Inf. Technol. **11**, 57–72 (2014)

22. Kutera, R., Łysik, Ł., Machura P.: Mobile and social technologies in marketing campaigns in Poland – a research study. In: Cunningham, P., Cunning ham, M. (eds.) eChallenges e-2013 Conference Proceedings. IIMC International Information Management Corporation Ltd. (2014)
23. Marsden, P.V., Wright, J.D.: Handbook of Survey research, p. 13. Bingley, Emerald (2010)

The Relative Advantage of Collaborative Virtual Environments and Two-Dimensional Websites in Multichannel Retail

Alex Zarifis[1(✉)] and Angelika Kokkinaki[2]

[1] University of Liverpool, Merseyside, Liverpool, L69 3BX, UK
A.Zarifis@liverpool.ac.uk
[2] University of Nicosia, Makedonitisa 46, 1700 Egkomi, Nicosia, Cyprus
kokkinaki.a@unic.ac.cy

Abstract. Collaborative Virtual Environments (CVE) have been with us for some years however their potential is unclear. This research attempts to achieve a better understanding of retail in CVEs from the consumer viewpoint by comparing this channel with the competing retail channels of 'bricks and mortar', or offline, and two dimensional navigation websites (2D websites), in order to identify their respective Relative Advantages (RA). Five categories of RA between retail channels were identified and explored using focus groups and interviews. These five categories explore distinct characteristics of each channel, consumer preferences over three stages of a purchase, differences between simple and complex products and lastly the role of trust. Participants showed a preference for offline and 2D in most situations however there was evidence that enjoyment, entertainment, sociable shopping, the ability to reinvent yourself, convenience and institutional trust were RA of CVEs in comparison to one of the other two channels.

Keywords: Collaborative virtual environments · E-commerce · Multichannel retail · Trust

1 Introduction

This research compares three retail channels between themselves in order to identify their respective Relative Advantage (RA). The first channel is 'brick and mortar' retail outlet that does not involve the internet. The second channel is the two-dimensional (2D) business to consumer website that displays information and offers navigation in two dimensions (2D websites). The third channel is the three dimensional online environment known as Collaborative Virtual Environment (CVE) [1]. The CVE used for this research is Second Life (SL) because it is widely used with over 20 million users [2] and it has all the functionalities such environments can offer. From the three channels the CVEs have not reached wide adoption and are therefore less understood in terms of their benefits from a consumer's perspective and hence their potential for retail. There are strong indications of CVEs potential for socialising and collaboration

© Springer International Publishing Switzerland 2015
W. Abramowicz (Ed.): BIS 2015, LNBIP 208, pp. 233–244, 2015.
DOI: 10.1007/978-3-319-19027-3_19

[3], and conducting exhibitions and conferences, [4]. Furthermore their relationship and RA of each channel in relation to the others is less clear. Each channel is in competition with the others in a multichannel environment [5]. The theory of the diffusion of innovation [6] suggests that an innovation, in this case the new retail channel, must offer a RA for consumers to adopt it. For time and energy efficiencies consumers often avoid measuring things against an agreed form of measurement and compare between alternatives. This is related to the concept of satisficing, combining satisfy with suffice, as opposed to maximisation [7].

The retail channels this research is comparing could be seen as cases that are being contrasted. Choudhury and Karahanna [8] was chosen as the foundation of this research and five objectives that captured the issues accurately were identified to guide this exploratory research. The first objective to explore was to assess whether a RA of CVEs compared to the 2D navigation Internet for e-commerce was the aspects of offline retail that it includes, that do not exist in the 2D websites. The second objective therefore explores the same topic between the other pair: Could a RA of CVEs for retail over offline retail be aspects of 2D navigation websites that it includes that are not included in the offline retail environment? The third objective was to explore whether consumers vary their intended usage of CVE across the different stages of the purchase process and whether this happens because the significance of the dimensions identified in Choudhury and Karahanna [8] vary across those stages. The fourth objective was to explore whether consumers' usage of CVE is different for simple and complex products. The consumer approaches a purchase of a simple and a complex product differently and it is therefore possible that some characteristics of CVEs are valued differently in these different processes. The fifth objective was to explore whether CVEs such as SL may have the RA of a higher degree of institutional trust compared to the 2D websites.

2   Multichannel Retail

Multichannel retail covers the activities involved in selling products and services to consumers using more than one channel [9]. This process has implications for many aspects of an organization such as legal, bookkeeping, enterprise systems, human resources, new product development, servicing and corporate marketing [10]. The most common retail channels are bricks and mortar, online 2D website, online virtual world, catalogue, call centre and television retail. The first three are explored extensively in this research. Today multichannel is considered the dominant approach to retail [11]. This can be seen in many industry sectors other than retail such as travel, banking, computer hardware, computer software and manufacturing [12]. It is important for an organization to coordinate the channels it uses [13] in order to achieve the best results, first for the customer and then for themselves.

Each channel has certain advantages and disadvantages. For a channel to survive in the highly competitive environment that exists today it must have some form of an advantage. Online retail, or B2C e-commerce, has certain advantages and disadvantage compared to other channels. The main advantage is convenience [14]. The main disadvantages are: Firstly requiring the necessary experience to use online stores [15] and secondly the lower level of trust in comparison to bricks and mortar shops.

Research in this area suggests it is important to take a customer and not a channel centred view [16]. This research follows this assertion. Research suggests that multi-channel consumers' behaviour and channel choices are more strongly influenced by psychographics than demographics [17]. There is evidence that consumers prefer different channels for different actions [18]. Searching for information about a purchase such as price, and making a purchase have differences and different channels may be preferred for each stage [19]. Each channel is found to have different utility [19]. Additional distinctions are examining and picking up the product that is being considered for purchase [20]. This research explores all of these stages.

3   Objectives to Be Explored

By analysing the existing literature and anecdotal evidence two high level categories of the relative advantage of CVEs seem to exist. The two categories are the RA of CVEs for retail that come from characteristics it draws from the 2D websites and the RA of CVEs that come from characteristics it draws from the 'bricks and mortar' environment. The first and second issues identified for further investigation that follow on from this are:

Objective 1:   A relative advantage of CVEs to the 2D Internet for e-commerce, may be the aspects of offline retail that it includes that do not exist in the 2D websites.

Objective 2:   A relative advantage of CVEs for retail compared to offline retail may be aspects of 2D e-commerce that it includes that are not included in the offline retail environment.

Choudhury and Karahanna [8] suggested that a consumer would adopt a new channel only if it was perceived to offer an advantage to existing channels. This argument is built on the theory of diffusion of innovation [6]. The third objective states that the 'variable' dimension of RA will vary across the 'variable' of stages of the purchasing process:

Objective 3:   Consumers may vary their intended usage of CVEs across the different stages of the purchase process because the significance of the dimensions may vary across those stages.

The nature of gathering information and making a purchase for a complex product in comparison to a simple product is different. Therefore, the nature of how the technology and the other aspects of a channel are used is different. Therefore the variables to compare are consumer usage, product complexity and purchase stages:

Objective 4:   Consumers' usage of CVEs may be different for simple and complex products.

Based on the literature on trust as it has been defined and modelled by McKnight et al. [22, 23] institutional trust has been identified as the relevant aspect. This is in agreement with Choudhury and Karahanna [8]. When considering institutional trust for CVEs it is important to clarify what the institution being considered is. For the purpose of this study the institution is SL, as opposed to CVEs in general:

Objective 5: CVEs such as SL may have the RA of a higher degree of institutional trust compared to the 2D websites.

4 Research Method

The methodology chosen was qualitative research starting with focus groups and carrying on with interviews. The focus groups had limited structure to allow themes to emerge. The data collection was progressively more focused utilizing the data collected so that insight could be gained on the themes that emerged. The interviews were therefore more structured while still allowing the participants to introduce issues they considered pertinent. The strength of qualitative methods lie in capturing what quantitative research is not strong at capturing. This is often referred to as why something social happened and the meaning as opposed to the causal effect [24]. Furthermore, its proponents argue that the lower level of, or lack of, abstraction from specific contexts makes it more effective at capturing the individuals point of view, the constraints of life and thus achieves richer descriptions [24]. The weakness of this qualitative method is its inability to achieve the generalizable results that quantitative methods can. Another important limitation is that the observer is active in the world being researched and, therefore, the level of objectivity of quantitative methods is not achievable.

5 Data Collection

Before the focus groups were carried out some trial focus groups were conducted. The trial focus group schedule was developed based on the literature. This was refined at each subsequent stage. After the trial five focus groups were carried out with a total of twenty six participants. Five or six participants were in each group and there was an even number of males and females. The participants were postgraduate students at an English university.

In depth semi-structured interviews in SL formed the second stage of the data collection. The recruitment was carried out 'in world' in order to ensure that participants had full SL membership and were active users and purchasers in SL for at least twelve months. Beyond the recruiting the interviews were also carried out in SL 'in-world' offices. This would act as an additional safeguard ensuring that the participants had the relevant experience. Their behaviour, level of skill and avatars appearance would be an indication of their experience. There were twelve participants, six female and six male, between 23 to 54 years of age and from various parts of the UK. Nine had a university education and the other three had graduated from high school.

6 Data Analysis

The transcribed data of the focus groups and interviews were entered separately into Nvivo 8 and the analysis was implemented using template analysis [21]. The template analysis method allows verification and, crucially, extension of previous research which is one of the aims of this research [25]. This method of analysis is compatible with the realist approach taken [25]. In this research it develops Choudhury and Kara-hanna [8] which provided the template foundation with some additions based on the broader literature review. The templates were the five objectives identified in from the literature.

For some objectives there were a number of issues within them so child nodes were created. All these nodes were created and populated within Nvivo. The level of support for each opinion was shown in the findings. The reason why an opinion was included was explained and its popularity was then indicated. This could be described as quantifying qualitative data [24].

7 Findings

The qualitative data analysis findings are presented firstly for the focus groups and then for the interviews. For the focus groups a short summary is given while the findings of the interviews are reported more extensively.

7.1 Focus Groups

There were some broad issues that could be identified. Firstly the participants showed an interest and ability to identify the features of each channel, assess what those meant to them, and compare them to the specific issue under question. Beyond this making the data richer this is a strong indication that this thought process happens naturally when making a purchase. This strengthens the argument that this research has a valuable contribution to make. While no conclusive findings could be claimed after the focus groups the richness of meaning drawn out in relation to the five objectives explored was another indication of the value of this research. Some features that were pervasive across all issues discussed had to be noted. These were the ability to socialize in CVEs, the richer more personal interaction and the value of the shopping assistant. All these issues were further explored in the interview stage.

7.2 Interviews

The responses from the interviews were useful but varied. On a small number of issues there was consensus, such as the unsatisfactory quality of the images and on the rest there were varying, often conflicting views. These conflicting views were often based on different motivations and expectations.

7.2.1 Results for Research Objective 1
A RA of CVEs to the 2D Internet for e-commerce may be the aspects of offline retail it includes that do not exist in the 2D Internet.

The first objective that was identified from the literature was significant because it proposed a RA of CVEs in the simplest form, offering a high level category of RAs that could include many more specific RAs. Many participants' responses were on this objective and subcategories were identified. These were primarily enjoyment, which will be discussed in more detail in the following section of the findings; social shopping, a richer and more emotive 3D environment, 'face to face' and the shopping assistant, and to a lesser extent location. For the 'social shopping' and 'richer more emotive environment', there was extensive evidence. An example of the former is: '…you can actually view the products with other people, even if they are just virtual representations…' (male, 26) and 'I was thinking it could be more sociable, you can't really go with friends around Amazon' (female, 22). As the second quote illustrates the ability to go shopping with their friends was considered an advantage. This could be especially useful when those friends do not live near them and would therefore not be able to shop together offline.

In the case of the latter, which channel was most emotive, all participants considered the real life to be first, which is understandable, and most considered CVEs such as SL to be more emotive than 2D websites. One participant stated in response to the question about which is more emotive: 'Virtual worlds but only because the 2D websites have no emotion at all' (female, 22). One other participant was more enthusiastic: 'Yes it is more interactive, I guess, you control your avatar, you walk into the shop you walk around, you can fly if you get bored. I think it is more fun' (female, 19). Those that did not consider SL to be more emotive than the 2D web pointed to the shortcomings of its current implementation: 'SL last at the moment because it is slow sometimes' (male, 28), and '…occasionally you find someone in a SL shop interested in helping you but you don't get their full attention. They usually have instant messaging going on, maybe music streaming' (male, 60). Since the reasons given were about the current implementation and not the fundamental nature of a CVE it suggests that if these issues were overcome those participants may also agree with the rest.

That last quote leads on to the third RA of CVEs compared to 2D websites for retail identified: That of 'face to face'. The most prominent of those interactions in relation to retail was interacting with a shopping assistant in a CVE. In addition to the shopping assistant there were positive comments about the ability to communicate to the person that created the product being purchased 'face to face'. The first point to clarify is that what is meant by 'face to face' here is virtual face to virtual face. More importantly it means communicating to a real person, in real time. In other words synchronous communication, possibly by voice and some, virtual, body language as opposed to the asynchronous forms mostly used on 2D website browsing, such as email. One participant said: 'SL can offer one on one contact with the seller where you would have to telephone someone in real life when shopping online, if you needed further information, one on one.' (male, 62). The same participant went further giving a reason why the one to one contact felt better: 'Second Life can bring intimacy to business relationships between companies and individuals who are physically maybe thousands of miles apart. By intimacy, I don't mean anything rude! Just as we are in the same 'room', now' (male, 62).

Lastly two more RAs compared to 2D websites that were less prominent but useful nevertheless were location and the nature of navigating a 3D environment and the way it influences your experience of information. Regarding the benefits of location one participant's response was very enlightening: '…unless you pay a lot of money for a shop in a high traffic area you don't make much I guess… I think there are some places where people shop as they would in Real Life…They wander and browse and hopefully buy… Maybe because all the better stuff, the stuff you pay a lot for is in the places where you pay more to rent… Maybe because Second Life is so huge that these high traffic places are successful because people want all the good stuff in one area. It's easier; all the shops with the fashionable stuff are in one place. You get seen there, I guess that maters to some people' (female, 24). That quote emphatically shows that parallels between the real world and CVEs for retail. This is not just stylistic or visual but functional. Regarding the related point of navigating a 3D environment: 'I like pretty shops, well designed architecture, well laid out. I dislike shops that have no navigation' (female, 55) and 'All depending on the shop really, most of them are set out nicely so you can see the sections nicely, men and women etc., and prices…' (male, 24). These typical quotes illustrate how the layout is both pleasurable and functional.

7.2.2   Results for Research Objective 2

A RA of CVE retail compared to offline retail may be aspects of 2D e-commerce it includes that are not included in the offline retail environment.

The second objective that was identified from the literature, like the first, proposes a RA of CVEs in retail in the simplest way, offering a high level category of RA. It is equally significant to the first but far simpler. The nature of the technology of CVEs operating on the Internet guarantees that they will contain some of the Internet's benefits compared to offline. It is therefore not controversial or likely to be disputed in any way since it is based directly on the functionality of the technology and not its implementation by a retailer. As we have seen when the issue in question results from an implementation of a technology the users' perceptions tend to vary. The data collected is nevertheless useful as it illustrated this point with empirical evidence. Unsurprisingly there was ample data on this objective. Three related RAs found were convenience, speed, 24–7 availability and global reach. Characteristic responses were: 'You could meet your friends… if they are in a different country' (female, 19) and 'the fact your sat at home and able to check out other places of interest without dragging family from shop to the next, so peaceful' (male, 38). Another RA in relation to offline was the ability to access additional information such as reviews and profiles: 'I could read the profile of the vendor… you can't do that in real life!' (female, 22).

7.2.3   Results for Research Objective 3

Consumers may vary their intended usage of CVEs across the different stages of the purchase process because the significance of the dimensions of RA may vary across those stages.

Since this research was qualitative, in regard to the third objective the purpose was not to conclusively identify the most popular channel for each stage. The purpose was to investigate whether people vary their usage, whether there was an outright winner and importantly the explanations the participants gave for their beliefs. That is the strength of qualitative research and why it was chosen. If people prefer different channels for each stage this would be a strong indication of the benefit of retailers utilizing a multichannel approach. The third objective was illustrated by participants primarily in response to question eight which asked about the stages in the purchasing process.

Secondly, following on from that, the question asked the participants to map the stages of their purchasing process, as they understood it, onto the three retail channels in question. There was evidence that participants had an evaluation of each channel's advantages and disadvantages and chose the one they would use for a given task accordingly. They often did not have an outright favourite for all the stages. What could be considered surprising is that no participant chose the same channel for all stages.

This was shown firstly in individual responses. For example 'online shopping is the best as you can really research the goods you buy' (male, 60), 'for browsing the 3D could be more fun' (female, 22), and 'for the purchase stage I prefer to go to the shop because you get what you buy instantly' (female, 22). Beyond the individual comments there were some aggregate patterns for each stage. Some considered SL to be good for payment: 'The payment in SL is probably the easiest; just two clicks, no filling in your card number and so on. It is slightly quicker and easier' (female, 26). Many considered SL as the best channel for after sales service: 'For after sales service, hm. If I wanted to return something I would like to send an email or a letter. If I wanted help to solve a problem, or help to show me how to use it then I would prefer something more direct like the phone or SL.' (female, 26). This quote illustrates how the ability to communicate with a real person, in real time 'face to face', which was identified as an RA in the first issue, plays a significant role here. The related point, of being able to contact the creator of the product directly was also stated as a benefit again: '... the fact that there's no hours waiting for customer support at a call centre, the ability to contact the creator directly.' (male, 26). Contacting the creator of the product directly is of course not a standard characteristic of retail in a CVE channel but a special case. In response to the questions about the purchasing stages some participants believed it depended on the type of product. This indicates the significance of the fourth objective which will be discussed now.

7.2.4   Results for Research Objective 4

Consumers' usage of CVEs may be different for simple and complex products.

Regarding the fourth issue identified from the literature, most participants considered the two dimensional websites and offline as best for simple products: 'If I knew what I wanted, so let's say the new Dan Brown book and I just had to buy it then the 2D Internet. 'It is the most practical' (male, 28). For complex products overall most participants considered the offline world as the best. Some considered two dimensional websites better because you can get more information in a shorter space of time: '... 2D, purely because of the increased amount of data that can be viewed in a reasonable amount of time' (male, 26). The other reason given for preferring 2D was that they preferred to absorb information in text form: 'when you are using the 2D websites at the moment all the information comes up...' (female, 19). There is of course information in two dimensional text in Second Life but there is usually some navigation involved before it can be consumed. Some participants championed the benefits of comparison websites: 'you can use comparison websites from independent people' (female, 21).

Those that considered CVEs to be better than two dimensional, believed this primarily because of the shopping assistant once again: 'SL can offer one on one contact with the seller where you would have to telephone someone in RL when shopping online, if you needed further information' (male, 62) and 'I do not know anything about laptops, so if there was someone there...' (female, 22).

7.2.5   Results for Research Objective 5

CVEs such as SL may have the RA of a higher degree of Institutional Trust compared to the 2D websites.

There were four types of responses. The most common was to group the two dimensional and three dimensional together because the underlying technology was the same: 'I don't think it would make a difference, because if you think about it logically it is just a different interface' (female, 22). There were some that trusted two dimensional websites, the most common reason being that it was more established: 'the two dimensional because it is tried and tested everyone knows it is safe, while as this is quite new, it does not have a reputation' (female, 21), and the ability to read feedback. There were some that preferred the CVEs sighting the payment system: 'the company you buy the products off don't see your bank details, so that would be fine' (female, 19). Some highlighted how SL the 'institution' influenced 'institutional trust' positively: '...I was very unsure when I first started purchasing on SL, now unless there is a problem that has been highlighted with the grid, I generally don't have a problem with SL...' (female, 22). What the participant was referring to here was that Linden Labs informs users about retailers that are not trustworthy. This illustrates how the fact that SL is owned by Linden Labs and has the potential to cultivate greater institutional trust. That logic was the reason why this issue was identified as an area to investigate.

Despite the varying opinions about how trustworthy SL was for retail there was only one specific problem mentioned and that was specific to virtual products: 'In most cases on SL, what is sold is no transfer, so you cannot be refunded for anything. You cannot exchange anything etc.' (female, 21). Despite this being a valid concern this is inevitable at the moment.

7.2.6   Emergent Issues

The emergent themes identified one additional RA of CVEs for retail, that of enjoyment or fun: 'I suppose it is a little bit like a game so it is more enjoyable than just clicking on something like Amazon and buying something, without browsing or doing anything else' (female, 22). Research suggests [26], enjoyment is a factor in online purchases. Further research, [27, 28] identified enjoyment as a possible construct for consumer behaviour in CVEs. Along similar lines to that it has been suggested [29], that 'entertainment' can be enabled in retail by CVEs and further research [30], suggests retail in CVEs may be more experience orientated as opposed to either customer or product orientated. The second research [30], did not have empirical evidence to support this and encouraged that to be done in their conclusion. The constructs put forward by Kim and Forsythe [29] and Bourlakis et al. [30] are similar to that found by this research. This research found empirical evidence to support previous findings about the role of enjoyment in this context [27, 28, 31] and beyond that found evidence that enjoyment is a RA of CVEs compared to the other channels. The model used, as put forward by

Choudhury and Karahanna [8] stated that for the purpose of assessing the RA of channels for retail:

Relative Advantage (RA) = Convenience + Trust + Efficacy of Information
Those with an awareness of TAM would immediately identify that:

Convenience is similar to Perceived Ease of Use (PEOU)
Efficacy of Information is similar to Perceived Usefulness (PU)

And since: TAM = PEOU + PU, Choudhury and Karahanna [8] could be rewritten as: RA = TAM + Trust.

Many models of TAM include 'enjoyment' as a variable, Perceived Enjoyment, (PE). This is related to PEOU: 'PE has been theorised and empirically validated as either an antecedent or a consequence of PEOU', [32]. Hence 'enjoyment' is proposed as an addition to this model in this context:

Choudhury and Karahanna [8] model for assessing the RA of channels:
RA = Convenience + Efficacy of Information + Trust

Similar and logically compatible with:

RA = TAM (PEOU(... + PE) + PU) + Trust (Institutional Trust)

Therefore a potential model based on this research for assessing the RA of channels:

RA of retail channels = Convenience + Efficacy of Information + Enjoyment + Institutional Trust

8   Conclusion

The main contribution of this research was analysing five areas of RA that should be considered when comparing these three retail channels. The first objective was that the RA of CVEs to the two dimensional Internet for retail were aspects of the offline world that it included. These are primarily enjoyment social shopping, a richer and more emotive three dimensional environment, 'face to face' and the shopping assistant, and to a lesser extent location and navigation. The second objective was that the advantages of CVEs compared to the offline world for retail were aspects of the two dimensional Internet that it included. These were found to be convenience, speed, twenty four-seven availability and global reach, and additional information such as reviews and profiles. The third objective was that consumers would vary their use of CVEs across different stages of the purchase process because the RA would vary across each stage. Some considered SL to be good for payment after sales service. The fourth objective was that consumer's usage of CVEs may vary for simple and complex products. Most participants considered 2D websites and offline as the best for simple products. For complex products overall most participants considered the offline retail as the best. Some considered 2D websites better because you get more information and you can use comparison websites. Those that considered CVEs to be better than two dimensional, once again valued the

ability to negotiate with a real person such as the shop assistant which they found especially beneficial for complex products. Lastly the fifth objective is that CVEs such as
SL may have a higher degree of institutional trust compared to the 2D websites. There
were four types of responses. The first and most common were to group the two dimensional and three dimensional together because the underlying technology is the same.
There were some that trusted the 2D websites more than CVEs the most common reason
being that it was a more established technology. A second reason for this was because
they valued the ability to provide feedback that many 2D websites offer. Those that
preferred the three dimensional retail of CVEs sighting the payment system and some
highlighted how SL, the 'institution' influenced institutional trust positively.

The range of preferences in the retail channels in general, in relation to the purchasing
stages and product complexity and the lack of outright 'winner' with complete RA indicate the value of a multichannel approach for retail. The value of a multichannel
approach has been identified [5, 16, 34], but it is investigated and proved here for three
reasons. Firstly this research illustrates that CVEs have a position alongside the other
channels. Secondly this research illustrates how simple and complex products influence
the consumers' choice of channel. Hence it is useful to analyse the three channels separately for simple and complex products. Therefore we propose six dimensions should
be investigated (three channels by two types of products). Thirdly this research gives
some indications about what the nature of that position should be. A secondary contribution was the extension of the Choudhury and Karahanna [8] model: Relative
Advantage of a retail channel = Information Efficacy + Convenience + Institutional
Trust + Enjoyment.

References

1. Benford, S., Greenhalgh, C., Rodden, T., Pycock, J.: Collaborative virtual environments.
   Commun. ACM **44**(7), 79–85 (2001). European Central Bank: Virtual Currency Schemes.
   Frankfurt am Main: European Central Bank (2012)
2. Domina, T., Lee, S.-E., MacGillivray, M.: Understanding factors affecting consumer
   intention to shop in a virtual world. J. Retail. Consum. Serv. **19**, 613–620 (2012)
3. Cagnina, M.R., Poian, M.: Beyond e-business models: The road to virtual worlds. Electron.
   Commer. Res. **9**, 49–75 (2009)
4. SL Actions (2014). www.slactions.org/2012/?page_id=2. Accessed on 15 August 2014
5. Telzrow, M., Meyer, B., Lenz, H.: Multi-channel consumer perceptions. J. Electron. Commer.
   Res. **8**, 18–31 (2007)
6. Rogers, E.M.: Diffusion of Innovations. Free Press, New York (1995)
7. Simon, H.A.: Rational choice and the structure of the environment. Psychol. Rev. **63**, 129–
   138 (1956)
8. Choudhury, V., Karahanna, E.: The relative advantage of electronic channels: a
   multidimensional view. Manag. Inf. Syst. Q. **32**, 179–200 (2008)
9. Levy, M., Weitz, B.: B.A. Retailing Management. McGraw Hill, New York (2009)
10. Rangaswamy, A., Van Bruggen, G.H.: Opportunities and challenges in multichannel
    marketing: an introduction to the special issue. J. Interact. Mark. **19**, 5–11 (2005)
11. Zhang, J., Farris, P.W., Irwin, J.W., Kushwaha, T., Steenburgh, T.J., Weitz, B.A.: Crafting
    integrated multichannel retail strategies. J. Interact. Mark. **24**, 168–180 (2010)

12. Kumar, V., Venkatesan, R.: Who are the multichannel shoppers and how do they perform?: correlates of the multichannel shopping behaviour. J. Interact. Mark. **19**, 44–62 (2005)

13. Yan, R., Guo, P., Wang, J., Amrouche, N.: Product distribution and coordination strategies in a multi-channel context. Journal of Retailing and Consumer Services **18**, 19–26 (2011)

14. Chang, Y., McFarland, D.: Managing a well-integrated multichannel retail strategy. Int. J. Retail Distrib. Manag. **32**, 147–156 (1999)

15. Bellman, S., Lohse, G.L., Johnson, E.J.: Predictors of online buying behaviour. Commun. ACM **42**, 32–48 (1999)

16. Schoenbachler, D.D., Gordon, G.L.: Multi-channel shopping: understanding what drives channel choice. J. Consum. Mark. **19**, 42–53 (2002)

17. Konus, U., Verhoef, P.C., Neslin, S.A.: Multichannel shopper segments and their covariates. J. Retail. **84**, 398–413 (2008)

18. Schroder, H., Zaharia, S.: Linking multi-channel customer behavior with shopping motives: an empirical investigation of a German retailer. J. Retail. Consum. Serv. **15**, 452–468 (2012)

19. Noble, S.M., Griffith, D.A., Weinberger, M.G.: Consumer derived utilitarian value and channel utilization in a multi-channel retail context. J. Bus. Res. **58**, 1643–1651 (2005)

20. Berman, B., Thelen, S.: Managing a well-integrated multichannel retail strategy. Int. J. Retail Distrib. Manag. **32**, 147–156 (2004)

21. King, N.: Using templates in thematic analysis of text. In: Cassel, C., Symon, G. (eds.) Essential Guide to Qualitative Methods in Organizational Research, pp 256–270. Sage, London (2004)

22. McKnight, D.H., Cummings, L.L., Chervany, N.L.: Initial trust in new organizational relationships. Acad. Manag. Rev. **23**(3), 473–490 (1998)

23. McKnight, D.H., Choudhury, V., Kacmar, C.: Developing and validating trust measures for e-commerce: an integrative model typology. Inf. Syst. Res. **13**(3), 334–359 (2002)

24. Denzin, N.K., Lincoln, Y.S.: Handbook of Qualitative Research. Sage, Thousand Oaks (2003)

25. Miles, M.B., Huberman, A.M.: Qualitative Data Analysis. Sage, Thousand Oaks (1994)

26. Koufaris, M.: Applying the technology acceptance model and flow theory to online consumer behaviour. Inf. Syst. Res. **13**(2), 205–223 (2002)

27. Holsapple, C.W., Wu, J.: User acceptance of virtual worlds: the hedonic framework. The Database Adv. Inf. Syst. **38**(4), 86–89 (2007)

28. Guo, Y., Barnes, S.: Why people buy virtual items in virtual worlds with real money. The Database Adv. Inf. Syst. **38**(4), 69–76 (2007)

29. Kim, J., Forsythe, S.: Adoption of virtual try-on technology for online apparel shopping. J. Interact. Mark. **2**(2), 45–59 (2008)

30. Bourlakis, M., Papagiannidis, S., Li, F.: Retail spatial evolution: paving the way from traditional to metaverse retailing. Electron. Commer. Res. **9**(1–2), 135–148 (2009)

31. Lee, K.C., Chung, N.: Empirical analysis of consumer reaction to virtual reality shopping mall. Comput. Hum. Behav. **24**(1), 88–104 (2008)

32. Sun, H., Zhang, P.: Causal relationships between perceived enjoyment and perceived ease of use: an alternative approach. J. Assoc. Inf. Syst. **7**(9), 618–646 (2006)

33. Tang, F.F., Xing, X.: Will the growth of multi-channel retailing diminish the pricing efficiency of the web? J. Retail. **77**(3), 319–333 (2001)

Enterprise Architecture
and Business-IT-Alignment

Combining Business Processes and Cloud Services: A Marketplace for Processlets

Danillo Sprovieri[✉] and Sandro Vogler

Institute for Information Systems, School of Business University of Applied Sciences and Arts
Northwestern Switzerland, Olten, Switzerland
{danillo.sprovieri,sandro.vogler}@students.fhnw.ch

Abstract. *Context:* Business agility has been approached on BPM-level using process reuse and on execution-level using cloud computing. *Objective:* We combine these levels linking business processes repositories and cloud-based services. The goal is to provide a framework for BPaaS marketplaces by introducing Processlets as cloud-based implementations of business processes. *Method:* We design the marketplace's components and functionalities based on existent literature and survey relevant technologies. *Results:* We identified the features of the marketplace's components and discuss suitable technologies for implementation. *Conclusion:* Our framework fosters reuse of business processes and mitigates the problem of a scattered cloud market.

Keywords: Business process as a service (BPaaS) · Cloud computing · Marketplace · BPM · SOA · Business-IT alignment

1 Introduction

Business agility has become a major factor for many companies due to market dynamics, changing business conditions, and new regulations [1]. These challenges are addressed on two different levels: On the execution level, where agility is achieved by applying Service Oriented Architecture (SOA) principles and on the level of Business Process Management (BPM), where Business Process (BP) reuse has been proposed to foster agility [2].

The implementation of SOA principles in cloud services[1] helps to achieve agility and increased flexibility [3], as it allows to react quickly and easily to changes [4]. However, cloud adaption is significantly more complex due to system integration and the management of multiple cloud providers, than traditional engineering for BPs [5]. Furthermore, the increasing amount of heterogeneous cloud services poses the problem of selecting and composing a set of services, which fits to organization's needs [6].

---

[1] The term *Web service* is used ambiguously: Often, it refers to a WSDL-based service. However, an increasing number of providers are offering RESTful Web services to make them more accessible [14]. For this reason, we use the more general term *Cloud services* to refer to any kind of Web services that exploit the potential of clouds, regardless of their architecture.

© Springer International Publishing Switzerland 2015
W. Abramowicz (Ed.): BIS 2015, LNBIP 208, pp. 247–259, 2015.
DOI: 10.1007/978-3-319-19027-3_20

In order achieve agility on BPM level, organizations need to design and adapt BPs frequently. However, designing BP models is a highly complex, time consuming and error prone task. BP model reuse reduces modeling time and errors, increases model quality and flexibility [7]. To support such reuse, BP repositories have been developed and largely adapted by enterprises [8].

Efforts towards agility on the levels of BPM and SOA are mutually complementary [9]. This paper aims to combine the efforts towards agility on both described levels by focusing on cloud services and BP repositories. We propose a framework for a marketplace that supports organizations in finding cloud-enacted BPs. A marketplace is a reasonable approach, because it mitigates the problem of the scattered cloud service market by offering a broad palette of services at a single location. Moreover, organizations cannot be experts in all supporting processes they need to perform (e.g. accounting principles, legally required processes, etc.). A marketplace supports this issue by offering processes that are needed in various organizations. Our framework enables reuse of BPs, which decreases the time to model and deploy a BP, and enables enacting these processes using cloud services. To design the marketplace, we perform a literature review about cloud service marketplaces. Next, we specify the necessary components and their functionalities. We survey existing technologies, which can be used to implement the discovery and repository functionality of the marketplace. We classify, analyze and compare the characteristics of these technologies and discuss their suitability for our framework. In this paper, we use the term Business Processes as a Service (BPaaS) for BPs, which are (partly or entirely) enacted by cloud services.

The remainder of this paper is organized as follows: In Sect. 2, we position our work with respect to previous research. In Sect. 3, we outline the contribution of the marketplace, describe the necessary components, and give in Sect. 4 the results of a survey on the suitability of existing approaches for implementation. In Sect. 5, we discuss our framework and summarize the identified research challenges.

2 Related Work

We review related work based on existing solutions for cloud marketplaces and business process marketplaces. Most cited papers for cloud marketplaces are the 4CaaSt marketplace [10] and the CCMarketplace [11]. 4CaaSt presents a marketplace framework for trading XaaS solutions and provides an overview of necessary components. The work acknowledges the need for both business resolution and technical resolution. The CCMarketplace presents an architecture for a cloud service marketplace. The research focuses on describing the specification of the requirements of the marketplace. Neither paper does consider integration of BPs in their framework.

Besides scientific literature, there are existing marketplaces for cloud services. IBM provides a marketplace[2] for cloud services, which differs from our framework in the sense that the cloud services are only provided by IBM itself. The same is true for the

---

[2] http://www.ibm.com/marketplace/cloud/ [accessed: January 20, 2015].

SAP marketplace.[3] Thus, these marketplaces lack any competition and do not allow users to contribute their own solutions. Therefore, these marketplaces merely provide pluggable software components. Cisco[4] allows third party cloud service providers to offer their services. The provided services are independent of each other and there are no functionalities supporting their combination. Also, the offerings are not related to BPs. Oracle's cloud marketplace[5] provides applications as well as consulting services delivered by third party organizations. The applications can be integrated into different Oracle products. This approach differs from our framework, since there is no arbitrary combination of services supported. Moreover, the Oracle cloud marketplace does not incorporate BPs.

We performed a market survey due to the lack of scientific literature about BPaaS marketplaces. We only found IBM's marketplace to offer BPaaS. In this context, BPaaS refers to software solutions for specific purposes. The marketplace offers a description and additional material, such as case studies. This does not match our definition of BPaaS, as it does not allow the customer to understand how the process is defined.

Some papers approach similar issues we identified: [12] approaches the problem of service composition by introducing a business-IT mapping model to trace implementation of business process activities and implementing services. This supports agile reaction upon service changes. While this approach is not focusing on cloud services, we adapt the basic idea of a linkage between business processes activities and services. However, we consider complete BPs instead of process fragments.

Service-oriented requirements engineering is proposed in [13] to align business needs with services. While their work focuses on the engineering process of SOA services, our framework provides a holistic view and approaches the problem from a business perspective.

3 Marketplace Framework

We propose a framework to foster business agility by enabling reuse of BPs and cloud services. Using a marketplace, we provide a one-stop-shop for organizations to buy deployable BPs. The BPs are enacted using cloud services. The marketplace provides a large set of BPs (in the sense of a process library). To connect these BPs to cloud services, we introduce Servicelets, which are executable implementations of BPs using cloud services. Each BP can be enacted by different sets of cloud services and can thus have several Servicelets. Servicelets connect activities with cloud services using execution environments. In the simplest case, the services can interact with each other directly and the execution environment consists merely of the configuration of these services. In a more complex example, a workflow management system (WfMS) executing WS-BPEL code can be used; or, since most cloud services have RESTful interfaces, an execution language like BPEL for REST [14] can be utilized. If components like a WfMS are

---

[3] https://websmp106.sap-ag.de/public/home [accessed: January 20, 2015].

[4] https://marketplace.cisco.com/cloud [accessed: January 20, 2015].

[5] https://cloud.oracle.com/marketplace/ [accessed: January 20, 2015].

needed, they must be part of the Servicelet themselves to ensure the Servicelet is self-containing.

Three different providers populate the proposed marketplace with products: business process providers, cloud services providers and Servicelet providers. These providers should be considered as roles, as e.g. a certain provider might offer a cloud service and a Servicelet that uses the service he provides. This open architecture allows any stakeholder to add new products to the marketplace. It is also possible that a customer enhances a Servicelet and offers the new version for sale to profit from the invested effort. This ensures, besides the competition among the providers, that the marketplaces products are continuously updated and enhanced. The framework does not impose a specific granularity of the BPs on the BP provider. It is up to the provider, to decide what granularity is appropriate.

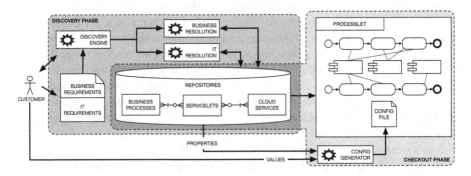

Fig. 1. Overview of involved components during the purchase process of the marketplace.

In Fig. 1, we give an overview of the components involved in the purchase process: The customer accesses the marketplace via a Web-interface, where the customer selects BPs and corresponding Servicelets. The discovery phase is assisted by a business-resolution engine and IT-resolution engine, which cover two aspects of the query: business-related and IT-related aspects. For selecting BPs, only business-related aspects are relevant (e.g. process aim, category, country, and price). Except for price, all of these aspects can be matched against the metadata of a BP, while the price depends on the selected Servicelets. The IT-related aspects act as a filter on the available Servicelets for each BP, as it excludes Servicelets, which do not conform to the IT-aspects (e.g. data storage must be provided by a European resident company).

The customer can select several BPs and must choose a Servicelet for each. The discovery engine supports this process by sorting the Servicelets according to the customer's settings (e.g. minimizes total running cost or minimizes number of services). A trivial approach is to compare all possible Servicelets and evaluate their cost or number of services respectively. In addition, the marketplace can give recommendations for Servicelets based on the analytics of previous purchases and based on the customer's context (e.g. country-specific processes or Servicelets).

After the customer has completed the BP and Servicelet selection, checkout phase takes place, where the customer enters a binding contract with the providers of the services, the Servicelets and the business processes. In this phase, all Servicelets are compiled into a container, which we call Processlet. A Processlet is a deployable BP, which contains at least one Servicelet. The Processlet contains the execution environment that allows enacting all selected BPs. Along with the Processlet, the customer gets access to the full documentation about the BP (e.g. a more detailed process model, instructions on how to perform the process, etc.). The execution environment or the service itself might allow the customer to modify the service or the Servicelet. For this reason, Servicelets are accompanied with config files. An example of a configurable option is the GUI of a cloud service, where the customer can place his own logo. During the checkout phase, all configuration files of selected Servicelets are compiled into the Processlet. Finally, the Processlet is deployed and the customer gets full access to the BP documentation.

Our framework supports both structured and unstructured BPs: For structured BPs, the Servicelet can comprise e.g. a WfMS, whereas for unstructured (knowledge-intensive) processes, the Servicelets consist of loosely coupled cloud tools supporting different activities. An example for an unstructured process is an innovation process, which is assisted by (cloud-based) mind-mapping and collaboration tools that use the same cloud storage.

As multiple providers and different cloud storages are involved, interoperability becomes an issue. Our framework relies on cloud services itself to provide the necessary adaptors (either by compatible API or by providing middleware). Also, the framework is not limited to Servicelets completely enacted by cloud services: Servicelets can require services, which are not cloud based, e.g. a process requiring a desktop word processor to fill a template. In this case, the customer must provide this application in order to enact the process. The Servicelet-provider will try to maximize the compatibility with the word processor by e.g. providing the necessary template from a website to download in the word processor's format.

3.1 Components of the Marketplace

The proposed framework for a marketplace requires the following components. Note that the purchase process illustrated in Fig. 1 does not depict all components.

The *discovery engine* provides the user interface and the search functionalities. The user interface is either a Web GUI or an API, which allows other components to access the marketplace. Its core feature is to present the content of the repositories to the user and assist in discovering suitable BPs and Servicelets. Recommendation features are assisted by the information gathered by the *analytic engine*. Therefore, the discovery engine is able to suggest suitable BPs to the customer based on his context and similar previous purchases. As the customers might not be domain experts, the search methods must be intuitively and not require extensive knowledge about how the process is structured. However, as there are processes with highly structured naming conventions (e.g. public administration, process libraries) a search based on metadata must also be provided. The discovery engine must be able to distinguish between business-requirements and IT-requirements.

Both *business resolution* and *IT resolution* are engines that further process these requirements and provide the backend for the discovery engine. Business resolution retrieves BPs matching the defined business requirements and interacts with the *business process repository*. IT resolution ensures retrieved Servicelets comply with the defined IT-requirements. The two resolution engines make sure Processlets comply with the defined business- and IT- requirements.

Three repositories form the storage facilities: The *business process repository* stores all information about the process in form of metadata: graphical representation (e.g. in BPMN), process documentation, business objective, textual description, versioning, author, price, categories, etc.). The *service repository* holds information about the cloud service offerings. This comprises service description, pricing models, options (such as SLA-levels), APIs, provider, license, use cases, etc. The *Servicelet repository* stores combinations of BPs and cloud services along with an execution environment and configuration files for the cloud services it uses. Moreover, it contains variants, such as different SLA levels of the same service.

The *config generator* is responsible for combining the configurations of all Service-lets into the config file of the Processlet. As each Servicelet contains configurations for all used cloud services, it is possible that two selected Servicelets use the same service and configure it differently. The config generator must ensure there are no contradictions while compiling the Processlet. If such contradiction cannot be resolved (not even manually), two separate instances of the service are necessary.

The *management component* is geared towards the needs of the marketplace operator and the providers. It offers tools for developer and providers to add new products to the marketplace. This comprises graphical modelling support, execution language editor, as well as an upload-functionality for additional material, such as a user guide. It also supports providers to continuously update their products. Whenever a BP or cloud service (e.g. an updated API) is updated, Servicelet providers that utilize the updated component are notified. This facilitates a quick adaption of Servicelets.

The *user management* stores the customers' context, such as country, industry sector, IT infrastructure and size of company to derive requirements, which assists the discovery engine. In addition, this component is responsible for user authentication. The *analytics engine* is a mining component that continuously improves recommendations by clustering products and profiling customers. Besides recommendation during the discovery phase, push notification can be sent to the customers if better suiting products or updates are available.

3.2 Discussion About the Benefits

We describe four cases and demonstrate how the framework can be applied to support these scenarios.

SMEs. Typically, SMEs do not have an in-depth knowledge on all tasks they need to perform (lacking expertise in supporting processes). An example are mandatory codes in administration, such as legal requirements for data security of tax related documents [15]. Hence, European companies using a cloud-based CRM solution need to make sure the providers comply with the national data protection law (safe harbor rule).

For this reason, all enacting services must adhere to these laws. The marketplace tags BPs (or sets thereof) and/or certain cloud-implementation as compliant. Based on the customers' context (e.g. country) and the specifications, the marketplace only shows suitable BPs and Servicelets that comply with applicable laws.

Startup Company. It has been shown that startup companies can benefit from cloud solutions, due to features like dynamic scalability and "pay-as-you-go". Startups do lack the necessary infrastructure or financial resources to implement the services and BPs themselves. Competition between cloud providers enables startups to implement these services for limited costs [16]. By offering cloud-based solutions and fostering the competition between providers, startup companies can quickly implement processes with limited financial efforts. Furthermore, the startup company profits from the notification of updated Processlets.

Established Corporations. Inter-organizational alignment of BPs has become common for companies [17]. Examples are integrated supply chains. Since the buyer wants to be able to change the supplier if necessary, new suppliers need to be integrated quickly. Integration requires not only technical compatibility, but also forces the partnering company to implement the same inter-organizational BPs. On one hand, the marketplace helps to find easy-to-implement Processlets to adapt the own business to a dominating party. On the other hand, the dominating party can supply the BPs together with a cloud implementation (Processlet) to increase the amount of possible business partners.

Public Administration. Public administrations are known to have similar BPs [18]. Courts, for instance, do often have similar processes to handle an appeal. Public administration is a good example for BP reuse, which is fostered by the marketplace.

4   Suitability of Existing Technologies for Implementation

In this Section, we analyze existing technologies that are able to support the key components of our framework: *discovery* and *business process repository*. We give a survey to give indication how existing technologies can be used for implementation.

4.1   Discovery

The classification scheme for retrieval methods has been created by combining and extending [19, 20]. Three types of discovery methods are described in [19]: search, navigation and query. Querying is the most powerful method as it uses the models directly, and does not rely on metadata. Therefore, we focus on query-based methods for BP retrieval and present our classification in Table 1.

Query-based methods are divided using three criteria. *Type of search* describes the properties of the algorithm. It can take three values: *graph-based* search checks for structure and similarity patterns; *behavior semantics* tests for satisfaction of process semantics (e.g. task A is executed before task B); *operation semantics* assess operational meaning of the process (e.g. how to buy a flight ticket). *Query type* defines what type of input the algorithm accepts. This can be a query string or a graphical model. *Implementation* describes for which kind of models the method has been implemented.

We investigate query languages that use models as input, as textual input requires expert knowledge, which does not apply to the marketplace's users. In addition, the users do not have domain knowledge about how the processes are structured, which makes methods using graph structure unsuitable. From these facts, we derived the following requirements: *R1.1*: The query should be model-based. *R1.2*: Search should be performed using operation semantics (preferred) or behavior semantics. *R1.3*: An implementation for process models should exist.

Only two methods satisfy all requirements: BPMN-VQL (BPMN Visual Query Language) and VMQL (Visual Model Query Language). BPMN-VQL is based on semantic annotations as well as ontology inference. In our scenario, the ontology approach is very useful, since it allows searching for semantically equivalent terms. VMQL allows modeling of restrictions on pattern matches; model queries are expressed as annotated models. A disadvantage of this approach is that very detailed knowledge (e.g. the exact name of an activity to match) is needed.

Table 1. Classification of query-based retrieval methods for business processes.

	Type of search	Query type	Implementation
aFSA [21]	Graph structure	Textual	Generic Process Model
APQL [22]	Behavior semantics	Textual	Generic Process Model
BPMN VQL [23]	Behavior semantics & Graph structure	Model	BPMN
BPMN-Q [24]	Graph structure	Model	BPMN
BP-QL [25]	Graph structure	Model	BPMN
GMQL [26]	Graph structure	Textual	General
IPM-PQL [27]	Graph structure	Textual	Generic Process Model
LM + SM [28]	Operation semantics & Graph structure	Textual	Generic Process Model
OCL [29]	Graph structure & Behavior semantics	Textual	UML
Operational Similarity [30]	Operation semantics & Graph structure	Textual	Generic Process Model
PPSL [31]	Behavior semantics	Model	UML
vGMQL [32]	Graph structure	Model	General
Vistrails [33]	Graph structure	Model	Workflow Model
VMQL [34]	Behavior semantics	Model	UML & BPMN
Wise [35]	Operation semantics & Graph structure	Textual	Workflow Hierarchy

Some solutions using operation semantics provide unique features worth mentioning: Wise focuses on presentation of the result. It builds a hierarchy of workflow fragments and presents the workflow with the sub-workflows extended, which match the keywords. LM + SM improves retrieval by combining textual information with the structure of the process. It retrieves text features (received from title, documentation, activity names, etc.) and clusters them based on topic similarity. In a second step, the clusters are partitioned using structural features (control flows). For our framework, this is beneficial, as it allows the user a "fuzzy" search, since the algorithm is able to find processes using different wording. The algorithm of "searching business process repositories using operational similarity" (Operational Similarity) creates taxonomies from labels in the model and allows searching using natural language processing.

None of the methods satisfies *R1.2* using operation semantics for search. We suggest to combine the ontology-based approach of BPMN-VQL with the advanced analytical features of LM + SM.

4.2 Business Process Repositories

The classification scheme relies on criteria in [36, 19], which have been extended to reflect the requirements of the framework. We do only consider repositories, which does not include process libraries such as SCOR or the MIT Process Handbook. The classification is presented in Table 2: *Supported functions* explain the features that the business repositories support. *Retrieval method* describes the language that is used to retrieve the BPs. *Output format* lists in which language the BPs can be exported.

The criteria *supported functions* is encoded using the following values: 1: Supports classification of processes into categories; 2: Presents BP models in graphical form; 3: Describes satisfied business objectives by the process; 4: Stores metadata for BPs; 5: Supports create, update and delete functions; 6: Supports navigation; 7: Supports search.

We consider the listed *supported functions* to be required for such a repository. Furthermore, the repository should support a retrieval method, which satisfies all mentioned requirements (R1.1 – R1.3).

Literature review has revealed no business repository supporting execution using cloud services. None of the surveyed repositories supports all functions neither a suitable retrieval method.

Apromore covers most of the aspects needed, lacking a description of business objectives for the process as well as a suitable query language. It is designed to convert the stored models to various process model notations (e.g. BPMN, EPC and BPEL). This is certainly a useful feature but not necessarily required for the marketplace. Fragmento stores process fragments to enable their reuse. The process fragment collection stores fragments in BPEL. This storage of process fragments could possibly lead to easier integration of new services. SBPR is an ontology-based repository, which supports semantic querying. It enriches BP models by annotating artifacts with entities using predefined ontologies. This feature comes very close to search using operational semantics; however, it is more general. Similarly, the Querying Framework stores information about several aspects (activity, control flow, data, goal, resource, and authorization) in an

Table 2. Classification of business process repositories.

	Supported functions	Retrieval method	Output format
Apromore [37]	1, 2, 4, 5, 6, 7	BPMN-Q & PQL	XML
BPCM repository [38]	3, 5, 6, 7	not specified[a]	XML
BPEL repository [39]	4, 5	OCL	BPEL
BPMN repository [40]	4, 5	not specified[a]	XML
Fragmento [41]	1, 4, 5, 7	SQL	BPEL
IPM [27]	1, 4, 5, 6, 7	IPM-PQL	IPM-EPDL
Osiris [42]	2, 4, 5	not specified[a]	XML
Prosero [43]	1, 2, 4, 5, 7	not supported[a]	BPEL
RepoX [44]	2, 4, 5, 6	SQL	XML
SBPR [45]	4, 5, 7	WSML	WSML
ServicePot [46]	4, 5, 7	XML	XML
Querying Framework [47]	4, 5, 6	WSML	WSML

[a]Query functionality not explicitly mentioned.

ontology repository. ServicePot focuses on choreographies and not on process diagrams. It extends the UDDI registry, which allows an easy integration of service discovery. IPM follows a more holistic approach and integrates process analysis and optimization, process automation and control, and process-oriented integration. Furthermore, it contains a process knowledge repository to foster knowledge discovery.

5   Conclusions and Future Work

In this paper, we presented a framework for a BPaaS marketplace enacting BPs by cloud services. The marketplace provides a one-stop-shop for organizations' needs mitigating the problem of the scattered service market. We introduced the concept of Processlet for a deployable set of cloud services, which enact one or more BPs. The marketplace fosters reuse of existing BPs with the corresponding Servicelet and reduces modeling time and errors. In addition, the marketplace offers updates of offered BPs, cloud services and Servicelets and enables organizations to more quickly adapt their BPs. In four cases, we demonstrated the advantages of the concepts of Processlet and Servicelet in context. Furthermore, we analyzed suitable technologies for implementation of core components of the marketplace.

Besides discussed advantages of our framework, we encountered some drawbacks: Any modification of BPs or cloud service APIs can cause compatibility issues. It should be evaluated how providers of BPs and cloud services can be assisted to mitigate this issue. In addition, more research work is needed in order to define how business- and IT-requirements can be formalized to use them to support the discovery process.

As part of our future work, we will elaborate the method how to apply the framework and perform validations in order to test usability and sensibility; these validations will facilitate transferring the method to industry and real-world application. The method will also describe the roles involved (e.g. analyst, engineer, business expert etc.). Furthermore, we plan to propose a method for using a goal modeling language to support the discovery process and to present an implementation for the business- and IT-resolution engines. In addition, a more in-depth analysis of the ecosystem and the participants of the marketplace should be performed and it should be analyzed whether storing BP fragments instead of complete processes is beneficial. Finally, the service composition task should be assisted or automated. Such approach could also include run-time service composition.

Acknowledgement. We would like to thank Marcela Ruiz and Barbara Re for valuable discussions and their profound remarks, which led to considerable improvement of the paper.

References

1. Goodhue, D., Che, D., Boudreau, M.C., Davis, A.R.: Addressing business agility challenges with enterprise systems. MIS Q. Exec. **2**, 73–88 (2009)
2. Krafzig, D., Banke, K., Slama, D.: Enterprise SOA: Service-oriented Architecture Best Practices. Prentice Hall Professional, Upper Saddle River (2005)
3. Ren, M., Lyytinen, K.: Building enterprise architecture agility and sustenance with SOA. Commun. Assoc. Inf. Syst. **22**, 75–86 (2008)
4. Grivas, S.G., Kumar, T.U., Wache, H.: Cloud broker: bringing intelligence into the cloud. In: 3rd IEEE International Conference on Cloud Computing, pp. 544–545 (2010)
5. Avram, M.G.: Advantages and challenges of adopting cloud computing from an enterprise perspective. Procedia Technol. **12**, 529–534 (2014)
6. Sheng, Q.Z., Qiao, X., Vasilakos, A.V., Szabo, C., Bourne, S., Xu, X.: Web services composition: a decade's overview. Inf. Sci. (Ny) **280**, 218–238 (2014)
7. Fellmann, M., Koschmider, A.: Analysis of business process model reuse literature: are research concepts empirically validated? In: Modellierung, pp. 185–192 (2014)
8. Dijkman, R., La Rosa, M., Reijers, H.A.: Managing large collections of business process models—current techniques and challenges. Comput. Ind. **63**, 91–97 (2012)
9. Kapuruge, M., Han, J., Colman, A.: Service Orchestration As Organization. Elsevier, Waltham (2014)
10. Menychtas, A., Vogel, J., Giessmann, A., Gatzioura, A., Garcia Gomez, S., Moulos, V., Junker, F., Müller, M., Kyriazis, D., Stanoevska-Slabeva, K., Varvarigou, T.: 4CaaSt marketplace: an advanced business environment for trading cloud services. Futur. Gener. Comput. Syst. **14**, 104–120 (2014)
11. Li, H., Jeng, J.: CCMarketplace: a marketplace model for a hybrid cloud. In: Conference of the Center for Advanced Studies on Collaborative Research, pp. 174–183 (2010)

12. Buchwald, S., Bauer, T., Reichert, M.: Bridging the Gap between business process models and service composition specifications. In: Lee, J., Ma, S-P.M., Liu, A. (eds.) Service Life Cycle Tools and Technologies: Methods, Trends and Advances, pp. 124–153. IGI Global (2011). doi:10.4018/978-1-61350-159-7

13. Adam, S., Riegel, N., Doerr, J., Uenalan, O., Kerkow, D.: From business processes to software services and vice versa. J. Softw. Evol. Process. **24**, 237–258 (2012)

14. Pautasso, C.: RESTful web service composition with BPEL for REST. Data Knowl. Eng. **68**, 851–866 (2009)

15. Stephens, D.O.: Protecting personal privacy in the global business environment. Inf. Manag. J. **41**, 56–59 (2007)

16. Smith, R.: Computing in the cloud. Res. Technol. Manag. **52**, 65–68 (2009)

17. Vathanophas, V.: Business process approach towards an inter-organizational enterprise system. Bus. Process Manag. J. **13**, 433–450 (2007)

18. Janssen, M., Joha, A.: Motives for establishing shared service centers in public administrations. Int. J. Inf. Manage. **26**, 102–115 (2006)

19. Yan, Z., Dijkman, R., Grefen, P.: Business process model repositories – framework and survey. Inf. Softw. Technol. **54**, 380–395 (2012)

20. Wang, J., Jin, T., Wong, R.K., Wen, L.: Querying business process model repositories. World Wide Web. **17**, 427–454 (2013)

21. Mahleko, B., Wombacher, A.: Indexing business processes based on annotated finite state automata. In: IEEE International Conference on Web Services, pp. 303–311 (2006)

22. ter Hofstede, A.H., Ouyang, C., La Rosa, M., Song, L., Wang, J., Polyvyanyy, A.: APQL: a process-model query language. In: Song, M., Wynn, M.T., Liu, J. (eds.) AP-BPM 2013. LNBIP, vol. 159, pp. 23–38. Springer, Heidelberg (2013)

23. Di Francescomarino, C., Tonella, P.: Crosscutting concern documentation by visual query of business processes. In: Ardagna, D., Mecella, M., Yang, J. (eds.) Business Process Management Workshops SE - 3, pp. 18–31. Springer, Berlin Heidelberg (2009)

24. Awad, A.: BPMN-Q: a language to query business processes. In: Workshop on Enterprise Modelling and Information Systems Architectures, pp. 115–128 (2007)

25. Beeri, C., Eyal, A., Kamenkovich, S., Milo, T.: Querying business processes with BP-QL. Inf. Syst. **33**, 477–507 (2008)

26. Delfmann, P., Steinhorst, M., Dietrich, H.-A., Becker, J.: The generic model query language GMQL. Inf. Syst. **47**, 129–177 (2015)

27. Choi, I., Kim, K., Jang, M.: An XML-based process repository and process query language for integrated process management. Knowl. Process Manag. **14**, 303–316 (2007)

28. Qiao, M., Akkiraju, R., Rembert, A.J.: Towards efficient business process clustering and retrieval: combining language modeling and structure matching. In: Rinderle-Ma, S., Toumani, F., Wolf, K. (eds.) BPM 2011. LNCS, vol. 6896, pp. 199–214. Springer, Heidelberg (2011)

29. Object Management Group: Object Constraint Language Version 2.4 (2014). http://www.omg.org/spec/OCL/2.4/PDF

30. Lincoln, M., Gal, A.: Searching business process repositories using operational similarity. In: Meersman, R., Dillon, T., Herrero, P., Kumar, A., Reichert, M., Qing, L., Ooi, B.-C., Damiani, E., Schmidt, D.C., White, J., Hauswirth, M., Hitzler, P., Mohania, M. (eds.) OTM 2011, Part I. LNCS, vol. 7044, pp. 2–19. Springer, Heidelberg (2011)

31. Forster, A., Engels, G., Schattkowsky, T., Van Der Straeten, R.: Verification of business process quality constraints based on visual process patterns. In: First Joint IEEE/IFIP Symposium on Theoretical Aspects of Software Engineering, pp. 197–208. IEEE (2007)

32. Steinhorst, M., Delfmann, P., Becker, J.: vGMQL-introducing a visual notation for the generic model query language GMQL. In: PoEM (2013)
33. Scheidegger, C., Vo, H., Koop, D.: Querying and re-using workflows with VsTrails. In: ACM SIGMOD International Conference on Management of Data, pp. 1251–1254 (2008)
34. Störrle, H., Acretoaie, V.: Querying business process models with VMQL. In: 5th ACM SIGCHI Annual International Workshop on Behaviour Modelling - Foundations and Applications – BMFA 2013, pp. 1–10. ACM Press, New York (2013)
35. Shao, Q., Sun, P., Chen, Y.: WISE: a workflow information search engine. In: IEEE 25th International Conference on Data Engineering, pp. 1491–1494. IEEE (2009)
36. Bhagya, R.P.T., Vasanthapriyan, S., Jayaweera, P.: Survey on requirements and approaches of business process repositories. Int. J. Sci. Res. Publ. **4**, 1–4 (2014)
37. La Rosa, M., Reijers, H.A., van der Aalst, W.M.P., Dijkman, R.M., Mendling, J., Dumas, M., García-Bañuelos, L.: APROMORE: an advanced process model repository. Expert Syst. Appl. **38**, 7029–7040 (2011)
38. Gao, Shang, Krogstie, John: A repository architecture for business process characterizing models. In: van Bommel, Patrick, Hoppenbrouwers, Stijn, Overbeek, Sietse, Proper, Erik, Barjis, Joseph (eds.) PoEM 2010. LNBIP, vol. 68, pp. 162–176. Springer, Heidelberg (2010)
39. Vanhatalo, J., Koehler, J., Leymann, F.: Repository for business processes and arbitrary associated metadata. In: BPM Demo Session, pp. 25–31 (2006)
40. Theling, T., Zwicker, J.: An architecture for collaborative scenarios applying a common BPMN-repository. In: 5th IFIP WG 6.1 International Conference, pp. 169–180 (2005)
41. Schumm, D., Karastoyanova, D., Leymann, F., Strauch, S.: Fragmento: advanced process fragment library. In: Pokorny, J., Repa, V., Richta, K., Wojtkowski, W., Linger, H., Barry, C., Lang, M. (eds.) Proceedings of the 19th International Conference on Information Systems Development, pp. 659–670. Springer, New York (2010)
42. Weber, R., Schuler, C., Neukomm, P.: Web service composition with O'GRAPE and OSIRIS. In: 29th International Conference on Very Large Data Bases, pp. 1081–1084 (2003)
43. Elhadad, M., Balaban, M., Sturm, A.: Effective business process outsourcing: the Prosero approach. IBIS (London 1859) **6**, 8–31 (2007)
44. Song, M., Miller, J., Arpinar, I.: Repox: an xml repository for workflow designs and specifications (2001)
45. Ma, Z., Wetzstein, B., Anicic, D., Heymans, S., Leymann, F.: Semantic business process repository. In: CEUR Workshop Proceedings, pp. 92–100 (2007)
46. Ali, M., de Angelis, G., Polini, A.: ServicePot – an extensible registry for choreography governance. In: IEEE 7th International Symposium on Service Oriented System Engineering, pp. 113–124 (2013)
47. Markovic, I., Pereira, A.: Towards a formal framework for reuse in business process modeling. In: Business Process Management Workshops, pp. 484–495 (2008)

The Nature and a Process for Development
of Enterprise Architecture Principles

Kurt Sandkuhl[1,3], Daniel Simon[2], Matthias Wißotzki[1(✉)], and Christoph Starke[1]

[1] Chair of Business Information Systems, University of Rostock, 18051
Rostock, Germany
{kurt.sandkuhl,matthias.wissotzki,christoph.starke}
@uni-rostock.de
[2] Scape Consulting GmbH, Cologne, Germany
daniel.simon@scape-consulting.de
[3] ITMO University, Kronverkskiy Pr. 49, 197101 St. Petersburg, Russia

Abstract. Enterprise architecture management (EAM) is expected to contribute
to, e.g., strategic planning and business-IT-alignment by capturing the essential
structures of an enterprise. Enterprise architecture principles (EAPs) are among
the subjects in EAM research that have received increasing attention during the
last years, but still are not fully covered regarding their characteristics and their
development and use in practice. The aim of this paper is to contribute to the EAM
field by investigating the nature of EAPs and by proposing and validating a
development process for EAPs. The main contributions of the paper are (a) an
analysis of the characteristics of EAPs, (b) an initial development process for
EAPs, and (c) the results of expert interviews for validating this process.

Keywords: Enterprise architecture · Principles · Application scenarios · Process

1 Introduction

Responsiveness to market changes, adaptability of business models, and continuous
business-IT-alignment are among the acknowledged factors for competitiveness on a
globalized market. Enterprise architecture management (EAM) is expected to contribute
to the above challenges by capturing the essential structures of an enterprise at and across
different architectural levels (e.g., business, data, application, technology). In the last
decade, EAM has attracted a lot of interest in the industry and a lot of activity in the
research community. This is confirmed by surveys of practitioners regarding the use and
importance of EA (see, e.g., [1]) and by literature studies investigating the research field
(e.g., [2]).

EA principles (EAPs) are among the subjects in EAM research that have received
increasing attention during the past years (e.g., [3]), but still are not fully covered
regarding their characteristics and their development and use in practice. The aim of this
paper is to contribute to the EAM field by investigating the nature of EAPs and by
proposing and validating an EAP development process. The main contributions of the

© Springer International Publishing Switzerland 2015
W. Abramowicz (Ed.): BIS 2015, LNBIP 208, pp. 260–272, 2015.
DOI: 10.1007/978-3-319-19027-3_21

paper are (a) an analysis of the characteristics of EAPs, (b) an initial development process for EAPs, and (c) the results of expert interviews for validating this process.

The research work in this paper started from the following research question: *What characteristics and steps should a process for developing EAPs have?* The research approach underlying this work is an abductive approach, i.e., a combination of (a) a deduction from the body of knowledge in terms of what theoretical basis applies to a development process for EAPs and (b) the definition of such a development process and an induction from work with experts in the field to evaluate to what extent our development process would work in practice.

The remainder of the paper is structured as follows: Sect. 2 briefly summarizes relevant background for our work on EAPs, based on a literature analysis in this area. Section 3 investigates the nature of EAPs and derives requirements for a development process by applying the Cynefin framework. Based on the background work and the results of Sect. 3, Sect. 4 proposes an initial EAP development process. Section 5 evaluates the process by means of expert interviews and discusses threats to validity of the findings. Summary and outlook on future work are presented in Sect. 6.

2 Research Background

Enterprise architecture (EA) denotes the fundamental conception or representation of an enterprise—as embodied in its main elements and relationships—in an appropriate model. EAM provides an approach for a systematic development of an enterprise's architecture in line with its goals by performing planning, transforming, and monitoring functions. The reasons for applying EAM are manifold, such as supporting the alignment of IT to business goals or the reduction of complexity.

To support EAM tasks, EA principles (EAPs) can be used. In general, EAPs (such as, e.g., "reuse is preferable to buy" or "processes are standardized") provide a stable instrument to support the transition from a current to a target architecture in line with goals and strategies [3]. As such, they are considered an important element in the management of an EA (cf. [4–6]). A literature analysis in the field of EAM, however, returned diverse definitions of the term "EA principle."

The IEEE standard 1471-2000 refers to EAPs as elements which guide the design and evolution of EAs from a current state to a target state (cf. [7]). Schekkerman [8] defines EAPs as general underlying rules and guidelines for the use and provision of business and IT resources in the enterprise and, in general, as a basis for business and IT decision making (cf. [8]). Aier et al. [4] define EAPs as restrictions to design freedom for projects transforming an EA. In 2009, Stelzer [5] conducted a systematic literature review that identified twelve articles dealing with research on EAPs. Based on that, Stelzer suggests EAPs to be considered as "fundamental propositions that guide the description, construction, and evaluation" of EAs (cf. [5]). The numerous approaches, however, reveal that there is not yet a consistent definition of EAPs. This can also be seen in the various synonyms used for EAPs, such as guidelines, laws, rules, and policies (cf. [5, 9]). Fischer et al. [10] extend on this and also survey the meta-models underlying the perception of EAPs in the literature.

All in all, however, based on our literature analysis, we start off with the following definition of EAPs that is to be substantiated in the next parts of this paper: *EAPs can be described as an enterprise-specific and abstract, yet simple collection of statements, which generally provide a framework for decision making and thus support the transformation process of an enterprise from a current to a target EA.*

3   Nature of EA Principles

As has been outlined in Sect. 2, there are different perceptions of what EAPs are. This section takes an application-oriented view on the nature of EAPs. Based on the identification and characterization of scenarios for EAP's application, it sheds light on whether EAPs can thus be of a law-like character ("rules") or whether they should rather be considered a guiding instrument ("guide rails"). This finally provides the basis for deriving requirements with respect to a process for developing EAPs.

3.1   Application Scenarios for EA Principles

We ground the identification of application scenarios for EAPs in the "Architecture Development Method" (ADM) provided by "The Open Group Architecture Framework" (TOGAF) [11]. We believe this is a reasonable approach given the fact that the ADM represents a generic, but comprehensive method for architecture development or, say, management of the architecture lifecycle going from vision via architecture definition and transition planning to implementation (and beyond). The ADM, as such, needs to be integrated with several other practices in the enterprise. Phases at different points in the ADM are of relevance depending on the practice to interact with, and thus the context to operate in. These contexts are what possible scenarios for the use of EAPs shall be based on.

This approach is largely in line with the general classes of EAM engagement defined by TOGAF itself, such as "supporting business strategy," "architectural portfolio management of the landscape," "architectural definition of [...] change initiatives," or "architectural portfolio management of projects." The latter one, for example, comes with a focus on the ADM's transition planning phases.

To become more specific, but abstract from the architectural focus prevalent in the classes' labels, we took on a view that is rather process-oriented and focuses on the actual practice or activities at hand in which EAM shall play a role. We aggregated different phases of the ADM accordingly. This let us derive the following application scenarios for EAPs:

- Strategy making,
- Business transformation/reorganization,
- IT transformation/landscape management,
- Project portfolio planning/management, and
- IT solution projects.

3.2 Characterizing the Application Scenarios Using the Cynefin Framework

The Cynefin framework is designed to support sense and decision making in varied contexts. Specifically, it sorts "the issues facing leaders into five contexts defined by the nature of the relationship between cause and effect": simple, complicated, complex, chaotic, and disorder. While the former four require "leaders to diagnose situations and to act in contextually appropriate ways," the latter one applies in case it is unclear which of these four contexts is predominant. As such, the framework can "help executives sense which context they are in so that they can not only make better decisions but also avoid the problems that arise when their preferred management style causes them to make mistakes" [12]. In other words, this means tailoring one's approach to fit the complexity of the current circumstances.

The framework's contexts can be characterized as follows. Simple contexts come with clear cause-and-effect relationships, based on which the "right" answer is self-evident. As such, they allow for following a sense-categorize-respond approach: leaders assess the facts of the situation at hand, categorize these facts, and then base their actions on established practice. In contrast, complicated contexts are the domain of experts since they contain multiple "right" answers that, although there are clear cause-and-effect relationships, not everyone can see. Here, the approach is to sense, then analyze, and finally respond. Both simple and complicated are ordered contexts.

With the complex context, "disorder" begins. Here, no "right" answer exists; the situation can be characterized by flux and unpredictability. That is why the recommended approach is to allow patterns to emerge and thus probe first, then sense, and finally respond. In the chaotic context, relationships between cause and effect are impossible to determine. So it is advisable to look for what works rather than seeking "right" answers. This means first acting, then sensing, and finally responding.

Against the background of this framework, the identified application scenarios can be characterized as follows. *Strategy making* is made a challenging undertaking by the "constant change in the external environment and evolving organizational structures" [13]. Given these dynamics, it seems unlikely that there are distinct and predictable relationships between cause and effect that allow definitely "right" answers to be made upfront through "simple" categorizations or analyses. Of course, deliberate strategy making requires thorough analyses, but it has been widely acknowledged that strategies sometimes also need to build on patterns that emerge over time.

Aiming at an elaboration and/or realization of a crafted strategy, *business transformation*—as a subtype of enterprise transformation—may be considered a shift within a defined enterprise that is a "response to radical changes in the economic, market, or social environment," a "fundamental alteration of context," and/or a "step change in performance" [14]. As such, it also comes with a number of challenges. Among these are, in addition to the dynamics propagated from the overall strategic level, the variety of stakeholders (with different cultural backgrounds) and the organizational interdependencies that a transformation initiative has to cope with. Even experts might thus not foresee a priori what will work; in consequence, we might consider this a "complex" context as well.

IT transformation programs are likely to have similar properties. In addition to the structural challenges as given by the variety and number of stakeholders, they usually also face a considerable "organizational dynamic complexity" [15]. In particular, this includes the following factors: unclear or ill-defined requirements; degree and frequency of change in scope, goals, budget, or duration during program execution; degree and frequency of requirements change during program execution; unclear, ill-defined, and/or unrealistic scope, goals, target, budget, and/or duration. Stakeholder response may also suddenly change "from support to attack" [16]. Again, given these dynamics and uncertainty, it may not work to just apply general rules but it may be necessary now and then to let patterns emerge and then respond in an adequate way.

Project portfolio planning decides on whether or which projects are to be executed to realize a defined target landscape (i.e., as part of wider transformation efforts) and/or meet certain business demands. Decisions here may be hard to make due to, for example, multiple and potentially conflicting objectives, diverse uncertainties and risks (e.g., technological, legal, cost-related) with respect to the projects in scope, and the interdependencies between projects [17]. Again, it might thus be difficult to foresee a priori what will be the "best" decisions in terms of the projects to be executed; in fact, some of the uncertainties may be reducible only later in the course of a certain project. At least it requires some expert knowledge to come to well-grounded decisions in the selection process.

Finally, a similar line of reasoning applies to *IT solution projects*, for which there is multiple evidence that these are not necessarily ordered contexts. At this micro architectural level, despite a more narrow scope, one is still likely to face a certain volatility of requirements [18] and a considerable number of organizational and technological elements with various interdependencies [19]. Although projects may not involve a stakeholder variety as great as prevalent in (wider) transformation programs [15], it is not uncommon to have objectives that are unclear and/or change over time and stakeholders that come with competing agendas [20]. In essence, these factors can be categorized into structural ("difficulty in managing and keeping track of the huge number of interconnected tasks and activities"), technical ("interconnection between multiple interdependent solution options"), directional ("ambiguity related to multiple potential interpretations of goals and objectives"), and temporal complexity ("uncertainty regarding future constraints, the expectation of change and possibly even concern regarding the future existence") (cf. [21, 22]).

For each of these scenarios, there may of course be differences between particular instances of that scenario in practice (e.g., not every project has the same complexity, but each project has to be assessed individually in terms of its complexity). Although the identified characteristics can thus not be generalized, they can well be considered an indication of how the scenarios may tend to be and will hence be applicable to a large number of cases in practice.

Given the identified dynamics and uncertainties in these scenarios, we argue that an instrument that provides orientation and defines a uniform corridor within which to act is what is likely to be of help here, while something that restricts the freedom to act through definite and detailed rules seems rather counterproductive.

3.3   Implications for the Nature of EA Principles and a Development Process

With application scenarios that are at least of a complicated nature and usually require some sort of analysis, it seems reasonable to conclude that EAPs need to be considered, formulated, and applied as *"guide rails"* rather than imposed as detailed "laws." This essential property, in turn, implies specific requirements in terms of how to manage EAPs, or make selected qualities of EAPs as present in the literature (e.g., [11]) especially crucial:

1. *Goal orientation (R1):* Given that EAPs as "guide rails" leave some room for interpretation and/or detailed design, it seems more essential than in a "rule-based" decision-making context to properly ground the identification of EAPs in motivational elements such as strategic goals, objectives, or also values [3] to make all relevant stakeholder understand the underlying rationale or, in other words, the contribution to the achievement of defined ends.
2. *Meaningful (but still generic) description (R2):* In addition to the motivation for each principle, it is crucial to develop a thorough description of what the principle actually means (beyond its title/name) and of how it is to apply [11]. It seems highly advisable to develop such a description in close coordination with relevant stakeholders from both the strategic and operational level to avoid any ambiguities and facilitate the actual use of that "guidance" in practice. However, this elaboration should not foreclose decisions for any potential situation at hand and thus embrace a too much detailed level, which, given the character of the identified application scenarios, may actually be a rather pointless endeavor.
3. *Proper communication (R3):* As opposed to rules (or "laws"), specified at a higher level of detail and leaving less options for action, "guide rails" are likely to pose higher requirements in terms of communication. It therefore seems essential to thoroughly communicate and thus explain defined EAPs to all relevant stakeholders.
4. *Process anchoring (R4):* Although EAPs are not to be used as laws, their application should well be obligatory. This may be facilitated by integration into methods and/or process models for the relevant application scenarios (if existing), e.g., by making principles a dedicated part in project quality gates.
5. *Regular measurement/controlling (R5):* Unlike rules or laws, principles may not simply be assessed in terms of whether they are followed are not (that is, in a "black-or-white" manner); however, it may well be assessed if one stays within the "guide rails." Further, based on their immediate link to strategic ends (R1), principles should be measured on a regular basis in terms of their overall fulfillment and their effect on the achievement of strategic goals.

4   Process for Development of EA Principles

When presenting our proposal for developing EAPs, we will refer to the requirements identified in Sect. 3.3 by indicating for each process part what requirement is addressed or supported (e.g., "(R1)"; see Fig. 1). The process has primarily been derived from the PDCA cycle and can roughly be divided into three phases: *planning, implementation,* and *control*. The process itself is thus used for the *identification/formulation, implementation, monitoring,* and *adjustment* of EAPs. The result of the process—the EAPs—primarily serve as basis for consistent [23] decision making (see Sect. 3).

The building block design of the process structure is in line with, e.g., [24] and starts with the preparation phase, which initially gathers the drivers and parameters for EAP development. This is followed by an elaboration of the drivers and problems underlying the EAP development. The subsequent process block then deals with the generation of "solutions" in terms of EAPs. These principles are finally communicated, anchored, and regularly monitored during application. After each process block, results should be approved by a decision-making body (e.g., an architecture board). Among others, this is to identify possible deficiencies or inconsistencies that may need to be resolved before proceeding to the next steps. In the following, the individual sub-processes are explained in more detail.

(1) The starting point of the process is the *preparation* phase. First, the drivers—such as goals, trends, problems, or challenges the organization faces (e.g., increase agility in processes, reduce complexity)—have to be identified (R1). Without clear drivers, principles may not have any benefit. Based on that initial grounding, required resources are planned for each process block (and its individual sub-processes) and a schedule is set up. That is, not only resources for the initial development, but also those for the implementation and governance (esp. R3, R5) should be planned. This goes hand in hand with the definition of relevant terms (e.g., principle, instruction/rule) and the envisioned application scenarios, since a common understanding is the prerequisite for the successful development and adequate use of EAPs (cf. R2, R3).

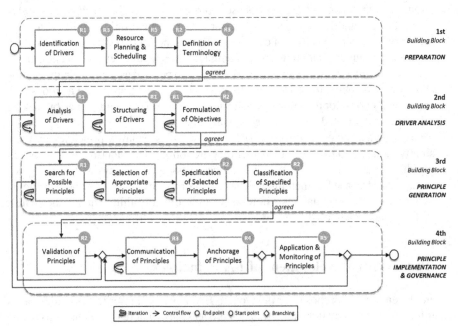

Fig. 1. EAP development process.

(2) The phase of *driver analysis* comprises a detailed elaboration of the drivers identified in the previous phase to facilitate an effective search for solutions (i.e., EAPs) in the next phase. The starting points here may initially be rather abstract drivers (e.g., values, visions, challenges) that are incrementally converted into specific objectives. This is done by studying the drivers in terms of, e.g., their individual elements/facets, their environment and context (e.g., financial, technical), their causes (in case of drivers in the form of problems), and any inter-relationships (R1). The latter leads over to the structuring and systematization of the identified drivers (R1). The final step of this phase then is the formulation of concrete objectives to be addressed by EAPs (R1, R2).

(3) Once driver analysis has been completed and approved, the actual *generation of principles* occurs. In general, problems/challenges, goals, and objectives are transformed into principles in this process block. It thus starts with the identification of possible principles on the basis of the identified drivers (R1) and with all relevant stakeholder groups, for example in workshops or with brainstorming techniques. Next, appropriate principles are selected out of the generated list based on predefined criteria (e.g., cost-benefit analysis). As part of this, the generated ideas are initially checked in terms of whether they are actually principles (or rather actions or requirements). Subsequently, the selected principles are specified in further detail and provided with a proper definition/description, rationale, and an outline of implications (R2) [11]. While the rationale part emphasizes the benefits of adhering to the principle, the implications provide an overview of the key tasks, resources, and potential costs required for the application of the principle. To account for possible conflicts during later EAP application (e.g., "access to all relevant data" vs. "data security"), the specified EAPs may also be prioritized here. Finally, the EAPs are classified according to, e.g., their architectural levels (R2) to facilitate their accessibility and maintenance.

(4) The final part of the process is the *implementation and governance* of the defined EAPs. This process block is substantial for the entire process, because validation (R2), communication (R3), anchorage (R4), application, and regular monitoring (R5) of the principles takes place in here. At the beginning of this process block, the previously classified EAPs are validated. This validation is performed with the help of selected stakeholders (R2) and using quality criteria such as understandability and consistency [11]. If certain quality requirements are not met, it may be necessary to return to the process block of principle generation. Once the principles have been successfully validated, they should be communicated in the next step, e.g., by means of texts or also graphics/models (e.g., using the ArchiMate language [24]) (R3). This is because principles can only be effectively applied if they have also been properly explained to the intended users. Subsequently, principles are formally anchored in the organization. For this purpose, the EAPs and their application are integrated into established methods and processes (e.g., project phase models, project prioritization processes) (R4) and, in addition, measures/key figures may be defined. This leads over to the final sub-process of actual principle application and concurrent monitoring. The latter is to control whether the principles are properly applied and whether the principles themselves still meet the business needs and continually contribute to the achievement of objectives (or solving of problems) (R5). If analyses indicate that drivers may have changed or that principles are not effective, it may be necessary to return to previous phases, e.g., to driver analysis.

5   Evaluation of the Process for EAP Development

The process for EAP development presented in Sect. 4 was evaluated from two perspectives. One perspective was to make sure that all requirements identified in Sect. 3.3 were met. This perspective is already covered as part of Sect. 4. The second perspective is to check suitability for the problem at hand and pertinence for practical use. This perspective is covered in this section. More specifically, we used expert interviews to validate the EAP process. As the development of the process was primarily based on academic literature, we selected experts from industrial practice to balance theory and practice. The interviews were guided by two questions:

1. From the experts' perspective, does the process cover all aspects of developing/ managing EAPs in a way suitable for industrial practice?
2. What shortcomings and improvement potential do the experts see?

From a methodical perspective, the aim of the interviews was to collect qualitative data which would be analyzed with Mayring's approach to qualitative content analysis (QCA) [26]. The use of QCA for evaluation purposes is unconventional, as qualitative research designs are usually used for theory building prior to designing artifacts. However, in our case the intention was to gain additional knowledge about potential domain-specific refinement needs, while at the same time evaluating the process. Two interviews were conducted in the time frame of one week: one by phone, the other one face to face. The interviews were recorded and transcribed.

Mayring's approach includes 6 steps. Step 1 is to decide what material to analyze, which obviously consists of the transcribed recordings of the two expert interviews. Both experts are practitioners in the EAM area: one from a larger tool vendor, the other from a medium-sized solution provider with focus on the utility sector. Step 2 is to make explicit how the data collection was arranged and prepared. The experts were selected from existing contacts of the researchers involved. The interviewees received information about the purpose of the interview, important terms (e.g., EAP, policy, rule), the proposed development process as such, and the interview questions. Step 3 is to make explicit how the transcription of the material has to be done. The material was analyzed step by step following rules of procedure devising the material into content analytical units. Step 4 concerns the subject-reference of the analysis, i.e., it has to be made sure that the analysis is connected to the concrete subject of the investigation. Subject-reference was implemented by (a) defining the research questions and their sub-questions in the interview guidelines and (b) using the subjects of these sub-questions as categories during the analysis.

Step 5 recommends theory-guided analysis of the data, which is supposed to balance fuzziness of qualitative analysis with theoretical stringency. We took the state-of-the-art into account during both formulation of the sub-questions and analysis of the material. Step 6 defines the analysis technique, which in our case was content summary. This attempts to reduce the material and create a manageable corpus which still reflects the original material. For this purpose, the text was paraphrased, afterwards generalized, and then abstracted and reduced.

Table 1. Result of the 2nd reduction step (excerpt)

#	Category	Statement
C 2	Purpose of principles	Support of goal achievement; Support of decision making; Support of planning; Support of problem solving; Serve as guidance means
C 4	Develop-ment of principles	Workshops; Discussion and evaluation of principles with important decision makers; Include department heads; Verify management support; Derive from best practices; Derive from manifestos; Brainstorming;
		Variant 1: Development and validation of principles separated in two different processes with different participants; *Variant 2:* Development and validation of principles integrated into the same processes with the same participants; Combination of workshop and creativity techniques
C 11	Quality require-ments	Limited number; Short name with clear meaning; Precise wording; No contradiction between different principles; Use of the same syntax for all principles; No formal process for quality assurance; General applicability; Guiding nature; Persistence; Relevance; Possibility to check whether it is used; Adaptability; Possibility of interpretation how to implement a principle

For brevity reasons, Table 1 presents only a small part of the result of step 6. Out of 15 categories represented in the interview questions, we selected "purpose of EA principles," "development process," and "quality criteria" for presentation in this paper. The purposes identified in the interviews support our argumentation in Sect. 3 that principles rather are guidelines than rules or laws. The qualitative content analysis showed that the purpose of EAPs is support of goal achievement, decision making, planning, and many more activities. The category "development of principles" reflects the blocks contained in our process proposal, i.e., all statements of the experts can be assigned to process blocks and all process blocks are represented in the statements. Furthermore, the interviewees also confirmed the suitability of the different process steps for development of EAPs. This is not explicitly reflected in the (paraphrased and reduced) statements, since the general and introductory part of the interview was not subject to content analysis. On the other hand, the statements show variations of the process that we did not anticipate. For example, one expert stated that EAP validation could happen in different ways, including evaluation by all potential future users. This would mean that validation and communication could potentially happen in parallel. Finally, the quality criteria found in the analysis are manifold and also point at some improvement potential to be addressed in future work. In fact, statements such as "possibility to check whether it is used" call for more concrete activities and guidelines to be included (esp. in the later process steps).

Early identification of threats to validity and actions to mitigate the threats can minimize the effect on the findings. Common threats to empirical studies are discussed in, e.g., [27]. The threats to validity can be divided into four categories: construct validity, internal validity, external validity, and conclusion validity. With respect to construct validity, the results are highly dependent on the people being interviewed. Only persons experienced in EAM and the use of principles in this field will be able to judge appropriateness of the proposed development process. To obtain a high quality of the sample, only experts having worked in this area for a long time and hence having the required background were selected. Another common risk in studies is that the presence of a researcher influences the outcome. Since the selected experts in the study and the research group performing the study have been collaborating for a while, this is not perceived as a significant risk. Furthermore, there is a risk that the questions of the interviewer may be misunderstood or the data may be misinterpreted. In order to minimize this risk, the interview guidelines were double-checked by another researcher and the interviews were recorded, which allowed the researcher to listen to the interview again if portions seemed unclear.

Regarding the internal validity, a common risk is that the data collected in the interviews did not completely capture the view of the experts regarding the development process. However, this threat was reduced by breaking down the research question in sub-questions covering the different aspects of the process. Thus, this threat to validity is considered being under control. A potential threat of the study regarding external validity is that the actual interviews have been conducted with only two experts. It will be part of the future work to conduct a study with more participants. With respect to conclusion validity, interpretation of data is most critical. To minimize this threat, the study design includes capturing the relevant aspects by different interview questions, i.e., to include triangulation. Another risk could be that the interpretation of the data depends on the researcher and is not traceable. To reduce the risk, data analysis was performed according to Mayring's approach.

In summary, actions have been taken to mitigate the risks identified, which from our perspective results in an appropriate confidence level regarding construct and internal validity. Future work (i.e., an extension of the study) will contribute to increasing the confidence level regarding external and also conclusion validity.

6   Summary and Future Work

Based on findings from a literature analysis and grounded in requirements derived from the Cynefin framework, this paper proposes a development process for EAPs. The main purpose of this development process is to provide a recommendation for systematic principle development in industrial practice of EAM. Furthermore, the approach is supposed to contribute to research in the field of EAM by addressing a so far less researched question. Although a first validation of the EAP development process was performed, the process so far more has to be considered as an initial version ready for application in industrial practice than as a highly mature process specification. Future work will have to investigate further details of the process (including involved roles and deliverables for each step) and identify improvement potentials. Evaluation of the process so far is based on two expert interviews only which confirmed the suitability of the process for the defined purpose and its applicability. For the future, we envision the application of the development process in industrial practice and a systematic evaluation of the findings (e.g., in the context of case studies). Finally, from a conceptual perspective, we consider a further specialization of the process for different drivers of EAP development, as different drivers on a more fine granular process level may require slightly different activities.

References

1. Niemi, E.: Enterprise architecture benefits: perceptions from literature and practice. In: Proceedings of the 7th IBIMA Conference on Internet & Information Systems in the Digital Age, pp. 14–16, Brescia, Italy (2006)
2. Simon, D., Fischbach, K., Schoder, D.: An exploration of enterprise architecture research. Commun. Assoc. Inf. Syst. **32**, 1–72 (2013)
3. Greefhorst, D., Proper, E.: Architecture Principles: The Cornerstones of Enterprise Architecture. Springer, Berlin (2011)
4. Aier, S., Fischer, C., Winter, R.: Construction and evaluation of a meta-model for enterprise architecture design principles. In: Proceedings of the 10th International Conference on Wirtschaftsinformatik WI 2011, pp. 637–644 (2011)
5. Stelzer, D.: Enterprise architecture principles: literature review and research directions. In: Dan, A., Gittler, F., Toumani, F. (eds.) ICSOC/ServiceWave 2009. LNCS, vol. 6275, pp. 12–21. Springer, Heidelberg (2010)
6. O'pt Land, M., Proper, E., Waage, M., Cloo, J., Steghuis, C.: Enterprise Architecture: Creating Value by Informed Governance. Springer, Berlin (2009)
7. Aier, S.: The role of organizational culture for grounding, management, guidance and effectiveness of enterprise architecture principles. IseB **12**(1), 43–70 (2014)
8. Schekkerman, J.: Enterprise Architecture Good Practices Guide: How to Manage the Enterprise Architecture Practice. Trafford Publishing, Victoria (2008)
9. van Bommel, P., Buitenhuis, P., Hoppenbrouwers, S., Proper, E.: Architecture principles –a regulative perspective on enterprise architecture. In: Proceedings of the 2nd International EMISA Workshop, pp. 47–60 (2007)
10. Fischer, C., Winter, R., Aier, S.: What is an enterprise architecture principle? In: Lee, R. (ed.) Computer and Information Science 2010. SCI, vol. 317, pp. 193–205. Springer, Heidelberg (2010)
11. The Open Group: TOGAF Version 9.1. Van Haren Publishing, The Netherlands (2011)
12. Snowden, D.J., Boone, M.E.: A leader's framework for decision making. Harvard Bus. Rev. **85**(1), 68–76 (2007)
13. Hammer, R.J., Edwards, J.S., Tapinos, E.: Examining the strategy development process through the lens of complex adaptive systems theory. J. Oper. Res. Soc. **63**(7), 909–919 (2012)
14. Purchase, V., Parry, G., Valerdi, R., Nightingale, D., Mills, J.: enterprise transformation: why are we interested, what is it and what are the challenges? J. Enterp. Transform. **1**(1), 14–33 (2011)
15. Piccinini, E., Gregory, R., Muntermann, J.: Complexity in IS programs: a Delphi study. In: Proceedings of the Twenty Second European Conference on Information Systems (2014)

16. Schefe, N., Timbrell, G.: A conceptualization of complexity in IS-driven organizational transformations. In: Proceedings of the 34th International Conference on Information Systems (2013)
17. Ghasemzadeh, F., Archer, N.P.: Project Portfolio selection through decision support. Decis. Support Syst. **29**, 73–88 (2000)
18. Jurison, J.: Software project management: the manager's view. Commun. Assoc. Inf. Syst. 2 (1999)
19. Baccarini, D.: The concept of project complexity: a review. Int. J. Project Manage. **14**(4), 201–204 (1996)
20. Ameen, M., Jacob, M.: Complexity in projects: a study of practitioners' understanding of complexity in relation to existing theoretical models. Umea University, Umea School of Business and Economics, Student thesis (2009)
21. Remington, K., Pollack, J.: Tools for Complex Projects. Gower Publishing, Aldershot (2007)
22. Xia, W., Lee, G.: Complexity of information systems development projects. J. Manage. Inf. Syst. **22**(1), 45–83 (2005)
23. Lindström, A.: On the syntax and semantics of architectural principles. In: Proceedings of the 39th Annual Hawaii International Conference (HICSS) (2006)
24. Wißotzki, M.: The capability management process: finding your way into capability engineering. In: Simon, D., Schmidt, C. (eds.) Business Architecture Management – Architecting the Business for Consistency and Alignment, pp. 77–105. Springer, Heidelberg (2015)
25. The Open Group: ArchiMate 2.1. Van Haren Publishing, The Netherlands (2013)
26. Mayring, P.: Qualitative content analysis. Forum Qual. Soc. Res. **1**(2), 38–48 (2000). Art. 20, http://nbn-resolving.de/urn:nbn:de:0114-fqs0002204
27. Wohlin, C., Runeson, P., Host, M., Ohlsson, C., Regnell, B., Wesslén, A.: Experimentation in Software Engineering: an Introduction. Kluver Academic Publishers, MA (2000)

Interrelations Between Enterprise Modeling Focal Areas and Business and IT Alignment Domains

Julia Kaidalova[1(✉)], Elżbieta Lewańska[2], Ulf Seigerroth[1], and Nikolay Shilov[3,4]

[1] School of Engineering, Jönköping University,
P.O. Box 1026 55111 Jönköping, Sweden
{julia.kaidalova,ulf.seigerroth}@jth.hj.se
[2] Department of Information Systems, Poznań University of Economic,
Al. Niepodleglosci 10, 61-875 Poznań, Poland
e.lewanska@kie.ue.poznan.pl
[3] St. Petersburg Institute for Informatics and Automation of the Russian Academy of Sciences,
39, 14 Line, 199178 St. Petersburg, Russia
nick@iias.spb.su
[4] ITMO University, 49, Kronverkskiy pr., 197101 St. Petersburg, Russia

Abstract. Efficient support of business needs, processes and strategies by information technology is a key for successful enterprise functioning. The challenge of Business and IT Alignment (BITA) has been acknowledged and actively discussed by academics and practitioners during more than two decades. On one hand, in order to achieve BITA it is required to analyse an enterprise from multiple perspectives. On the other hand, it is also required to deal with multiple points of views of involved stakeholders and create a shared understanding between them. In relation to both of these needs EM is considered as a useful practice, as it allows representation of various focal areas of an enterprise and facilitates consensus-driven discussion between involved stakeholders. Various focal areas of an enterprise are linked to various domains that BITA include, for example, time and function focal areas from the Zachman Framework contribute to analysis of business domain of BITA. This and other interrelations are investigated in this study, which is done by conceptual positioning of Zachman's focal areas on Information System strategy triangle and Generic Framework for Information Management.

Keywords: Business and IT alignment · Enterprise modeling · Enterprise modeling focal area · BITA domain

1 Introduction

IT is a key facilitator for a successful functioning of the most of today's enterprises. Through IT we are able to change the way companies organize their business processes, communicate with their customers and deliver their services (Silvius 2009). IT is often considered as a necessary mean for realization of business strategies and processes. On the other hand, IT also creates new possibilities including new strategic directions and new business models for enterprises. The highly dynamic nature of requirements from

© Springer International Publishing Switzerland 2015
W. Abramowicz (Ed.): BIS 2015, LNBIP 208, pp. 273–284, 2015.
DOI: 10.1007/978-3-319-19027-3_22

the business side raises the need for continuous reconsideration of supporting IT. The quest of finding efficient IT support that satisfies business needs has been addressed in the literature as Business and IT Alignment (BITA) (Luftman 2003; Chan and Reich 2007). In early studies BITA implied alignment of the business plan and the IT plan, or alternatively the business strategy and the IT strategy. Later consideration of BITA implicated the fit between business needs and information system priorities. These meanings expanded over time and currently research recognizes many facets of alignment between business and IT. In general it is possible to differentiate between four dimensions of BITA: (1) strategic/intellectual, (2) structural, (3) social, and (4) cultural (Chan and Reich 2007). Various frameworks have been presented that conceptualize BITA emphasizing different dimensions of alignment. For example, Strategic Alignment Model (Henderson and Venkatraman 1993) focuses on strategic dimension of business and IT. It also implies structural alignment dimension to some extent. BITA in itself also implies various domains. Existing BITA frameworks define various sets of domains to consider. For example, Strategic Alignment Model includes four related key domains of strategic choice, namely business strategy, organisational infrastructure and processes, IT strategy, and IT infrastructure and processes (ibid).

The concept of BITA often implies enterprise transformation, i.e. an actions for taking an enterprise from one state to an improved state (Seigerroth 2011). In order to achieve BITA many enterprises need to transform on the fly and perform changes in their operations reactively, while others have the possibility to be more proactive in planning, design and implementation of changes. Regardless of the type of change (reactive or proactive), the importance of two issues becomes apparent: (1) planning and agreeing on a future state of the enterprise (TO-BE state), including vision and strategy, and (2) analysis of the current praxis of the enterprise (AS-IS state) and making sure that stakeholders have a common understanding about it. Dealing with both of these issues in a successful way call for representation of various enterprise focal areas (perspectives), since clear representation of various perspectives of an enterprise enable their careful analysis and alignment. In this context Enterprise Modeling (EM) is an acknowledged and widely used practice that enables alignment of different perspectives and thus facilitates BITA (Seigerroth 2011). EM facilitates the creation of a number of integrated models, which capture and represent different focal areas of an enterprise, for example, business processes, business rules, concepts, information, data, vision, goals, actors etc. (Stirna and Persson 2009). Enterprise models contain representations of various focal areas of an enterprise; therefore they allow to slice an enterprise in a multidimensional way and to integrate these multiple dimensions into a coherent structure. Therefore EM has established a strong position in relation to BITA, which is acknowledged by several scholars (e.g., Seigerroth 2011; Chan and Reich 2007; Wegmann et al. 2005; Christiner et al. 2012; Gregor et al. 2007).

There are many different EM frameworks available today, each contain a sets of focal areas for viewing an enterprise in a comprehensive way. For example, The Zachman Framework defines six focal areas that in combination can represent an enterprise: data, function, network, people, time, and motivation (Zachman 1987; Lillehagen and Krogstie 2008). The CIMOSA framework defines the following four focal areas: function view for describing the expected behavior and functionality of the enterprise;

an information view for describing the integrated information objects of the enterprise; a resource view for describing the resource objects of the enterprise; an organisation view for describing the structure of the enterprise (Zelm 1995; Lillehagen and Krogstie 2008). There is though quite little research done regarding how various EM focal areas are related to BITA, and particularly how different BITA domains can be covered by modelling of various focal areas (Seigerroth 2011). Several studies investigate this contribution with a focus on EA, where enterprise models are regarded as instruments to handle EA (c.f., e.g., Plazaola et al. 2007; Aier and Winter 2009; Saat et al. 2010; Engelsman et al. 2011). Considering the same problem but with focus on EM would be beneficial, as it could contribute to the support of EM practitioners who apply EM to achieve BITA, which currently receives a scant attention in existing EM guidelines. What is more, existing studies have not yet proposed a conceptual description regarding the role of EM in relation to BITA (Kaidalova et al. 2014). Such conceptual description could provide a ground for more structured and purposeful usage of EM in order to achieve BITA, which will allow going beyond general recognition of EM potential of EM for BITA. Therefore, positioning different EM focal areas in relation to BITA domains would contribute to progress of BITA in that regard. Thus, the research question of this work is the following:

What EM focal areas are required to deal with BITA domains?

The purpose of this paper is therefore to investigate what EM focal areas that require attention in order to cover the facets/focal areas that BITA comprises. The research process was organized in the following way. First, a number of enterprise modeling frameworks as well as BITA frameworks were identified and analysed. Based on this analysis, the authors selected two BITA frameworks and seven enterprise modelling frameworks for further investigation that is presented in Sect. 2. To start with, the Zachman Framework was selected as a reference point to present EM focal areas. The selection criteria are described in Sect. 2 below. EM focal areas defined in The Zachman Framework were then mapped to BITA domains used in both selected frameworks. This mapping has been made independently by two experts and later compared and discussed until the mutual agreement been reached.

The rest of the paper is structured in the following way: Sect. 2 describes the theoretical foundation that served as a basis for this study. Results of the study are presented and discussed in Sect. 3. Finally, conclusion is presented in Sect. 4.

2   Theoretical Foundation

In this section the theoretical foundation for the study is presented. The idea of this study is to conceptually position EM in the context of BITA. Particularly, to investigate how BITA domains are covered and assisted by EM focal areas. However, as it has been mentioned above in Sect. 1, a set of BITA domains and a set of EM focal areas might vary between different frameworks. Therefore, it is important to describe chosen frameworks that will be used in the study and, what is more, to define what is meant by EM focal area and BITA domain to set the terminology for the paper. The meaning of EM focal area is closely connected to the concept of method component as part of a methods perspective, which has been introduced by Goldkuhl et al. (1998). A focal area is a viewpoint from which an enterprise can be investigated and modeled (processes, goals, problems, information demand etc.). In this paper the meaning of BITA domain is similar to how it is used by Chan and Reich (2007) where they differentiate between a BITA dimension and a BITA domain. A BITA domain is a bounded area that an enterprise structure contains and that together with other domains show the constitution of business and IT architecture.

In Sub-sect. 2.1 below relevant BITA frameworks are introduced. After this several frameworks that propose a set of EM focal areas are described in Sub-sect. 2.2.

2.1   Business and IT Alignment Frameworks

Maes et al. (2000) have proposed a unified framework for business and IT alignment. It is built on fundaments of two complementary frameworks: generic framework for information management, that is an extension to the model proposed by Venkatraman (1993), and Integrated Architecture Framework (Maes et al. 2000).

Generic Framework for Information Management was designed as a tool for management support. The framework consists of three areas and three levels, as depicted on Fig. 1 below. In the framework technological aspects are divided into two parts: (1) Information and communication, i.e., software components for interpreting information, communication and supporting knowledge processes, and (2) Technology, i.e., infrastructure: hardware and middleware. Third area represents business expertise. In the rows of the framework, different levels of analysis (representing three areas of management), are depicted: (1) strategy is about making decisions about scope, capabilities and governance; (2) structure relates to architecture of the organisation; (3) operations apply to processes and skills. Authors argue that middle column (Information and communication) and middle row (Structure) are considered as dependent on the adjacent columns and rows while they are of high importance for achieving BITA.

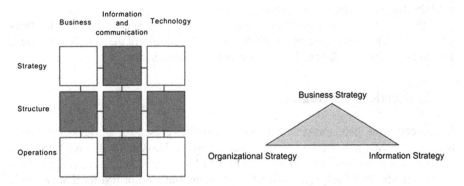

Fig. 1. On the left: Generic framework for information management. Source: (Maes et al. 2000). On the right: Information Systems Strategy Triangle. Source: (Pearlson and Saunders 2010)

The framework covers the most important areas in the enterprise but gives only a static view as it was designed as a management tool. Architecture of business and IT in the enterprise on different stages is not included in it. Thus, Maes et al. (2000), complement the Generic Framework with their **Integrated Architecture Framework (IAF)**. IAF is aimed "to support the integrated architectural design of business and IT" (Maes et al. 2000) and consist of three dimensions: design phases, architecture areas and special viewpoints, which serve as a base for model customization according to different aspects of on organisation. As a result, a **Unified framework for business – IT alignment** was created. It covers all areas important for BITA, on different modelling stages.

Although the Unified Framework for BITA is more complex and covers both management and architectural related areas, in the analysis conducted in this paper, only the generic framework for information management was included. An additional dimension of architecture phases does not provide any additional information for analysis.

Another approach is adopted by Pearlson and Saunders (2010) in the framework **Information Systems Strategy Triangle** (see Fig. 1 above). The framework focuses on relationship between information, business and organisational strategy, and also how Information System (IS) strategy can influence other strategies in a company (sometimes can be schematically positioned inside the strategic triangle). As all decisions regarding organisation structure, personnel hiring, infrastructure, etc. are made on business level, business strategy drives both organisational and information strategies. However, changes in IS strategy must be in line with the appropriate changes in business and organisational strategies. Business strategy is an upper strategy that shapes information strategy and organisational strategies. Authors describe business strategy as a plan expressing where a business aims to go and how it expects to get there. Organisational strategy defines how business is organized, it is organisation's design as well as all method and means used for coordination and control. IS Strategy must be complementary to the business strategy and should support business goals. However, IS strategy can trigger changes in organisation (in both business and organisational strategy). IS strategy is described as the plan the organisation uses in providing information systems and services.

2.2 Enterprise Modeling Frameworks and Included Focal Areas

A number of Enterprise Modeling Frameworks have been developed. It is worth noticing that in the literature the same framework is often addressed as both EM framework and Enterprise Architecture (EA) framework. This is mainly due to the fact that EA frameworks are used to model and plan various states of the enterprise (Abraham and Aier 2012). Therefore available EM and EA frameworks can be used as parts of a coupled domain of Enterprise Architecture Modeling (Engelsman et al. 2011). In this sub-section several frameworks are described in order to present diversity in approaches to Enterprise Modelling.

The Zachman Framework for Enterprise Architecture (Zachman 1987; Zachman International 2014) was developed in the 1980s at IBM by John Zachman, and is widely used ever since.

The Zachman Framework uses two dimensions for classifying elements of an enterprise: focal areas and levels of abstraction (or perspectives) (see Fig. 2 above). Focal areas are defined according to six basic questions: (1) data (what?) – data needed for the enterprise to operate, (2) function (how?) – concerned with the operation of the enterprise, (3) network (where?) - concerned with the geographical distribution of the enterprise's activities, (4) people (who?) - concerned with the people who do the work, allocation of work and the people-to-people relationships, (5) time (when?) – to design the event-to-event relationships that establish the performance criteria, (6) motivation (why?) – the descriptive representations that depict the motivation of the enterprise, which typically focuses on the objectives and goals. Specified perspectives are: Contextual (Scope), Conceptual (Business Model), Logical (System Model), Physical (Technology Model), Detailed Representation, and Functioning Enterprise. The goal of the framework was to present main areas in the enterprise on different levels of abstraction, according to the specific purpose. This means, that for each column a basic model is defined and these models are further detailed by the perspectives defined in the rows. The Zachman Framework has been developed as a basic tool, understandable for both technical and non-technical experts. It is comprehensive and claims to cover all areas of an enterprise. It was designed as a tool for planning and problem solving, and it is independent of any specific tool or methodology.

	Why	How	What	Who	Where	When
Scope	Goal List	Process List	Material List	Organizational Unit and Role List	Geographical Locations List	Event List
Business Model	Goal Relation-ship	Process Model	Entity Relation-ship Model	Organizational Uni and Role Relation-ship Model	Locations Model	Event Model
System Model	Rules Diagram	Process Diagram	Data Model Diagram	Role Relation-ship Diagram	Locations Diagram	Even Diagram
Technology Model	Rules Specifi-cation	Process Function Specifi-cation	Data Entity Specifi-cation	Role Specifica-tion	Location Specifica-tion	Event Specfica-tion
Detailed Representation	Rules Details	Process Details	Data Details	Role Details	Location Details	Event Details
Functioning Enterprise	Working Strategy	Working Function	Usable Data	Functioning Organiza-tion	Usable Network	

Fig. 2. The Zachman Framework. Based on (Zachman International 2014)

GERAM (Generalized Enterprise Reference Architecture and Methodology) (IFIP-IFAC 1999) consists of several interrelated components, each focused on different area of EM. The one focused on enterprise architecture and modelling is **GERA (Generic Enterprise Reference Architecture)**, which goal is to identify concepts of enterprise integration (Bernus and Nemes 1994; IFIP-IFAC 1999). GERA classifies concepts into three groups: (1) Human oriented concepts - capabilities, skills, know-how, competencies, roles; these aspects are related to organisation itself (e.g., decision level, responsibilities, authorities, etc.), communication (between humans as well as humans and IT systems); (2) Process oriented concepts – operations and life-cycle phases; (3) Technology oriented concepts – infrastructure used to support business processes; includes models of resources, IT systems, communication systems, etc. In GERA, four focal areas have been identified: Resource, Organisation, Information, and Function.

ARIS (Architecture of Integrated Information Systems) uses a set of different views to model all organisational aspects (Scheer 2000). Models for different views can be described using different languages (e.g., EPC diagrams, UML) and are integrated through the control view model (so called ARIS house) (Scheer and Nüttgens 2000). These views can be interpreted as EM focal areas.

The function view includes all activities performed by organisation, its inputs and outputs, goals. Also, IT systems are included in the function view due to the fact that they support activities performed in the enterprise. Organisation view focuses on organisation structure. It includes human actors as well as IT systems – it groups actors according to the responsibility to process specific work object. Data view covers various aspects of data processing, exchanged messages. Output view represents all physical and nonphysical input and output. Central part of the ARIS house, that integrates all previously mentioned views is Control view/process view.

CIMOSA (Computer Integrated Manufacturing Open System Architecture) was designed as a tool to analyse the evolving requirements of an enterprise and model functions and systems that match those requirements (Lillehagen and Krogstie 2008). The framework consists of three dimensions: (1) dimension of genericity; (2) dimension of model and (3) dimension of view. Dimension of view consists of four focal areas: function view (expected behavior and functionality of the enterprise), information view (integrated information objects of the enterprise), resource view (resource objects of the enterprise), organisation (structure of the enterprise).

Urbaczewski and Mrdalj (2006) propose a comparison of a number of the frameworks (The Zachman Framework, DoDAF, FEAF, TEAF and TOGAF). The comparison is based on the Zachman's views. They argue that Zachman's set of views is the most comprehensive and therefore allows to represent stakeholder perspectives in full. According to Urbaczewski and Mrdalj (2006) comparison, The Zachman framework is the most comprehensive one and covers most focal areas (e.g., it is the only one that directly covers time and motivation aspects). Without doubt, Enterprise Modelling frameworks vary significantly. Each framework has slightly different scope and goal, uses different set of views and focal areas. After our analysis of EM frameworks and the performed comparison of another set of frameworks, we are leaning towards the same observation – Zachman framework provides a comprehensive set of enterprise focal areas. Moreover, The Zachman Framework is well defined in the literature, while some aspect of the other frameworks are described in a very general way and might be subject for different interpretations. Thus, The Zachman Framework was chosen for further analysis.

3　Results and Discussion

This section presents how EM focal areas (Zachman 1987) can be positioned within the domains of the chosen BITA frameworks: first the Generic framework for information management in Sub-sect. 3.1 and then IS Strategy triangle in Sub-sect. 3.2. The positioning is then discussed in Sub-sect. 3.3.

3.1　Analysis of the Generic Framework for Information Management

The positioning of Zachman's six focal areas within the domains of the Generic Framework for Information Management is presented below in Fig. 3 below.

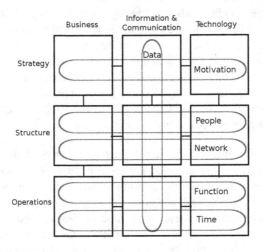

Fig. 3.　The positioning of EM focal areas (Zachman 1987) within the BITA domains of Generic Framework for Information Management (Maes et al. 2000)

EM within data focal area provides a support for dealing with Information & Communication domain of BITA, since it provides various kinds of information that are fundamental for enterprise functioning. It does not have direct connection to Technology domain, which has to do with infrastructure of the enterprise in terms of hardware and middleware. EM of the areas function and time are able to facilitate dealing with the operations domain, as together these two focal areas are able to describe business processes of the enterprise and the way it operates. EM of people and network provides a strong support for the structure BITA domain, as it allows describing the hierarchy and disposition of business units and employees within it. EM of motivation-related issues contributes to the clear picture regarding an enterprise strategy, as it gives an understanding regarding visions and goals of an enterprise.

3.2   Analysis of the IS Strategy Triangle

The positioning of Zachman's six focal areas within the domains of the IS strategy triangle is presented below in Fig. 4. An important point here is that the domain of the IS strategy triangle are considered to imply not only strategies, but also operational issues to a certain extent.

Fig. 4. The positioning of EM focal areas (Zachman 1987) within the BITA domains of IS Strategy triangle (Pearlson and Saunders 2010)

EM within the focal area of data can facilitate dealing with the domain of Information Strategy, as it enables analysis of various information needed to make an enterprise operational. EM of relations within networks and between people is able to contribute to analysis of organisation strategy, as it gives a clear picture of how responsibilities are distributed in an enterprise between employees and units, the hierarchy of units that form an organisational structure and the disposition of this structure. EM within the areas of time and function provides a clear picture of business processes within an enterprise, and thus plays an important role in dealing with the domain of business strategy. EM within the area of motivation is able to contribute to the domain of business strategy, as it represents vision and goals of an enterprise that have a decisive role in business strategy.

3.3   Discussion

According to the presented positioning above it is possible to see that the EM framework introduced by Zachman (1987) is able to cover all domains of analysed BITA frameworks. It expectedly confirms the status of this framework of being able to provide a comprehensive view of an enterprise and to serve as an enterprise ontology. However, it is possible to see that some domains of BITA frameworks are covered by more than one focal area, whereas others domains receive less support. For example, the technology dimension in the Generic Framework for Information Management is covered by several focal areas in an intersecting manner. There is no focal area that could deal with technology and business domains explicitly. The Zachman Framework covers business domain only partially, as 5 out of 6 focal areas are related to it. However, this might cause problems and misunderstanding due to the fact that none of those areas is explicitly designated to describe business-related issues. The same problem applies to the technology domain. However, from EM point of view, it is more important to handle and define how different systems in organisation works, which is covered by information & communication domain, than separately analyse organisation's infrastructure.

As it has been mentioned earlier in Sect. 1, apart from BITA domains, it is possible to differentiate between four dimensions of BITA: strategic/intellectual, structural, social and cultural. Each BITA framework has its own focus and puts an emphasis on certain BITA dimensions. It is equally important to deal with all four dimensions, but currently strategic and structural dimensions of BITA receive more attention than social and cultural (Chan and Reich 2007). The chosen BITA frameworks provide rather minor support for dealing with these two dimensions, particularly Generic Framework for Information Management includes Information & Communication domain that is to a certain extent related to these alignment dimensions. It is therefore interesting to investigate which of the existing BITA frameworks allow to deal with cultural and social alignment dimensions, which calls for a comprehensive state-of-the-art study in the BITA domain.

As it has been shown, Zachman framework proposes a set of focal areas that covers the domains of both of the analysed frameworks completely. However, in some cases in order to minimize the usage of resources it can be good to model an enterprise in a "good enough" way. In that case it might be suitable to decrease the number of modelled focal areas. Possible way to do it would be to unite people and network focal areas into an organisational structure, and unite function and time into business processes. By doing so the total number of focal areas to be modelled would decrease from six to four: motivation, data, organisational structure and business processes. This would be still a sufficient set of focal areas to deal with various BITA domains. Also, this approach is visible in other EM frameworks described in Sub-sect. 2.2. Most frameworks, even the ones created based on The Zachman Framework, have fewer, but more general focal areas.

Analysis of the other EM frameworks is planned as a future work. It would be beneficial to investigate existing EM and EA frameworks (as for example, Archimate framework) and the architectural domains that they define, with a specific attention to the distinction and similarity between EM and EA frameworks. This calls for investigation of the state-of-the-art in the general area of Enterprise Architecture Modeling. In addition, the presented conceptual positioning should be refined empirically.

4 Conclusions

This paper investigated how EM can facilitate BITA achievement. Particularly, the analysis has been done taking into account six EM focal areas presented by Zachman framework, and two sets of BITA domains proposed by Generic framework for information management and IS Strategy triangle. The presented positioning indicates that a set of focal areas proposed by Zachman is indeed comprehensive and can be used to deal with both sets of BITA domains. It is also possible to say that the set of Zachman focal areas can be reformed and still cover all of the considered BITA domains – people and network focal areas can be considered as constituents of organisational structure focal area; function and time can be considered as constituents of business processes. In case of such unification the sufficient set of EM focal areas to cover all BITA domains would include four focal areas instead of six. Using a set of fewer number, but more generic EM focal areas can be often observed in other existing EM frameworks. The analysed positioning does not address the EM contribution into dealing with the social and cultural dimensions of BITA, as the chosen BITA frameworks emphasize strategic and structural dimensions. This is an interesting aspect for future work, since EM is often used to develop a common understanding about the current multi-dimensional praxis and an agreement on future vision and strategies within an enterprise, and thus it can facilitate dealing with social and cultural dimensions of BITA.

References

Abraham, R., Aier, S.: Architectural coordination of transformation: implications from game theory. In: Rahman, H., Mesquita, A., Ramos, I., Pernici, B. (eds.) MCIS 2012. LNBIP, vol. 129, pp. 82–96. Springer, Heidelberg (2012)

Aier, S., Winter, R.: Virtual decoupling for IT/Business alignment – conceptual foundations, architecture design and implementation example. Bus. Inf. Syst. Eng. 1(2), 150–163 (2009)

Bernus, P., Nemes, L.: A framework to define a generic enterprise reference architecture and methodology. In: ICARV 1994, Singapore, pp. 88–92 (1994)

Chan, Y.E., Reich, B.H.: IT alignment: what have we learned? J. Inf. Technol. 22, 297–315 (2007)

Christiner, F., Lantow, B., Sandkuhl, K., Wißotzki, M.: Multi-dimensional visualization in enterprise modeling. In: Abramowicz, W., Domingue, J., Węcel, K. (eds.) BIS Workshops 2012. LNBIP, vol. 127, pp. 139–152. Springer, Heidelberg (2012)

Engelsman, W., Quartel, D., van Jonkers, H., Sinderen, M.: Extending enterprise architecture modelling with business goals and requirements. Enterp. Inf. Syst. 5(1), 9–36 (2011)

Goldkuhl, G., Lind, M., Seigerroth, U.: Method integration: the need for a learning perspective. IEEE Proc. Softw. 145(4), 113–118 (1998)

Gregor, S., Hart, D., Martin, N.: Enterprise architectures: enablers of business strategy and IS/IT alignment in government. Inf. Technol. People 20(2), 96–120 (2007)

Henderson, J.C., Venkatraman, N.: Strategic alignment: leveraging information technology for transforming organizations. IBM Syst. J. 32(1), 4–16 (1993)

IFIP-IFAC Task Force on Architectures for Enterprise Integration. GERAM: Generalised Enterprise Reference Architecture and Methodology (1999). http://www.ict.griffith.edu.au/~bernus/taskforce/geram/versions/geram1-6-3/v1.6.3.html

Kaidalova, J., Seigerroth, U., Bukowska, E., Shilov, N.: Enterprise modeling for business and IT alignment: challenges and recommendations. Int. J. IT Bus. Alignment Gov. 5(2), 44–69 (2014). ISSN 1947-9611

Lillehagen, F., Krogstie, J.: Active Knowledge Modeling of Enterprises. Springer, Heidelberg (2008). ISBN 978-3-540-79415-8

Luftman, J.: Assessing IT-Business Alignment. Inf. Syst. Manag. 20(4), 9–15 (2003)

Maes, R., Rijsenbrij, D., Truijens, O., Goedvolk, H.: Redefining business – IT alignment through a unified framework. PrimaVera Working Paper 2000-19 (2000)

Pearlson, K.E., Saunders, C.S.: Managing and Using Information Systems, A Strategic Approach, 4th edn. Wiley, Chichester (2010). ISBN 978-0-470-34381-4

Plazaola, L., Flores, J., Silva, E., Vargas, N., Ekstedt, M.: An approach to associate strategic business-IT alignment assessment to enterprise architecture. In: Fifth Conference on Systems Engineering 2007 (CSER 2007). Stevens Institute of Technology, New Jersey (2007)

Saat, J., Franke, U., Lagerstrom, R., Ekstedt, M.: Enterprise architecture meta models for IT/business alignment situations. In: 14th IEEE International Enterprise Distributed Object Computing Conference (2010)

Scheer, A.-W., Nüttgens, M.: ARIS architecture and reference models for business process management. In: van der Aalst, W.M., Desel, J., Oberweis, A. (eds.) Business Process Management. LNCS, vol. 1806, p. 376. Springer, Heidelberg (2000)

Scheer, A.-W.: ARIS – Business Process Modeling, 3rd edn. Springer, Heidelberg (2000). ISBN 978-3-642-63009-5

Seigerroth, U.: Enterprise modelling and enterprise architecture: the constituents of transformation and alignment of business and IT. Int. J. IT/Bus. Alignment Gov. (IJITBAG) 2, 16–34 (2011). ISSN 1947-9611

Silvius, A.J.G.: Business and IT alignment: what we know and what we don't know. In: The Proceedings of International Conference on Information Management and Engineering, pp. 558–563. IEEE (2009)

Stirna, J., Persson, A.: Anti-patterns as a means of focusing on critical quality aspects in enterprise modeling. In: Halpin, T., Krogstie, J., Nurcan, S., Proper, E., Schmidt, R., Soffer, P., Ukor, R. (eds.) Enterprise, Business-Process and Information Systems Modeling. LNBIP, vol. 29, pp. 407–418. Springer, Heidelberg (2009)

Urbaczewski, L., Mrdalj, S.: A comparison of enterprise architecture frameworks. Issues Inf. Syst. 7(2), 18–26 (2006)

Wegmann, A., Regev, G., Loison, B.: Business and IT alignment with SEAM. In: Glinz, M., Lutz, R. (eds.) The Proceedings of the 14th International Requirements Engineering Conference, pp. 291–302. IEEE Computer Society, Los Alamitos (2005)

Zachman International home page (2014). https://www.zachman.com/about-the-zachman-framework. Accessed 12 Dec 2014

Zachman, J.A.: A framework for information systems architecture. IBM Syst. J. 26(3), 276–291 (1987)

Zelm, M.: CIMOSA: A primer on key concepts, purpose, and business value. Technical report, Stuttgart, Germany (1995)

Specific BIS Applications

Multi-dimensional Performance Measurement in Complex Networks – Model and Integration for Logistics Chain Management

Lewin Boehlke[✉], André Ludwig, and Michael Gloeckner

University of Leipzig, Grimmaische Str.12, 04109 Leipzig, Germany
{boehlke,ludwig,gloeckner}@wifa.uni-leipzig.de

Abstract. Global demands and emerging markets resulted in a rise of complexity in logistics over the last years. Furthermore, producing companies outsourced their logistics for reasons of flexibility. Hence, clients consult solution specialists for customer-focused end-to-end processes. The 4th party logistics business model offers an overall service management throughout an entire logistics chain by planning and controlling inherent processes. But because physical provisioning is outsourced to 3rd party providers, performance measurement is essential in order to meet service levels. Measuring sub providers poses difficulties because of different process definitions and communication standards. A relief is the growing amount of information during the process lifecycle. In this paper we present an integrated, strategically aligned provider rating model, applicable throughout the entire logistics process and capable of summing up operational information into ratios that are further aggregated into key performance indicators. Thereafter, we prove its applicability through a prototypical implementation.

Keywords: Logistics chain controlling · Performance measurement · 4PL · KPI · Process evaluation

1 Introduction

Competition in a networked business world has evolved from competing between single companies towards competition between logistics chains. Thus, the effort for coordination and information management between partners and its resulting complexity has increased dramatically [1]. Within this context, the Fourth Party Logistics (4PL) represents a promising model for integrating business partners by designing, controlling and managing end-to-end logistics chains without running own physical resources [2]. A 4PL Provider (4PLP),[1] offers basic logistical functions such as transport, handling or storage and integrates them with additional services. Thus, value added and customer-oriented services can be offered to its customers. These additional services include e.g. refinement, consulting, financial settlement or human resource tasks. For the design of

---

[1] Gudehus uses the Term 4PL in order to describe the provider [2]. However, we differentiate between 4PL (Business Model) and 4PLP (Provider for this Business Model).

© Springer International Publishing Switzerland 2015
W. Abramowicz (Ed.): BIS 2015, LNBIP 208, pp. 287–298, 2015.
DOI: 10.1007/978-3-319-19027-3_23

these services, the 4PLP establishes long-term relationships with sub-providers. This business model is a modification known in contract logistics [2].

For logistics chain design and controlling, the 4PLP uses information systems in order to cope with the growing amount of information during the process execution and to enable efficient utilization of these information after process termination. Operational and strategic challenges occur within this context, such as efficiency and effectiveness [3]. Operational challenges subsume the synchronization between flow of material and information, mining and provision of information and incurred information preparation for fast decision-making. For example, process alternatives need to be found in order to adjust a process instance within a short time frame. Strategic challenges arise when correctly planning a future logistics chain is based on inherited and surrounding information, as well as under consideration of the many depending and independent variables.

Especially, virtual companies offering value-added services represent a challenge as they can be hardly rated compared to single companies. This business model describes short-term unions, which offer transparently bundled services to a customer [4]. The derived challenge is the standardization of key performance indicators (KPIs) beyond company level and the applied, uniform measurement methods.

	Tasks	Use of KPI-System
Analysis	Customer requirements Service criteria Service provider pool Contract 4PL <-> Customer	Match customer requirements with provider pool Transform customer requirements standardized service definition language
Design	Process modelling Process variants Provider option comparison Process simulation	Time and volume information integration of service providers for simulation Faster generation of process variants Allow for provider option comparison
Implemen-tation	IT-System composition Define information flows	Integration support via knowledge about sub-provider systems
Operation	Execute processes Mining and processing of process quality	Assignment event <-> provider Find alternate providers in case of process instance failure
Retirement	Process quality evaluation Process comparison Pattern detection	KPIs saved via provider profile Standardized approach allowing for provider comparison and pattern detection

Fig. 1. 4PL Service Lifecycle and use cases of KPI-System

The 4PLP needs a solution for logistics chain design and management to cope with these challenges operationally and strategically. Various authors have proposed a number of solutions in the last years [5–7], some of them describing a phase-oriented process lifecycle, similar to Fig. 1. The figure shows the phases and corresponding tasks for the 4PLP during the 4PL Service Lifecycle. In the analysis phase, customer requirements and contract tasks are taken into concern. The design phase copes with process modelling and simulation, while considering customer needs of the analysis phase. During the design phase, planning and simulation are essential. One integrated simulation concept is shown by [5]. The authors present a solution for using mined operational information in process simulation during the design phase. The main task of simulation is to find an ideal configuration of the logistics process, limited by the service level agreements established between 4PLP and its customer. This approach misses a comprehensive service provider profile, which generates an overall impression during process design as stated by the authors themselves [5]. This lag of information is addressed by our approach.

In this paper, we present a KPI-based approach for rating and selecting sub-providers in order to design and manage a logistics network vertically, horizontally, operationally and strategically throughout the entire 4PL Service Lifecycle. As stated by use cases shown in Fig. 1 solutions are presented in each of the phases. More concrete, operational retrieved information are wrapped into key performance indicators and persisted in provider profiles. These information are used during the retirement phase, where an internal evaluation for performance measurement purposes of contracted providers takes place. The retrieved information and patterns are then used for designing future logistics chains in a more detailed manner.

The paper is organized as follows: First, logistical controlling approaches are presented briefly. Thereby shown, state of research does not adequately support the 4PLP. As a reasonable consequence, a new measuring concept is introduced thereafter, inheriting concrete ratios on an operational level. Next, a concept for aggregating these ratios is presented, which takes operational and strategic levels beyond company borders into account. Thereafter, the feasibility of this concept is shown via a prototype. We finally present a conclusion with key findings, appraise the paper in a critical manner and give an outlook of future work.

2   Related Work

One of the main aspects of controlling logistical chains is the performance measurement, which primarily focusses on financial targets [8]. Still, non-financial aspects as customer satisfaction and process duration are important [9]. Another author add technical aspects and take efficient use of operating equipment into account [10]. Ultimately, a multi-dimensional representation of a company or (parts of) the logistics chain is created, in order to measure success of a cooperation. *Werner* postulates the transition from traditional ratio systems into performance measurement approaches [11]. These include soft factors such as customer opinion or the usage of multiple perspectives (e.g. customer perspective or financial perspective) in the company rating. Focusing the customer, service qualities have to match customer needs delivered in an end-to-end process. These processes are evaluated with performance measurement methods. Also, inherited ratios with absolute measurable values have developed into non-monetary KPIs and thus satisfy modern-day performance measurement requirements. This approach clearly states, that the vertical structure of traditional ratio systems is not valid anymore, but needs to be replaced by KPIs arranged along the logistic process beyond company borders.

Werner presents an approach for logistical benchmarking, in which company parts such as purchase, production or delivery are matched with different types of ratios. With this being said, this model applies to producing companies. Still, this approach is suitable for the present paper, but needs to be customized in a strong manner, as multiple ratios in the model of *Werner* are not to be measured by the 4PLP. If not contractually defined and provided, the 4PLP cannot measure them (e.g. throughput per employee). Focusing strategic logistics process management, the 4PLP only takes aggregated ratios into account, as there is no need to know for e.g. the throughput by employee in assigned companies. *Werner* further distinguishes between polycentric (equal companies) and hierarchically structured networks [11]. The present paper concentrates on the 4PLP-Perspective and such only on hierarchically structured networks. In contrast to *Werner*, we focus on the logistics chain management function, which is given to the 4PLP by the customer, instead of focusing the production function of a company, contracting the 4PLP.

Gleich/Daxboeck illuminate the crux of the matter directly: Logistics controlling means including customers and suppliers, but not including the whole logistical chain for performance measurement purposes [10]. The authors claim that the difficulties are because of missing (IT-) standards, as well as data security and data protection concerning the interaction between 4PLP and subcontractors. Rating virtual companies with a traditional ratio system is limited, due to missing aggregation-concepts. Although the process notion for KPI-retrieval in logistics is privileged, vertical company-layers are necessary in order to link operational and strategic ratios and such, allow aggregation. Thus, we present a method that establishes measurement standards throughout the entire logistics chain.

Mutke et al. also present a model for the integrated simulation based on gained information during process execution [5]. However, this approach does not fulfill the requirements of performance measurement, which demands the linkage between operational and strategic ratios. Furthermore, the authors claim the importance of provider selection, which needs to be further developed [5]. As stated before, the approach also misses an overall impression of assigned providers, in order to correctly plan future logistics chains. This is where the present paper comes into account, extending the paper of *Mutke* et al., by creating an overall provider impression.

An integrated, operational approach is shown by Frauenhofer IML with DISMOD [12]. This approach uses retrieved information for real-time reaction on process problems within the operations phase (e.g. breakdown of a transport service). However, we focus on the strategic analysis of processes and process comparison in the retirement phase. Thus, this approach is not suitable due to the strategic perspective in the current paper.

The presented approaches illustrated, that there is a need for a model that supports the 4PLP with a strategic designed, horizontally- and vertically-integrated ratio system, within the whole 4PL-Service-Lifecycle. Hereinafter, an appropriate model is presented, which fills this academic void.

3   Rating Approach

The rating takes place in different dimensions, so that a comprehensive picture of the provider can be drawn. The 4PLP rates contracted sub-providers retrospectively during the retirement phase in the 4PL Service Lifecycle. As all process instances are terminated when the retirement phase starts, the process dimension is the first perspective to be considered. The present paper relies on the approach of [8], in which process specific ratios are provided by operational installed Auto-ID-technologies in an aggregated form and stored in provider models. But because the present paper also relies on *Weber* et al. with the above described restrictions due to the business model of the 4PLP, a multi-dimensional approach is preferred. Still, not all ratios in *Weber* et al. apply. Thus, applicable ratios are presented hereinafter.

In order to create an overall process judgment, all ratios are introduced as proportions [13]. It should be highlighted, that in contrast to other ratios, the throughput time only serves as a template for the achieved throughput time, which includes the work in progress (WIP). Adherence to schedules reflects target-performance comparison between actual service delivery-time and agreed process instance duration. Standard deviation σ allows for a range of tolerance in time. As shown in Table 1, a percentage rate is generated by each further needed ratio. Hence, an overall performance measure is possible to be calculated, in order to state the process performance of a subcontractor.

Table 1. Process ratios and their metrics

Ratio label	Metric	Unit
Adherence to schedules	$\frac{Day_{delivery} - Day_{inward}}{\sigma} * 100$	%
Delivery fulfillment	$\frac{\#deliveries_{ontime}}{\#deliveries_{total}} * 100$	%
Throughput time	$\frac{WIP}{Throughput}$	sec, min, hr, d, m, y
Achieved throughput time	$\frac{Throughput\ time}{agreed\ throughput\ time} * 100$	%
Downtimes	$\left(t_{total} - availability * t_{total}\right) * 100$	%

Beside process specific ratios, *financial ratios* form a second dimension. As in the first dimension, not all ratios of [8] apply to the 4PLP. Still, freight costs or storage costs depending on the offered service are to be determined. As presented in Table 2, financial ratios consist of a quotient, which determines costs per unit (e.g. average storage costs per square meter). *Billing accuracy* reflects the problem when service units are not precisely aligned in time. Thus, rounding appear so that there is a difference between charged and actual price (e.g. simplified work hours). *Failure costs* states the average number of errors throughout all process instances by multiplying its cardinality with the average time and costs per service.

Table 2. Finance ratios and their metrics

Ratio label	Metric	Unit
Avg. service costs	$\dfrac{price_{payed}}{serviceUnit}$	Costs/unit
Failure costs	$\#error_{service} * t(avg(error))$ $* cost_{serviceUnit}$	Used currency
Billing accuracy	$\dfrac{price_{real}}{price_{charged}}$ or $\dfrac{unit_{real}}{unit_{charged}}$	Used currency

Including soft-factors such as provider satisfaction, the *provider-relationship dimension* forms a third dimension as shown in Table 3. *Provider satisfaction* is a weighted mark representing the overall impression on the subcontractor, which is calculated via surveys. This dimension also includes *reaction times for requests*. The ratio takes into account that subcontractors might never respond to requests. If the maximum time of request is e.g. 30 days, a value >= 1 is created, as long as the subcontractor reacts within 30 days. The maximum time of request is to be neglected, if its value is <1. *Subjective evaluation* reflects the overall satisfaction with the subcontractor. Because its value is not objective, still influencing, a weight (w) is introduced to subjective evaluation (n), which equals over all subcontractors for comparison reasons. The values n and w are inserted by the 4PLP. Long term relations create steadiness, which is represented by *relationship period*. This steadiness is a positive factor, which also puts other provider-relationship ratios into perspective. If, for instance, a new provider is introduced to the system, this value enables the user to either meet expectations, when steadiness is needed or enabling new partnerships due to management guidelines and process requirements.

Table 3. Provider-relationships ratios and their metrics

Ratio label	Metric	Unit
Subjective evaluation	$n * w$	$n, w \in \mathbb{N}$, $1 \leq n \leq 6$; $1 \leq w \leq 2$
Subcontractor satisfaction	*Survey score n*	$n \in \mathbb{N}, 1 \leq n \leq 6$
Response time for request	$\dfrac{1+t_{maxResponseDays}}{1+(t_{response}-t_{request})}$	$t \in \mathbb{N}$
Relationship period	*#months*	

The three dimensions measure subcontractors in a direct form, in order to create an overall impression. This model is also capable of handling virtual companies and therefor allow an aggregation of provider performances within the process flow. Thus, the present performance measurement model provides the opportunity on rating atomic providers just as virtual companies. Ultimately, this model yet lags a comparable approach between subcontractors. Therefor it needs to be enhanced by aggregated values. A possible approach is presented hereinafter.

4 Aggregation Approach

Due to the standardized performance measurement approach for atomic subcontractors, a foundation is given in order to aggregate ratings for (parts of) the logistics chain. The presented three dimensions are to be enhanced by a dimensional score, creating a statement such as process fulfillment, as presented in Fig. 2.

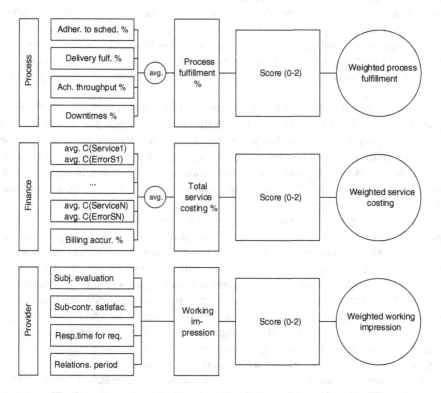

Fig. 2. Aggregation of ratios towards overall provider rating with KPIs

The common score of the process dimension, the *process fulfillment*, represents a statement on operational performance of the subcontractor compared to service level agreements. This includes speed and correctness of service delivery.

The financial dimension is aggregated to a *total service costing*. It is calculated with the average process instance costing. On a critical note, this costing is only comparable to similar services by other (virtual) providers, because average finishing process costs might be higher than storage costs. Furthermore, requested complex and nonrecurring services are not to be taken into account, but recurring services with multiple instances. Still, statements on main cost drivers within a process can be derived.

The provider relationship dimension is the third dimension, which needs to be aggregated. As presented above, it includes soft factors and not directly addable ratios. In an aggregated form, this dimension represents a *working impression* of subcontractors, weighted by his own satisfaction during the process operation. The working impression furthermore consists of the relationship duration modified through subjective evaluation and communication speed. With a questionnaire, which needs to be developed for the 4PLP, the subcontractor satisfaction is also taken into account. Because comprehensive ordinal scales ranging from 1 to 6 are used in the questionnaire, and also a dimensional score per dimension ranging from 0 to 2 is introduced, the system is user-friendly and easy to understand. The dimensional score allows the user to decide the amount of influence each dimension has. While zero means irrelevance, two means double relevance.

All three dimensions serve as aggregated statements, in such a way that information function of logistics controlling [14] is met. With the goal of providing information to the management, these KPIs are used in reports and the planning process as support for decision making.

Following the guidelines of balanced scorecard (BSC) development, a KPI-model has to focus on a few KPIs [15]. Even though the present approach does not try to serve as BSC, its guidelines such as linking of operational and strategic KPIs, involve the customer, multiple dimensions and concentrating on few KPIs are reused here. Furthermore, our approach is extendable with more dimensions. As stated in BSC-concept, *innovation* is mandatory [15], resulting in continuous service improvement and could be measured for example by number of patent fillings or publications.

The primary goal of the present concept is to support the 4PLP. Thus, it fulfills multiple functions. First, the *inform function* supports during planning, operation and retirement. As for planning, subcontractors can be chosen based on their performance in former processes. In operation, alternate solutions have to be found if a process instance is threatened to fail. During retirement its main goal is revealing performance patterns. Secondly, the *control function* is a central aspect of logistics controlling, which involves guiding, managing and regulation [14]. As described by *Pfohl,* these management functions are handled by the management itself, but the *coordination aspect* remains part of controlling. Hence, the present model serves as validator if strategic logistics process goals are satisfied with regard to subcontractors.

In summary, (virtual) subcontractors can be measured and matched this way. Still, the process for using this model is to be determined. Therefore, the following section describes the process with the aid of a prototypical implementation.

5   Prototypical Implementation

The prototypical implementation consists of different system parts and will be compared to a BI-architecture [16] hereinafter, as there are several similarities. As presented in Fig. 3, the prototype is integrated into the system landscape in between CEP and the simulation. The prototype itself is mostly subject matter of the retirement and planning-phase (subsuming analysis- and design-phase of the 4PLP-Service-Lifecycle).

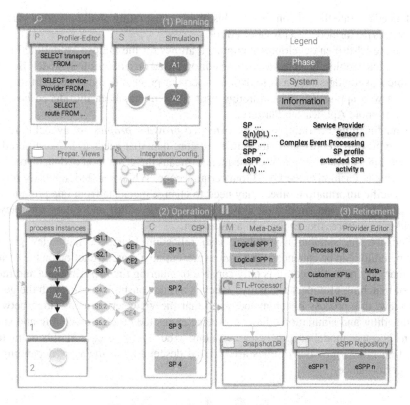

Fig. 3. System structure and integration

First, an ETL[2]-Process extracts data from external sources, transform data into the appropriate data-types and loads them into a data warehouse (DWH) [16]. All ETL processors have been developed by using Spring-Framework, allowing for *ReST-requests* on the complex event processing interface, in order to retrieve process specific data. Simultaneously, a second batch-load retrieves metadata of companies from another database. A third batch load purges operational determined process data with meta-information of the belonging subcontractor. Afterwards it loads this data into the now presented data-model.

As shown in Fig. 3, data is now stored into the snapshot-database, which corresponds the staging area in a BI-architecture and contains all batch-loaded data. The resulting data-structure (see Fig. 4) represents the data-structure of the DWH, as they are merged in the present prototype. A typical approach when modelling the DWH is the introduction of linked dimensions, so that a data-cube is created. This is where the present approach differs, as the central data-structure is the provider-profile, which is assembled with ratios mentioned in Fig. 4. Similar to a typical DWH, data is not overwritten but appended instead. After the enrichment of the DWH with operational- and meta-information, the

---

[2] Extract, transform, load.

4PLP needs to rate the subcontractor. This involves the rating of all non-process ratios. This step is performed via a provider-editor.

The model-driven development approach allows for the Eclipse Modelling Framework, which enables the generation of an editor, based on a descriptive model. Because multiple stakeholders such as simulation experts or process designers benefit from the editor, a web application with different user accounts has been introduced based on Eclipse Remote Application Platform.

After finishing the rating process, *extended provider profiles* are available for the planning process, representing the unfiltered core-data that needs to be pre-filtered for each working-purpose which, in consequence, means the generation of data marts. Data mart strive to customize the DWH to a certain scope. While simulation mostly rely on time-specific information, others may need to change metadata from one provider. As shown in Fig. 3, the simulation expert logs into the editor and filters for relevant providers and stores the response in a data-mart.

Because the simulation-prototype also relies on Eclipse-technologies [5], the existing editor can be enhanced with a view presenting a search-function on the created data mart. By now, this editor is only capable of entering time-specific data and metadata of matching providers manually, which is a time consuming task. With the generated view, this process is automated, such that the view serves as a bridge between profile editor and simulation editor. This accelerates the process modelling and simulation, because modelled processes can simply be linked to chosen subcontractors. Based on ratios, provider profiles are ordered descending in the view depending on their performance.

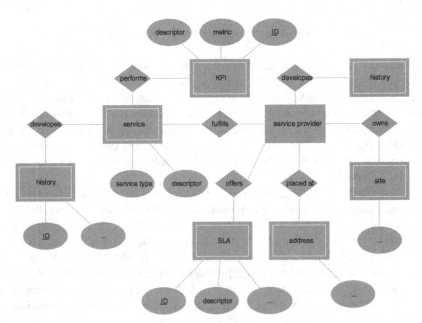

Fig. 4. Draft of provider and performance combination

6 Discussion and Future Work

The present paper described a rating approach for subcontractors from perspective of the 4PLP. For this purpose, techniques of logistics chain controlling have been introduced. With the use of KPIs, single companies or even virtual companies can be rated. Corresponding to the initial requirement of creating a strategically designed, horizontally- and vertically-integrated ratio system, we pointed out a valid approach and proved its use with a prototype implementation.

It has been further shown, that it is possible to integrate operational process data into the planning processes of future supply chain models. More precisely, operational process data is aggregated into KPIs and then resembled into a company profile. This profile is then used for comparing subcontractors in order to select participants for upcoming logistics processes. For an exemplary manner, the generated extended provider profiles were integrated in the process simulation task during the planning process.

However, certain requirements are to be met. First of all, the 4PLP needs to possess a huge information base, which requires a long term usage of this concept. In this regard, it is essential to continuously enter data into the system. Furthermore, the introduced dimensions are difficult to compare. For instance, a pure shipper might appear financially more reasonable than a value-added service provider, because processes of the latter are more diverse. Also, a new subcontractor might qualify himself as potential partner, but misses a performance measurement in the present system. In order to integrate new players into the system, further research has to be done.

Critically speaking, the presented KPI-Model should be further reviewed in order to test its acceptance in practice. Therefore, concept testing or expert interviews must approve applicability of the model, especially concerning the implemented ratios and dimensions. In this context, other dimensions such as an innovation dimension could be developed, giving the model a more strategic character. Considering ratios, the usage different KPIs should be analyzed. Also, the model uses subjective scale with a close range, which is to be tested for its overpowered influence. Lastly, the adaptability in other industries such as health or government must state the generality of the model. For this, scenarios and use cases have to be developed in order to test the model.

The present concept could furthermore be extended. One could for example filter potential subcontractors based on the customer specific service level agreements, such that only matching subcontractors are available during the planning phase. For this purpose, a matching algorithm needs to be developed, which adjusts customer requirements with service-offers in the system.

Finally, highlighting the similarity between the present prototype and a BI reference architecture, performance recognition patters need to be developed in the retirement phase, in order to enable risk assessment, which is crucial for strategic planning. This concept is well known as *data mining* [16], and allows for an explorative data analysis in huge amount of data.

References

1. Fawcett, S.E., Fawcett, S.A.: The firm as a value-added system. Int. J. Phys. Dist. Log. Manage **25**(5), 24–42 (1995)
2. Gudehus, T., Kotzab, H.: Comprehensive Logistics, 2nd edn. Springer, Heidelberg (2012)
3. Fugate, B.S., Mentzer, J.T., Stank, T.P.: Logistics performance: efficiency, effectiveness, and differentiation. J. Bus. Logist. **31**(1), 43–62 (2010)
4. Chesbrough, H.W., Teece, D.J.: Organizing for innovation. Harv. Bus. Rev. **74**(1), 65–73 (1996). http://search.ebscohost.com/login.aspx?direct=true&db=bth&AN=9601185343&site=ehost-live
5. Mutke, S., Roth, M., Ludwig, A., Franczyk, B.: Towards real-time data acquisition for simulation of logistics service systems. In: Pacino, D., Voß, S., Jensen, R.M. (eds.) ICCL 2013. LNCS, vol. 8197, pp. 242–256. Springer, Heidelberg (2013)
6. Augenstein, C., Ludwig, A., Franczyk, B.: Integration of service models - preliminary results for consistent logistics service management. In: 2012 Annual SRII Global Conference: SRII 2012: Driving Innovation for IT Enabled Services, pp. 100–109, San Jose, California, USA, Los Alamitos. IEEE Computer Society/IEEE, California/Piscataway, 24–27 July 2012
7. Kunkel, R., Ludwig, A., Franczyk, B.: Modellgetriebene ad-hoc Integration von Logistikdienstleistern - Integrationsansatz und Prototyp. In: Multikonferenz Wirtschaftsinformatik 2012: Tagungsband der MKWI 2012. Gito, Berlin (2012)
8. Weber, J., Wallenburg, J., et al.: Logistik-Controlling mit Kennzahlensystemen. http://www.whu.edu/fileadmin/data/Startseite/PDF_s_f%C3%BCr_Pressemitteilungen/2012_PDF_s/2012_PDF_s_Q2/WHU_Studie_2012_Logistikkenzahlen.pdf. Accessed 18 November 2014
9. Tonchia, S., Quagini, L.: Performance Measurement: Linking Balanced Scorecard to Business Intelligence. Springer, Heidelberg (2010)
10. Gleich, R., Daxböck, C.: Supply-Chain- und Logistikcontrolling - inkl. eBook: Instrumente, Kennzahlen, Best Practices. Haufe Lexware Verlag, München (2014)
11. Werner, H.: Kompakt Edition: Supply Chain Controlling: Grundlagen, Performance-Messung und Handlungsempfehlungen. Springer Gabler, Wiesbaden (2014)
12. Fraunhofer Institute for Material Flow and Logistics, DISMOD Cockpit - Fraunhofer IML. http://www.iml.fraunhofer.de/en/fields_of_activity/transport_logistics/products/dismod_cockpit.html. Accessed 16 March 2015
13. Werner, H.: Supply Chain Management: Grundlagen, Strategien, Instrumente und Controlling, 3rd edn. Gabler, Wiesbaden (2008)
14. Pfohl, H.-C.: Logistikmanagement: Konzeption und Funktionen, 2nd edn. Springer, Berlin (2004)
15. Biazzo, S., Garengo, P.: Performance Measurement with the Balanced Scorecard: A Practical Approach to Implementation Within SMEs. Springer, Heidelberg (2012)
16. Loshin, D.: BUSINESS intelligence: The Savvy Manager's Guide, 2nd edn. Morgan Kaufmann, Waltham (2012)

Engaging Facebook Users in Brand Pages: Different Posts of Marketing-Mix Information

Mathupayas Thongmak[✉]

MIS Department, Thammasat Business School,
Thammasat University, Bangkok, Thailand
mathupayas@tbs.tu.ac.th

Abstract. This research explores the posting quantity and posting types most likely to create brand engagement on Facebook brand-fan pages. Using content analysis, the study explores five posting types: product, price, place, promotion, and others, by analyzing 1,577 posts from 183 brand-fan pages. Findings suggest that high posting amount could increase brand popularity. Thus, a brand-fan page's content provider should add more price information, promotional information, other informational content, and emotional content on posted messages. This work amends the social media marketing literature by including the concept of marketing mix along with uses and gratifications (U&G). It also combines likes, comments, and shares to represent the post popularity. The study provides some managerial implications for effectively planning the content strategy to engage more fans or potential customers in brand-fan pages.

Keywords: Facebook · Brand · Fan pages · Posts · Marketing mix · Uses and gratifications

1 Introduction

In modern times, social media, especially social networking, has played a major role in corporate marketing. In February 2010, fifty percent of the total online users in the Asia-Pacific region visited a social networking site. Nowadays, the hub of customer activity is also located inside social media [1]. Social media marketing could increase customer engagement and brand awareness through viral marketing [2]. Viral marketing is spreading the messages about products or services through interactions among consumers, relevant users, and potential consumers [3, 4]. Positive attitude towards social media or social media attitudinal loyalty could lead to the higher purchase frequency, purchase quantity, and purchase expenditure of customers [5]. Facebook fan pages are popular to serve as a channel for social media marketing. They could be applied to create the brand community for sharing company, product or service information, communicating marketing messages, expanding network of customers, and getting feedbacks to build a strong brand [6, 7]. In this paper, these fan pages will be later called as brand-fan page. Brand community trust and affection create a positive influence on brand loyalty [8]. Companies owning the brand-fan pages need to draw customers' intention and sustain social interactions in the brand community regularly using relevant and interesting content [7]. Some research studied the social media

© Springer International Publishing Switzerland 2015
W. Abramowicz (Ed.): BIS 2015, LNBIP 208, pp. 299–308, 2015.
DOI: 10.1007/978-3-319-19027-3_24

marketing. Cvijikj and Michahelles explored the relationship between content characteristics, which were communicated through companies' Facebook brand-fan pages, and customers' online engagement level [2]. De Vries, Gensler, and Leeflang analyzed posts of 11 brand-fan pages in 6 product categories to specify drivers of the post popularity [9]. Chauhan and Pillai tried to understand the content strategy of Indian higher education institutes to build customer engagement in the brand community [7]. Swani, Milne, and Brown studied post strategies to promote electronic word of mouth (e-WOM) for B2B and B2C fan pages [10]. E-WOM is defined as "any positive or negative statement made by potential, actual, or former customers about a product or company, which is made available to a multitude of people and institutions via internet" [11]. Malhotra and Malhotra did a content analysis of 98 brand-fan pages [12]. Clemons, Barnett, and Appadurai indicate that e-WOM or viral marketing on social networks have not yet succeeded the high expectation set [13]. Understanding of how customers interact in social media channels and the guideline specifying what constitutes great content for fans are still lacking [2]. Shen and Bissell also emphasize the need of studies about likes and comments on wall posts [14]. Although some studies explore the content strategy for brand-fan pages, none of them studies content types considering the U&G and the classic marketing mix. Also, only little research has investigated on the complete set of the post popularity (likes, comments, and shares). Thus, the research questions are: (RQ1) How do the different wall posts (marketing-mix information versus others) affect the brand post popularity (the number of likes, comments, and shares)? (RQ2) How does the post popularity influence the brand engagement? This study applies PTAT to represent the brand engagement. People Talking About This (PTAT) is a popular metric that shows the holistic view of all interactions with a fan page, consisting of key components such as page likes, likes, comments, shares, RSVP's, check ins [15]. (RQ3) How does the brand engagement in terms of PTAT impact the increase in new fans (the number of new likes)?

2 Background

2.1 Facebook Fan Pages, E-WOM, and Brand Engagement

Facebook provides 5 tools for supporting companies to reach their market better: Facebook ads, Facebook fan pages, social plugins, Facebook applications, and sponsored stories [2]. Facebook ads are a method to acquire fans by letting people to hear about a brand from their friends who have already connected to the brand-fan page. Social plugins such as the activity feed, recommendations, comments and live stream could create Facebook social experiences for users. Facebook applications can be added to create a viral awareness campaign. Sponsored stories are a way to promote News Feed stories of a company and stories about people connecting with the brand-fan page, app, event to the fans' friends [16, 17]. Facebook fan pages enable companies to create brand communities with posts containing anecdotes, photos, videos, or other materials [9]. Brand community is an important tool to increase sales, improve the relationship with consumers, facilitate customers' opinion exchange about the brand, and involve customers in a form of WOM communications [2, 3, 15].

The WOM communications can be used as a powerful tool to achieve various marketing purposes such as launching new products or promoting brand images [2, 14]. The more positive content that user generated, the higher sales volume, sales share, and average customer lifetime value will be [4]. Companies can easily create their brand community using Facebook fan pages. A brand-fan page enables fans to engage more with the brand by posting content on the wall, commenting on existing posts, showing their interest towards existing posts by liking or sharing the posts on their walls. These interactions create WOM communications leading to the objectives of viral marketing [2].

2.2 Uses and Gratifications of Content: Marketing-Mix Information and Others

Uses and gratifications theory explains why people choose specific media and content [18]. Gratifications include informativeness, playfulness (or enjoyment), and interactivity [19]. Informativeness is the richness and availability of information [19, 20]. Past research indicates the impact of informativeness, playfulness, and interactivity on a user's intention or behavior [19]. The richness of user-generated content in terms of information about product or brand attributes has a large impact on consumers' purchase behaviour [21]. Information and entertainment seeking are also drivers to commit brand communities generated by consumers [22]. In the informativeness aspect, marketing-mix information is interesting because the 4Ps concept could help companies meet their customers' needs and expand their sales in other markets [23]. The marketing mix model consists of product, price, place, and promotion. It is later extended to various concepts such as adding personnel, physical assets, and procedures [24]. Product is a necessary component to satisfy customers [23, 25]. Product information includes product performance, brand/brand reputation, product design, product quality, and product features [26]. Price is various costs endured by customers to achieve products or services [25]. Price information includes cheaper or affordable price, price discount, special price, and credit/installment facilities [26]. Place covers both electronic channels and physical channels to distribute products or services [25]. Place information includes availability of product, sold by reputed dealer, customer service by dealer, special warranty by dealer, etc. [26] Promotion is benefits gained from buying products or services [25]. Promotion information includes advertisements, sales promotion, exchange offer, buy back offer, free benefits, etc. [26].

3 Hypotheses

3.1 Facebook Posts: Marketing-Mix Information Versus Others

The content strategy strongly affects fans' participation [2, 7, 9, 10, 12]. Informative and entertaining content of U&G theory is crucial to increase online brand community engagement [2, 27–29]. Creating informative value is important to make people passing along information through likes and wall posts [12]. The pursuit of information explains why individuals consume brand-related content [29]. Information seeking enables people to use social networks and contribute to the Facebook community [9, 30]. Entertainment affects the positive attitude of people towards ads and drives them to participate in brand-related content online [9, 29]. Facebook brand-fan page administrators should create brand-related information and entertainment content to achieve higher number of likes, comments, and interaction duration [2, 9]. Consumers often face lacking of product-related information to complete their purchase decisions [21, 31]. Getting discounts or coupons, purchasing products and services, and reading reviews and product rankings are top three activities of customers researching about products before deciding to buy them [1]. Product information and place information are high impact variables influencing consumer buying and decision-making behavior [26]. Thus, hypotheses are as follows:

H1: Posts which contain 4Ps information and other content drive a high number of (a) likes, (b) comments, and (c) shares.

H1: There is a correlation relationship between different posts and total (d) likes, (e) comments, and (f) shares.

3.2 Likes, Comments, Shares by Facebook Users

Generally, fans participate in a brand community by liking, commenting, or sharing the wall posts [2]. When a user likes, comments, or shares a brand post, it automatically generates e-WOM such as appearing in his/her friends' feeds [9, 32]. Consumers liking brand posts endorses the posts throughout their friend networks. People tend to make brand purchases, talk about their experiences, and gain emotional attachment, commitment, and loyalty with the brand [10]. Sharing of positive comments on a brand post positively affects the brand post popularity [9]. Sharing action gains superior influence than liking and commenting actions [15]. Thus, hypotheses are as follows:

H2: Post popularity in terms of the number of (a) likes, (b) comments, and (c) shares positively causes the highest PTAT.

3.3 People Talking About This

A single brand engagement such as one 'like' could refer the message or post to a number of friends in the user's network, creating a social contagion effect [10]. A strong fan engagement can push the brand forward and create more and more reach on each wall post. People Talking About This shows the history of a brand-fan page's awareness and interaction [15]. It includes stories such as sharing, liking or commenting on posts, answering a question, responding to an event, or claiming an offer [33]. Page size is the total fans liking a Facebook fan page. It can be applied to evaluate the page growth from time to time. Facebook advertising also applies a page size as one of the key success metrics [15]. Thus, hypothesis is as follows:

H3: The higher level of PTAT leads to the more new likes in Facebook brand-fan pages.

4 Study Design

This study focuses on the brand communities built on Facebook. Facebook was chosen because of its maximum number of online users, its availability, and its user-friendly features to collect the post details, brand engagement, and other related content [7]. Firstly, top 100 brand-fan pages with highest fans in February 2013 were explored [34]. Then, other brand-fan pages in the same categories were collected to fulfill the samples. Thus, 210 brand-fan pages in Thailand (both local and international brands) were selected from 9 categories: Fast-Moving Consumer Goods (FMCG), e-commerce, retail foods, telecommunications, electronics, fashion, finance, jewellery & watches, and retail, under conditions consisting of: (1) brands having concrete products or services, (2) top brands in each category, listed in [35]. Secondly, the elements to be collected are the page size (current fan likes), the highest PTAT, the new likes per week (which were shown in a brand-fan page's statistics in the period from May 2013 to June 2013), and the details of each post. The details of each post consisted of (1) the post types, (2) number of likes, (3) number of comments, and (4) number of shares. Five post types compose of product, price, place, promotion, and others. Product post is a post providing a detailed narrative about the product. Price post is a post giving the price information. Place post is a post providing information about distribution channels to acquire products or services. Promotion post is a post about discounting prices, redeeming points or gifts, giving coupons or gift cards, persuading fans to participate in brand events, and giving free products or premiums. Others post is a post about other brand-related information, brand-unrelated information, emotional-related message, activities unrelated to the brand, and funny message. Thirdly, the mentioned elements were investigated by two research assistants using content analysis. The definition of each post type was formally described to them. Because of the potentially large number of each brand-fan page's posts, the whole brand-fan pages were split into two sets and were separately assigned to the research assistants. Each research assistant had to collaboratively make judgments about the unclear post types and randomly verify the post elements collected by another coder. This process took 7 months data-gathering period from June 2013 to December 2013 to gather the whole brand posts.

5 Results

As mentioned earlier, 210 brand-fan pages were picked. However, 27 brand-fan pages were excluded because they were not available or had no posts on those pages. So, the remains 183 brand-fan pages were left. A number of posts (1,584 brand posts) were collected. Six posts which could be categorized into more than one post types and a post with missing information were removed. Finally, 1,577 posts were used in the data analysis phase. Of 183 brand-fan pages, the average fans of each brand-fan page were 648,084 per a FMCG brand-fan page, 217,323 per an e-commerce brand-fan page, 180,305 per a retail foods brand-fan page, 465,709 per a telecommunications brand-fan page, 1,452,821 per an electronics brand-fan page, 114,626 per a fashion brand-fan page, 131,342 per a finance brand-fan page, 69,953 per a jewelry & watches brand-fan page, and 208,149 per a retail brand-fan page. Table 1 shows the average posts and the maximum posts per brand for each category. The most active brand categories are FMCG and e-commerce that communicate 15-16 posts/week. Retail brands provide the content an average of 7 posts/week. Brands in other categories generate messages around 3-4 posts/week. Common post types are other posts, posts about promotion, and posts about products respectively. Posts about pricing and place are quite rare, except for FMCG, e-commerce, and retail food categories.

Table 1. Summary of posts in brand-fan pages

Brand Category	4 Ps Information Posts/Week								Other Posts/ Week		Total Posts/ Week	
	Avg. P1	Max P1	Avg. P2	Max P2	Avg. P3	Max P3	Avg. P4	Max P4	Avg. O	Max O	Avg.	Max
FMCG	1	8	0	2	0	2	4	13	11	36	16	40
E-Commerce	2	11	0	5	2	20	4	49	7	75	15	140
Retail Foods	1	4	1	4	0	2	3	15	3	12	7	30
Telecommunications	1	2	0	0	0	0	1	3	1	1	3	4
Electronics	1	5	0	1	0	1	1	5	1	4	4	10
Fashion	1	4	0	1	0	1	1	5	1	5	4	8
Finance	1	5	0	0	0	1	1	3	1	3	3	9
Jewellery & Watches	3	5	0	1	0	0	0	1	1	2	4	8
Retail	1	4	0	0	0	1	2	4	1	3	3	7

*Note: P1: Product, P2: Price, P3: Place, P4: Promotion, O: Others

Simple linear regression was applied to test the relationship between total posts within one week and the post popularity. As shown in Table 2, hypothesis H1a, H1b, and H1c are supported. Given the Bs of the effects of total posts and some small R-squared of the model, only about 6-9 per cent of the variation in the number of comments and shares is explained by the variation in the total posts. Although, the small R-squared shows that predictability represents a tiny fraction of the variance in the dependent variable, even a small R-squared can signal significant predictability in the economic context [36]. The study of De Vries, Gensler, and Leeflang did not support that informative and entertaining brand posts are more popular than non-informative brand posts [9]. Content had an insignificant impact on the number of likes and the number of comments [7]. Entertainment posts yielded no significant impact on the total likes [14]. Information-seeking and entertainment-seeking had no influence on Facebook group members' viral advertising pass-on [3] and the community commitment in marketer-generated brand communities (MGBC) [19]. Promotion-related posts had a significant effect on comments but not likes or shares [14]. However, this study presents different results, showing the importance of promotional posts. As shown in Table 3, the results support H1d, H1e, and H1f. Multiple regression analysis was also applied to test whether the post popularity significantly contributes to the brand engagement (the highest PTAT). The results reveal 2 drivers (the sum of likes and the sum of comments), explained 45.5 % of the variance ($R^2 = .455$, $F(2,180) = 75.245$, $p < .01$). It is found that the sum of likes significantly predicts PTAT ($\beta = .513$, $p < .001$) as same as the sum of comments ($\beta = .195$, $p < .05$), thus H2a – H2b are accepted. Lastly, the highest PTAT is found to have a significant effect on new fans, supporting H3 with R-squared of 0.221 ($\beta = .470$, $p < .001$).

Table 2. Results of regression analysis

Dependent Variable	Independent Variables	R^2	β
The Sum of Likes	Total Posts	.372	.610***
The Sum of Comments	Total Posts	.066	.257***
The Sum of Shares	Total Posts	.095	.308***

*$p < .05$ **$p < .01$ ***$p < .001$

Table 3. Pearson correlations of different posts and the post popularity ($n = 183$)

	Sum Likes	Sum Comments	Sum Shares
Product Information Posts	.102	-.014	-.035
Price Information Posts	.423***	.135	.186*
Place Information Posts	.121	.019	.017
Promotion Information Posts	.651***	.389***	.373***
Other Posts	.510***	.176*	.264***

*$p < .05$ **$p < .01$ ***$p < .001$

6 Limitations and Implications

Limitations of this study are focusing only on Facebook brand-fan pages in Thailand (including some well-known international companies), the limited posts in some post types, and some statistical results with the small r-squared. So, the application of results in different environments should be done with caution. However, most studies have been conducted in the Western context. So, this research reveals some interesting perspectives as follows. First, although FMCG, e-commerce, retail foods, electronics, fashions, and finance brands have the highest Thai fans, only FMCG, e-commerce, and retail food brands actively build their communities, showing by the low number of posts per a brand-fan page. So, there are rooms for other brands to join brand-fan pages and attract their customers' attention. Brands in other categories rather than FMCG, e-commerce, and retail should create more valuable information on their brand-fan pages. Second, more posts, both informational and entertaining post types, increase the post popularity by generating more likes, comments, and shares. Therefore, administrators of each brand-fan page should post informational and emotional messages regularly. Unlike the 4Ps information on the National Soccer League website which place information are the primary content [37], promotional and price information are quite popular among all posts of Facebook brand-fan pages in Thailand. However, the brand-fan pages' administrators should balance the posts of marketing-mix elements to support different needs of customers. Third, the number of likes and the number of comments drives the brand engagement (PTAT). Brand engagement can later increase a page size by increasing new likes. The number of likes and comments on posts by fans create the number of likes, comments, and shares in the Facebook brand-fan pages by non-fans. Adding more of those content types, especially prices, promotion, and others, could increase the number of fans and fans' engagement.

7 Conclusion and Further Research

Understanding what customers value in a social platform is an important to build social customer relationship strategy [1]. This study explores the effects of different information provided on brand-fan pages on the post popularity along with the customer engagement in the brand-fan pages. Findings reveal the importance of the post frequency and post types on the post popularity. The post popularity also drives more fans and non-fans' participating in the brand-fan pages. Dynamics of fans' engagement could be driven by emotional posts (such as funny posts) and informational post (the posts about price and promotion). Future research should gather more posts of each post type and add more brand-fan pages of each brand category from other countries such as Malaysia to generalize the results. Comparative study regarding the effective content strategy in different environments should be studied to help better customize the social customer relationship strategy. Pitfalls and risks of using brand-fan pages should be studied to provide the guideline for brand-fan pages' administrators in properly responding to customers. More complex measures rather than likes, comments, and shares should be added to achieve further market insights.

References

1. Heller Baird, C., Parasnis, G.: From social media to social customer relationship management. Strategy Leadersh. **39**(5), 30–37 (2011)
2. Cvijikj, I.P., Michahelles, F.: Online engagement factors on facebook brand pages. Soc. Netw. Anal. Min. **3**(4), 843–861 (2013)
3. Chu, S.-C.: Viral advertising in social media: participation in facebook groups and responses among college-aged users. J. Interact. Advertising **12**(1), 30–43 (2011)
4. Koch, S., Elçiseven, Ö.: A theoretical framework for assessing effects of user generated content on a company's marketing outcomes. In: International Conference on Information Resources Management (CONF-IRM), Vienna, Austria, 21–23 May 2012 (2012)
5. Ping, J.W., Goh, K.Y., Lin, Z., Goh, Q., Chih, A.: Does social media brand community membership translate to real sales? a critical evaluation of purchase behavior by fans and non-fans of a facebook fan page. In: The 20th European Conference on Information Systems, Barcelona, Spain (2012)
6. Hsu, Y.-L.: Facebook as international eMarketing strategy of Taiwan hotels. Int. J. Hospitality Manage. **31**(3), 972–980 (2012)
7. Chauhan, K., Pillai, A.: Role of content strategy in social media brand communities: a case of higher education institutes in India. J. Prod. Brand Manage. **22**(1), 40–51 (2013)
8. Hur, W.-M., Ahn, K.-H., Kim, M.: Building brand loyalty through managing brand community commitment. Manage. Decis. **49**(7), 1194–1213 (2011)
9. De Vries, L., Gensler, S., Leeflang, P.S.: Popularity of brand posts on brand fan pages: an investigation of the effects of social media marketing. J. Interact. Mark. **26**(2), 83–91 (2012)

10. Swani, K., Milne, G., Brown, B.P.: Spreading the word through likes on facebook: evaluating the message strategy effectiveness of fortune 500 companies. J. Res. Interact. Mark. **7**(4), 269–294 (2013)
11. Hennig-Thurau, T., Gwinner, K.P., Walsh, G., Gremler, D.D.: Electronic word-of-mouth via consumer-opinion platforms: what motivates consumers to articulate themselves on the internet? J. interact. mark. **18**(1), 38–52 (2004)
12. Malhotra, A., Malhotra, C.K., See, A.: How to create brand engagement on facebook. MIT Sloan Manage. Rev. **54**(2), 18–20 (2013)
13. Clemons, E.K., Barnett, S., Appadurai, A.: The future of advertising and the value of social network websites: some preliminary examinations. In: Proceedings of the Ninth International Conference on Electronic Commerce, pp 267–276. ACM (2007)
14. Shen, B., Bissell, K.: Social media, social me: a content analysis of beauty companies use of facebook in marketing and branding. J. Promot. Manage. **19**(5), 629–651 (2013)
15. Socialbakers.com a marketer's guide to facebook metrics (2014). http://www.socialbakers.com/resources/studies/a-marketers-guide-to-facebook-metrics
16. Facebook.com best practice guide marketing on facebook (2011). http://www.cde.uac.pt/uploads/artigo/8142aeea2f34b45a61994e03cbe18328ac72d10b.pdf. Accessed 10 March 2015
17. Facebook.com sponsored stories for marketplace. (20 October 2011). https://www.facebook.com/ads/stories/SponsoredStoriesGuide_Oct2011.pdf. Accessed 10 March 2015
18. Katz, E., Blumler, J.G., Gurevitch, M.: Utilization of Mass Communication by Individual. In: Hanson, J., Maxcy, D. (eds.) Sources Notable Selections in Mass Media, pp. 51–59. Dushkin/MaGraw-Hill, Guilford (1999)
19. Chiang, H.-S.: Continuous usage of social networking sites: the effect of innovation and gratification attributes. Online Inf. Rev. **37**(6), 851–871 (2013)
20. Chakraborty, G., Srivastava, P., Warren, D.L.: Understanding corporate B2B web sites' effectiveness from North American and European perspective. Ind. Mark. Manage. **34**(5), 420–429 (2005)
21. Goh, K.-Y., Heng, C.-S., Lin, Z.: Social media brand community and consumer behavior: quantifying the relative impact of user-and marketer-generated content. Inf. Syst. Res. **24**(1), 88–107 (2013)
22. Sung, Y., Kim, Y., Kwon, O., Moon, J.: An explorative study of Korean consumer participation in virtual brand communities in social network sites. J. Glob. Mark. **23**(5), 430–445 (2010)
23. Mostaani, M.: The Management of Consumer Cooperatives. Paygan Publications, Tehran (2005)
24. Jain, M.K.: An analysis of marketing mix: 7Ps or more. Asian J. Multi. Stud. **1**(4), 23–28 (2013)
25. Davies, W., Brush, K.E.: High-tech industry marketing: the elements of a sophisticated global strategy. Ind. Mark. Manage. **26**(1), 1–13 (1997)
26. Manickam, S., Sriram, B.: Modeling the impact of marketing information on consumer buying behavior in a matured marketing environment: an exploratory study of the middle east consumers. J. Promot. Manage. **19**(1), 1–16 (2013)
27. Dholakia, U.M., Bagozzi, R.P., Pearo, L.K.: A social influence model of consumer participation in network-and small-group-based virtual communities. Int. J. Res. Mark. **21**(3), 241–263 (2004)
28. Raacke, J., Bonds-Raacke, J.: MySpace and facebook: applying the uses and gratifications theory to exploring friend-networking sites. Cyberpsychology Behav. **11**(2), 169–174 (2008)

29. Muntinga, D.G., Moorman, M., Smit, E.G.: Introducing COBRAs. Int. J. Advertising **30**(1), 13–46 (2011)
30. Lin, K.-Y., Lu, H.-P.: Why people use social networking sites: an empirical study integrating network externalities and motivation theory. Comput. Hum. Behav. **27**(3), 1152–1161 (2011)
31. Kivetz, R., Simonson, I.: The effects of incomplete information on consumer choice. J. Mark. Res. **37**(4), 427–448 (2000)
32. Brown, S., Kozinets, R.V., Sherry Jr., J.F.: Teaching old brands new tricks: retro branding and the revival of brand meaning. J. Mark. **67**(3), 19–33 (2003)
33. Othman, I., Bidin, A., Hussain, H.: Facebook marketing strategy for small business in malaysia. In: 2013 International Conference on Informatics and Creative Multimedia (ICICM), pp 236–241. IEEE (2013)
34. Socialbakers.com, February 2013 social marketing report: Thailand (2013). http://www.socialbakers.com/blog/1515-February-2013-social-media-report-facebook-pages-in-thailand
35. Socialbakers.com pages in Thailand (2013) http://www.socialbakers.com/facebook-pages/brands/thailand/
36. Ferson, W.E., Sarkissian, S., Simin, T.T.: Spurious regressions in financial economics? J. Finance **58**(4), 1393–1414 (2003)
37. Carlson, J, Rosenberger III, P.J., Muthaly, S.: Goal!: an exploratory study of the information content in the australian national soccer league websites. Paper presented at the Australian and New Zealand marketing academy conference, Massey University Albany Campus, Auckland New Zealand, 1–5 December 2001

Context Variation for Service Self-contextualization in Cyber-Physical Systems

Alexander Smirnov[1,3], Kurt Sandkuhl[2,3], Nikolay Shilov[1,3(✉)],
and Nikolay Telsya[1,3]

[1] SPIIRAS, 14 Line 39, 199178 St. Petersburg, Russia
[2] University of Rostock, Albert-Einstein-Str. 22, 18059 Rostock, Germany
[3] ITMO University, Kronverkskiy Pr. 49, 197101 St. Petersburg, Russia
{smir,nick,teslya}@iias.spb.su, kurt.sandkuhl@uni-rostock.de

Abstract. Operation and configuration of Cyber-Physical Systems (CPSs) require approaches for managing the variability at design time and the dynamics at runtime caused by a multitude of component types and changing application environments. As a contribution to this area, this paper proposes to integrate concepts for variability management with approaches for self-organization in intelligent systems. Our approach exploits the idea of self-contextualization to autonomously adapt behaviors of multiple services to the current situation. More concrete, we put the "context" of CPS into the conceptual focus of our approach and propose context variants for use in self-contextualization of CPS. The main contributions of this paper are to identify challenges in variability management of CPS based on an industrial case, the integration of context variants into the reference model for self-contextualizing services and an initial validation using a case study.

Keywords: Cyber-physical systems · Self-organization · Self-contextualization · Context variation

1 Introduction

Cyber-Physical Systems (CPSs) tightly integrate physical, and IT (cyber) worlds based on interactions between these worlds in real time. Such systems rely on communication, computation and control infrastructures commonly consisting of several levels for the two worlds with various resources as sensors, actuators, computational resources, services, humans, etc. Operation and configuration of CPS require approaches for managing the variability at design time and the dynamics at runtime caused by a multitude of component types and changing application environments. This is a relatively new research field demanding for new approaches and techniques. As a contribution to this area, this paper proposes to integrate concepts from product line engineering for improved control of variability with approaches for self-organization in intelligent systems. More concrete, we put the "context" of CPS into the conceptual focus of our approach and propose context variants for use in self-contextualization of CPS.

The presented approach contributes to the concept of self-contextualization to autonomously adapt behaviors of multiple services to the context of the current situation in

W. Abramowicz (Ed.): BIS 2015, LNBIP 208, pp. 309–320, 2015.
DOI: 10.1007/978-3-319-19027-3_25

order to provide their services according to this context and to propose context-based decisions. For this reason the presented conceptual model enables context-awareness and context-adaptability of the service. Based on an application example of a CPS in transportation, we will illustrate selected challenges as starting point for our conceptual contribution on integrating variability management and self-organization. This contribution is an extension of previous work on self-contextualizing services with a focus on introducing context variants into the existing reference model. For this purpose, a certain degree of formality is required in CPS models, which will also be subject of the paper.

The main contributions of this paper are to identify challenges in variability management of CPS based on an industrial case, the integration of context variants into the reference model for self-contextualizing services and an initial validation using a case study. The remaining part of the paper is structured as follows: Sect. 2 will give a brief overview to background for this work including existing self-organization approaches and concepts for variability management. Section 3 presents an industrial case for motivating context variants and illustrating challenges regarding variability and self-organization. Section 4 introduces the concept of self-organization in CPS including an extended reference model for self-contextualizing services, context variants and a formalization of the extension. Section 5 illustrates the use of context variants via a case study based on mobile robot interaction. Finally, Sect. 6 summarizes the paper and discusses future work.

2 Background

This section summarizes the conceptual background for our work with focus on self-organisation and context computing (2.1) and variability management (2.2).

2.1 Self-organisation and Context Computing

Self-organising systems are characterised by their capacity to spontaneously (without external control) produce a new organisation in case of environmental changes. These systems are particularly robust, because they adapt to these changes, and are able to ensure their own survivability [1].

Examples for research projects investigating self-organization are the EC FP7 projects SOCIETIES[1] and SENSEI.[2] The vision of SOCIETIES (Self Orchestrating CommunIty ambient IntelligEnce Spaces) is to develop a complete integrated solution via a Community Smart Space which extends pervasive systems beyond the individual to dynamic communities of users. SENSEI aimed at integrating the physical with the digital world of the network of the future. It produced: (i) a scalable architectural framework; (ii) an open service interface and corresponding semantic specification; (iii) network island solutions consisting of a set of cross-optimized and energy aware protocol stacks; (iv) pan European test platform.

---

[1] http://www.ict-societies.eu/partners/.
[2] http://www.sensei-project.eu/.

Context-computing plays an important role to enable services adapting to situations in CPS [2, 3]. Context-computing is first introduced in 1994 by [4]. They consider the context as the information about located-object and the changes to object over time. Dey [5] defined context as *"Context is any information that can be used to characterize the situation of an entity"*. With increasing mobility of users, increased performance and functionality of mobile devises and sensors, and increasing amount of information available, context computing also gains of importance in order to integrate circumstances and situations of the users, what is often referred to as human related context. Many different approaches were developed for representing a context using a formal language (e.g., UML [6] or OWL [7]) or informal language (e.g., Dey's approach to use the Context Toolkit to define the context through a GUI [8]).

2.2   Variability Modelling

Discussion of the characteristics of self-organization in Sect. 2.1 indicates that capturing and representing variations in sub-systems, sensors or other elements of CPS including the relationships or dependencies to other components is an essential task in context computing. The area of variability modeling offers concepts how to deal with variability in complex systems, which might be applicable for CPS and will be briefly presented in this section.

Variability modeling offers an important contribution to managing the variety of the variants of systems by capturing and visualizing commonalities and dependencies between features and between the components providing feature implementations. Since more many years, systematic management of variants is frequently used in the area of technical systems and in software product lines [9]. Feature models are one of the variability modeling approaches often used in product lines and product families. The purpose of a feature model is to extract, structure and visualize the commonality and variability of a set of products. Commonalities are the properties of products that are shared among all the products in a set, which places these products in the same category or family. Variability are the elements of the products that differentiate and show configuration options, variation points and choices that are possible between variants of the product and aim at satisfying different customer requirements. Variability and commonality is modelled as features and organized into a hierarchy of features and sub-features in the feature model, which sometimes is called feature tree. Feature diagrams are used to visualize the hierarchy and other properties of a feature model; they express the relation between features with the relation types mandatory, optional, alternative, required and mutual-exclusive. The exact syntax of feature diagrams is explained in [9].

The original concept of feature models was introduced by Kang et al. as "Feature-Oriented Domain Analysis" (FODA) [10]. The original definitions, notations and concepts used by FODA have been extended and modified as other application fields for feature models were identified. Many examples of requirement abstraction, architecture specification, etc. have been put forward over the years [7, 9]. FODA was followed by FORM (Feature-Oriented Reuse Method) [11]. A significant step towards formalizing feature models was taken by Czarnecki and Eisenecker in Generative Programming [12]; further refinement of the feature model notation was published in [13].

Recent work on variability modelling also addresses the field of services and service line engineering. A method for service line engineering is proposed in [14] that bundles all variations of a Software-as-a-Service (SaaS) application based on a common core. [15] work in the same area and use variability models to derive customization and deployment information for individual SaaS tenants.

Unlike conventional distributed agent-based systems, the resources of CPS interact in both cyber and physical space. For this reason, the mechanisms developed for agent-based systems in most cases are not efficient in CPS [16]. This paper is trying to make a step further in the area via adding the context variability management as a mechanism enabling self-contextualization and increasing flexibility in service-based CPS.

3   Case Study from Transportation

The case study in this section is presented with the intention to provide a motivating case for work on context variation and self-contextualization. The case is based on an industrial project from transport and logistics industries. The logistics industry makes intensive use of modern information technology and CPS [17] for achieving high efficiency of processes and solutions in a globalized market. One of the world's largest truck manufacturers designed and implemented new transport related services based on vehicles (e.g. trucks, trailers, cranes, warehouse doorways, etc.) and IT systems (e.g. on-board units in vehicles, traffic control systems and fleet management systems). The aspect of the case relevant for this paper aims at using wireless sensor networks (WSN) in trailers for innovative logistics applications. Compared to the highly computerized trucks, most of today's trailers are poorly equipped with electronic systems.

The WSN is installed in the position lights of a trailer. Each light carries a sensor node able to network by ZigBee[3] with neighboring nodes and furnished with a radar sensor. This sensor can be used for various purposes, including protection of the goods on the trailer against theft, assistance to the driver of the truck (e.g. lane control, blind spot support) or surveillance of the trailer or its different compartments (e.g. by electronically sealing them). A gateway in the trailer is controlling the WSN in the position lights; the gateway communicates with the back-office of the owner of the trailer or the owner of the goods. Several services were developed within the project, which exploit the possibilities of combining sensor information and IT services. One of these services is additional protection of the trailer when parked against theft (see [18] for more details).

The WSN also can be used to communicate information from sensors positioned in the goods loaded on the trailer to the trailer's gateway and to the back-office. Such goods-related sensor information can be used for value-added services, like monitoring of refrigerated goods (temperature sensor in the cargo box), checking whether the goods have been securely steadied (using a motion sensor), or whether vessels with fluids remain upright or are in danger to topple over (using a piezoelectric sensor). For many of the services offered, observations acquired through the different sensors in the trailer have to be combined with information coming from other sources, like from resource

---

[3] http://www.zigbee.org.

planning systems. The technical solution developed in the industrial case consists of a knowledge base in the trailer's gateway, an orchestrated set of web-services using the knowledge base and a control system operating and monitoring the business services provided to the customers of the transportation company. When configuring the knowledge base for different business services and customers, rules defining relevant situations and required actions had to be defined. Configuring the system usually led to a large number of rules which often were nested or had to be evaluated in a defined order. The many variations hidden in these rules were not only difficult to grasp and control by the engineers involved but also caused performance issues.

A simple example shall be used to illustrate the need for controlling variability and at the same time high flexibility. Let us assume we have three sensor types built into the trailer (radar sensors in position lights, sensor to control backdoor, motion sensor for the cargo area) and three sensor types potentially in the cargo goods (temperature sensor, piezoelectric sensor, malfunction of transport box). Each of these sensor types might be evaluated on its own (e.g. "IF temperature_in_cargo too_high THEN lead truck to next service point"), in combination with another sensor (e.g. "IF temperature_in_cargo too_high AND goods_not_securely_steadied THEN stop truck to check trailer and goods") or in combination with another sensor and external information from information systems and back office (e.g. "IF temperature_in_cargo too_high AND goods_not_securely_steadied AND truck_close_to destination THEN reduce speed and continue to destination"). If only 3 out of these 6 sensor types can be combined there would be already 41 potential combinations. Furthermore, for the cargo sensor types, there will be a variable number of actual sensors in the trailer depending on what goods are loaded and it is not known in advance whether or not a customer booked a certain service. Also, new types of sensors have to be integrated into the overall systems as a basis for new services. This requires a high flexibility and configurability of the knowledge base and system behavior.

4   Context Variants for Service Self-contextualization

Our approach for increasing flexibility and controlling variability in CPS such as the one introduced in Sect. 3 consists of the principle of self-contextualization (Sect. 4.1) and a reference model developed for this purpose which is enhanced by integrating context variants for better variability control (Sect. 4.2).

4.1   Self-contextualization in CPS

The cyber and physical spaces of CPS are represented by sets of resources. The resources have some functionality in result of which they provide services. The services provided by one resource are consumed by other resources. Since the resources are numerous, mobile, and with a changeable composition, the CPSs belong to the class of variable systems with dynamic structures. Restriction to only planned resource interactions in such systems is only a theoretical option, in practice this basically is just impossible. Resource self-organization is the most efficient way to organize interactions and communications between the resources making up CPSs. In order to achieve the dynamics of the self-organizing system, its components have to be creative, knowledgeable, active, and social. The resources that are parts of a system permanently change their joint environment what results in a synergetic collaboration and leads to achieving a certain level of collective intelligence.

In order for distributed systems like CPSs to operate efficiently, they have to be provided with self-organization mechanisms. In a CPS such mechanisms concern self-organization of CPS' resources. The goal of the resource self-organization is support of humans in their decisions, activities, solution of the tasks, etc. At that, humans are the participants of the self-organization process, as well.

The process of self-organization of a network assumes creating and maintaining a logical network structure on top of a dynamically changing physical network topology. The autonomous and dynamic structuring of components, context information and resources is the essential work of self-organization [19]. The network is self-organized in the sense that it autonomously monitors available context in the network, provides the required context and any other necessary network service support to the requested services, and self-adapts when context changes.

Due to the nature of CPS, semantics is one of the necessary bases to ensure that several resources arrive at the same meaning regarding the situation and data/information/knowledge being communicated. Ontologies provide for a shared and common understanding of some domain that can be communicated across the multiple CPS' resources. They facilitate knowledge sharing and reuse in open and dynamic distributed systems and allow entities not designed to work together to interoperate [20].

The present research inherits the idea of ontology usage for modelling context in CPSs. According to [21], any information describing an entity's context falls into one of five categories for context information: individuality, activity, location, time, and relations. The individuality category contains properties and attributes describing the entity itself. The category activity covers all tasks this entity may be involved in. The context categories location and time provide the spatio-temporal coordinates of the respective entity. Finally, the relations category represents information about any possible relation the entity may establish with another entity.

The context is purposed to represent only relevant information and knowledge from the large amount of those. Relevance of information and knowledge is evaluated on a basis how they are related to a modelling of an ad hoc problem. Resource's context is described by location, time, resource individuality, and event. Resources perform some activity according to the roles they fulfil in the current context and depending on the type of event. On the other hand, the type of activity that a resource performs defines the type of event. For example, the event of a phone call defines the human activity as answer the phone. But, when a person raises the hand at the lecture time, this activity defines an event as, for instance, lecture interruption. This explains bi-directionality of 'defines' relationship between event and activity. The context is updated depending on the information from the service's environment and as a result of its activity. The ability of a system (service) to describe, use and adapt its behavior to its context is referred to as self-contextualization [22].

4.2   Reference Model of a Self-contextualizing Service

Figure 1 represents the developed conceptual model for service's self-contextualization mechanism based on the context variation concept. The model includes the following concepts:

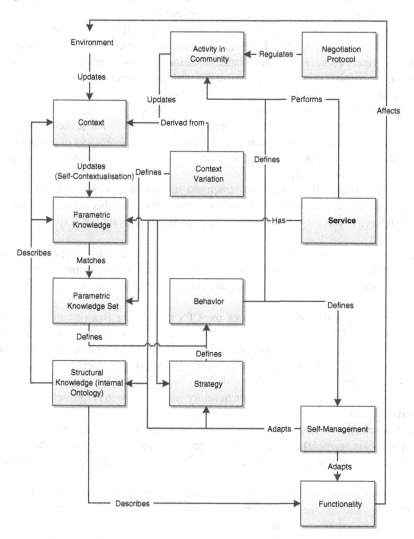

Fig. 1. Reference model of a self-contextualizing service

Service is used to represent CPS' resources. It is an acting unit of the self-organization process. The service has *structural and parametric knowledge* and predefined *strategies*. It performs some *activity in the community*.

Structural Knowledge is a conceptual description of problems to be solved by the *service*; the service's internal *ontology* represents this kind of knowledge. The internal ontology harmonizes with the common ontology. The structural knowledge describes the structure of the service's *parametric knowledge* and the structure and the terminology of the service's *context*. Depending on the situation the structural knowledge can be modified (adapted) by the *self-management* capability.

Parametric knowledge is the knowledge about the actual situation. This knowledge is the structural knowledge filled with the information characterizing this situation.

The service's *context* characterizes the situation of the service. It is represented by means of the service's internal *ontology* and is updated depending on the current event, location, and time. Information from the service's *environment* and results of its *activity in the community* define events and vice versa. The context updates the service's *parametric knowledge*, which in turn defines the service's *behavior*. The presented approach exploits the idea of self-organization to autonomously adapt *behaviors* of multiple services to the situation in order to provide their *services* according to this context and to propose context-aware decisions.

Context variation is a predefined structured sub-set of all potential contexts of an service for defining constraints in behavior of the service for this sub-set. *Parametric knowledge set* is the sub-set of parametric knowledge relevant for a specific context variant.

Strategy is a pre-defined plan of actions rules of action selection to change the service's own state and the state of the *environment* from the current to the preferred ones. The strategy defines the service's *behavior*. The service can modify its strategy through *self-management*.

Environment is the surroundings of the CPS the service is a part of, which may interact with the CPS. The environment produces events, which in turn affect the service's *context*. The service can affect the environment if it has appropriate *functionality* (e.g., a manipulator can change the location of a corresponding part).

Self-Management is a service's capability achieved through its *behavior* to modify (reconfigure) its internal *ontology*, *functionality*, and *strategy* in response to changes in the environment.

Behavior is the service's capability to perform certain actions (*activity in community* and/or *self-management*) in order to change the own state and the state of the *environment* from the current to the preferred ones. The *behavior* is defined by the service's *strategies*.

Functionality is a set of cyber-physical functions the service can perform. Via the functionality, the service can modify its *environment*. The service's functionality can be modified in certain extent via the *self-management* capability.

Activity in community is a capability of the service to communicate with other services and negotiate with them through the service's *behavior*. It is regulated by the *negotiation protocol*.

Negotiation protocol is a set of basic rules so that when services follow them, the system behaves as it supposed to. It defines the *activity in community* of the services.

5   Context Variation Usage for Mobile Robot Interaction

The objective of this section is to illustrate usage of context variation in a simple but practical case. Since the transportation case study introduced in Sect. 3 is an operational solution and not easily available for testing of new approaches, we had to select another case and decided to use the case of robots jointly solving a task. Two robots have to find an object and bring it to a storage (Fig. 2). Only one robot should handle this task. For this purpose, robots should interact to find the one who will bring the object to the storage. Originally, the robots do not know about the positions of other robots. Interaction process includes the object finding, distance measurement and information sharing between robots [23].

- The whole task can be split to the several subtasks:
- Task 1. Robots scan the area around. While scanning the robot controls the incoming data from the gyroscopic and ultrasonic sensors and records turn angle with measured distance.
- Task 2. Robots identify objects. The measurements from the previous step are processed. Processing determines objects that were captured by the ultrasonic field of vision. The result of parsing is the angle where the object was found and distance to this object. In Fig. 3 "Robot 1" finds two objects. First is on the angle 1.1, at distance 1.1; and second is on the angle 1.2 at distance 1.2. The same situation is for the "Robot 2".
- Task 3. Robots identify other robots. They share found objects through the smart space and compare the distances to objects. It can be concluded that since distance 1.1 is not equal to distance 2.1 and distance 1.2 is equal to distance 2.2 (Fig. 3), the angles 1.2 and 2.2 point to robots, and angles 1.1 and 2.1 point to the object. If there are more than two equal distances, robots should check that the object is not the other robot. For this purpose random robot moves to an object and tasks 1, 2, and 3 are repeated.
- Task 4. Each robot should interoperate with another robot and decide who will move towards the object. The robots choose the closest robot to the object, which then moves to the object.
- Task 5. Selected robot carries out the defined task with the object.

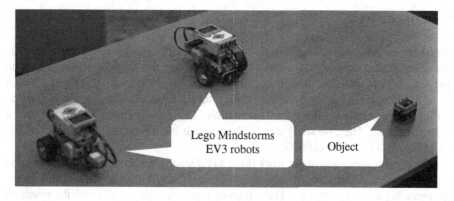

Fig. 2. Scenario implementation

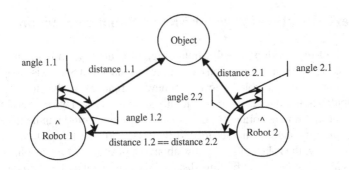

Fig. 3. Object finding scenario

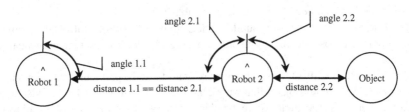

Fig. 4. Situation illustrating context variation 2

The context variation significantly simplifies the implementation of task 3. The variations can correspond to different situations with predefined solutions. Below, two examples of context variations are presented.

Context variation 1: distance 1.2 = distance 2.2; distance 1.1 > distance 2.1

This context variation corresponds to the situation presented in Fig. 3. In this situation robot 2 has to turn to angle 2.1 and move to the object.

Context variation 2: distance 1.2 not available; distance 1.1 = distance 2.1

This context variation corresponds to the situation presented in Fig. 4. In this situation robot 2 has to turn to angle 2.2 and move to the object.

6   Conclusions

Based on an industrial example from CPS in transportation, this paper motivated the introduction of context variants into modeling CPS and the need for self-organization and self-contextualization in such systems. We extended the existing reference model for self-contextualizing services accordingly and formalized the concept of context variant. As a proof-of-concept implementation, a case study demonstrated the use of context variants. The case study clearly showed that implementation of certain tasks is simplified using context variants.

The biggest limitation of our work is the so far small number of cases where context variants were used and the missing investigation of the impact of the changes in the reference model (see Sect. 4.2). The reference model was developed with the intention to serve as basis for implementing services for use in CPS. We found it valuable to conceptualize the overall behavior of CPS. Although we do not expect that the original purpose will be affected, this aspect still has to be investigated.

Future work will include conceptual and technical activities. From a technical perspective, implementation of context variants in the transportation CPS introduced in Sect. 3 is one of the planned activities. As the knowledge base used in the industrial case already supports the concept of aggregating different sensor types to "situations", we expect no principle problems when introducing context variants, as this concept to some extent also includes aggregation of elements. However, the flow of execution and the way to define customer services using the knowledge base and context variants will have to be changed. From a conceptual point of view, we plan to investigate effect of using context variants on the engineering process of CPS. The specification of variations probably has to be part of the design and specification of the overall CPS, which also will affect requirement elicitation.

Acknowledgment. The research was supported partly by projects funded by grants # 14-07-00378, # 14-07-00345, # 14-07-00363 of the Russian Foundation for Basic Research. This work was also partially financially supported by Government of Russian Federation, Grant 074-U01.

References

1. Serugendo, G.D.M., Gleizes, M.-P., Karageorgos, A.: Self-organisation and emergence in MAS: an overview. Informatica **30**, 45–54 (2006)
2. Preuveneers, D., Berbers, Y.: Internet of things: a context-awareness perspective. In: Yan, L., Zhang, Y., Yang, L.T., Ning, H. (eds.) The Internet of Things: From RFID to the Next-Generation Pervasive Networked Systems, pp. 287–307. Auerbach Publications, Taylor and Francis Group, New York (2008)
3. Zhang, D., Huang, H., Lai, C.-F., Liang, X., Zou, Q., Guo, M.: Survey on Context-Awareness in Ubiquitous Media. Multimedia Tools Appl. **67**, 1–33 (2011)
4. Schilit, B.N., Theimer, M.M.: Disseminating active map information to mobile hosts. Network, IEEE **8**(5), 22–32 (1994)
5. Dey, A.K.: Understanding and using context. Pers. Ubiquitous Comput. **5**(1), 4–7 (2001)
6. Henricksen, K., Indulska, J., Rakotonirainy, A.: Modeling context information in pervasive computing systems. In: mattern, f, Naghshineh, M. (eds.) PERVASIVE 2002. LNCS, vol. 2414, pp. 167–180. Springer, Heidelberg (2002)
7. Wang, X., Zhang, D., Gu, T., Pung, H.: Ontology based context modeling and reasoning using OWL. In: Proceedings of the Second IEEE Annual Conference on Pervasive Computing and Communications, pp. 18–22 (2004)
8. Dey, A.K., Salber, D., Abowd, G.D.: A conceptual framework and a toolkit for supporting the rapid prototyping of context-aware applications. In: Moran, T.P., Dourish, P. (eds.) Context-Aware Computing, Special Triple Issue of Human-Computer Interaction, vol. 16, pp. 229–241 (2001)
9. Thörn, C., Sandkuhl, K.: Feature modeling: managing variability in complex systems. In: Tolk, A., Jain, L.C. (eds.) Complex Systems in Knowledge-based Environments: Theory, Models and Applications. Studies in Computational Intelligence, vol. 168, pp. 129–162. Springer, Heidelberg (2009)

10. Kang, K., Cohen, S.G., Hess, J.A., Novak, W.E., Peterson, S.A.: Feature-oriented domain analysis (FODA) - feasibility study. Technical Report CMU/SEI-90-TR-21, Carnegie-Mellon University (1990)
11. Kang, K.C., Kim, S., Lee, J., Kim, K., Shin, E., Huh, M.: FORM: a feature-oriented reuse method with domain-specific reference architectures. Annals of Softw. Eng. **5**, 143–168 (1998)
12. Czarnecki, K., Eisenecker, U.: Generative Programming. Addison-Wesley, Reading (2000)
13. Riebisch, M.: Towards a more precise definition of feature model. In: Riebisch, M., Coplien, J.O., Streitferdt, D. (eds.) Modelling Variability for Object-Oriented Product Lines. BookOnDemand Publ. Co, Norderstedt (2003)
14. Walraven, S., Van Landuyt, D., Truyen, E., Handekyn, K., Joosen, W.: Efficient customization of multi-tenant software-as-a-service applications with service lines. J. Syst. Softw. **91**, 48–62 (2014)
15. Mietzner, R., Metzger, A., Leymann, F., Pohl, K.: Variability modeling to support customization and deployment of multi-tenant-aware software as a service applications. In: Proceedings of the 2009 ICSE Workshop on Principles of Engineering Service Oriented Systems, pp. 18–25. IEEE Computer Society (2009)
16. Lin, J., Sedigh, S., Miller, A.: A semantic agent framework for cyber-physical systems. In: Elçi, A., Koné, M.T., Orgun, M.A. (eds.) Semantic Agent Systems. SCI, vol. 344, pp. 189–213. Springer, Heidelberg (2011)
17. European Road Transport Research Advisory Council (2010) ERTRAC's Strategic Research Agenda: Towards a 50 % more efficient road transport system by 2030. ERTRAC, October 2010. http://www.ertrac.org/. Accessed 04 Jan 2011
18. Sandkuhl, K., Borchardt, U., Lantow, B., Stamer, D., Wißotzki, M.: Towards adaptive business models for intelligent information logistics in transportation. In: 11th International Conference, *BIR 2012*. Higher School of Economics, Nizhny Novgorod, Russia (2012)
19. Ambient Networks Phase 2. Integrated Design for Context, Network and Policy Management, Deliverable D10.-D1 (2006). http://www.ambient-networks.org/-Files/-deliverables/-D10-D.1_PU.pdf. Accessed 09 September 2014
20. Hong, J., Suh, E., Kim, E.: Context-aware systems: a literature review and classification. Expert Syst. Appl. **36**, 8509–8522 (2009)
21. Zimmermann, A., Lorenz, A., Oppermann, R.: An operational definition of context. In: Kokinov, B., Richardson, D.C., Roth-Berghofer, T.R., Vieu, L. (eds.) CONTEXT 2007. LNCS (LNAI), vol. 4635, pp. 558–571. Springer, Heidelberg (2007)
22. Raz, D., Juhola, A.T., Serrat-Fernandez, J., Galis, A.: Fast and Efficient Context-Aware Services. John Willey, New York (2006)
23. Teslya, N.: Smart space-based Lego® mindstorms EV3 robots interaction. In: Proceedings of the 16th Conference of Open Innovations Association FRUCT, pp. 195–200 (2014)

EPOs, an Early Warning System for Post-merger IS Integration

Erol Unkan and Barbara Thönssen[✉]

Hochschule Für Wirtschaft FHNW, 4600 Olten, Switzerland
Erol.Unkan@bluewin.ch, Barbara.Thoenssen@fhnw.ch

Abstract. Our approach allows for identifying and assessing indicators that signal the risk of failure in post-merger integration of information systems with the help of an early warning system. The applied research method is Design Science Research complemented by a real case study. The main artifact is the EPOs prototype we developed, applying semantic technologies, namely an ontology and an inference engine, for the automatic identification, validation and quantification of risk indicators. If a certain threshold is reached, a warning message is triggered including the ponderosity of the risk to fail. We evaluated our approach based on two real world scenarios.

Keywords: Risk · Post-merger integration · Early warning system · Semantic technology

1 Introduction

According to the Institute of Mergers, Acquisitions and Alliances (IMAA) [1] in 2014 alone in Europe about 14,000 announced mergers and acquisitions with about 1,400bil EUR value of transactions will be performed. However, most mergers & acquisitions (M&A) fail [2, 3] and they usually fail in the post-merger integration (PMI) phase [4]. One of the most critical factors for success in combining two or more organizations, while at the same time being highly susceptible to failure, is the integration of information systems (IS) [5–7]. This is not only very costly in various facets such as time, people and ultimately money, but also avoidable [3]. Furthermore, the problem concerns not only private companies, but also public organizations.

The work at hand focuses on IS PMI from a business perspective, i.e. on information system integration in the post-merger phase of M&A in academic organizations such as universities. M&A in universities are no rarity [8] and are as important as their private company counterparts. However, literature about risks in IS integration is scarce in general and even more about IS integration in academic organizations. Our work aims for closing the gap and we will show that it is possible to identify and assess risk indicators that signal the risk of failure in post-merger integration of information systems in academic organizations with the help of an Early Warning System.

The chosen method for our work is Design Science Research complemented by a study of a real case. The main artifact is the prototype of EPOs (Early Warning System for Post-Merger IS Integration) we developed, applying semantic technologies. The evaluation was carried out based on two real cases.

© Springer International Publishing Switzerland 2015
W. Abramowicz (Ed.): BIS 2015, LNBIP 208, pp. 321–333, 2015.
DOI: 10.1007/978-3-319-19027-3_26

The paper is structured as follows: In Sect 2 the motivating scenario and related research is provided. We continue with the chosen methodology in Sect 3. In Sect 4 we describe our approach in detail. In Sect 5 we describe the technical implementation of our approach. In Sect 6 we provide evaluation results and we close in Sect 7 with a conclusion and an outlook.

2 Motivating Scenario and Related Research

Mergers & Acquisitions is a collective description for a series of related legal corporate activities with the purpose of combining (sometimes only parts of) two or more organizations. In most discussions on this topic, the terms 'merger' and 'acquisition' are used interchangeably [9].

As statistics show, M&A increased from less than 5,000 to more than 50,000 deals per year world-wide [1] and a variety of research was done on why most mergers and acquisitions fail to deliver. According to different sources only about 20 % of all mergers really succeed [2]. A long term study over thirty years from Business Week in [3] and a study in Lajoux in [3] covering over seven thousand M&A show that only about 55 %-77 % deliver the announced financial promises and 61 % of deals made since 1998 decreased shareholder value [9]. The cost of failure is the destruction of shareholder value and possibly the destruction of the business [2, 4]. According to McGrath [4] 53 % of M&A fail in the post-merger integration phase, followed by 30 % in the prelude phase and only 17 % fail in deal negotiation, pre-change of control and change of control phase. There are many reasons discovered why M&A fail (amongst others by [3, 9]). Mehta [9] regards information systems and information technology (IT) integration problems as the main reasons for failure and states that the lack of attention attributed to the merging of IS and IT was one of the main reasons for failures in the 1980's merger wave.

Since it has shown that failure of M&A largely occurs in the post-merger phase and IS integration problems are considered critical to the realization of merger objectives and expected synergies, [9] our research focuses on this.

Little research has been done on M&A in academic organizations [10]. It represents only a tiny fraction of what is found in the business sector and mainly consists of studies that lack clear theoretical orientations [8]. Furthermore, main studies conducted in this field in the 1970's are focused on either the initiation or the negotiation stage [8], without much consideration for what happens after the merger, and may not represent today's world correctly. However, M&A in academic organizations are a common incident [8].

Risk assessment allows an organization to consider the extent to which potential risks have an impact on achieving objectives. When assessing risks, their probability is to be determined, i.e. how likely it is that a risk will occur [11] as well as their severity. These two classifications can be combined into one aspect which McGrath [4] calls the risk significance and measures it on a four-point scale: low risk, moderate risk, medium risk, high risk.

In addition a risk mitigation can be used to determine the priority of risk mitigation and is based on the risk significance and a mitigation effort. McGrath [4] measures this aspect on a three-point scale: low priority, medium priority, high priority.

When considering risks, qualitative and quantitative metrics that provide insight into potential events must be determined. Such metrics are called risk indicators [12] – also named Key Risk Indicator or KRI. We regard the measuring of KRI with appropriate metrics central element of an early warning system as according to Gilead [13, p. 109] "The central objective of early warning systems is to prevent surprises".

However, up to our knowledge there is few literature found discussing the risk of IS integration in M&A, and none focusing on the risk of IS integration in academic M&A. Lundqvist's [10] work is the most extensive found on IS integration in public organizations but no risk catalogue can be found that covers a complete list of relevant risks, respectively risks factors, the risk indicators and their metrics including probability and impact.

3 Methodology

For our work we chose Design Science Research (DSR) [14]. We consider DSR appropriate as design activities are a central element in applied disciplines, particularly in the information technology field. Design Science is knowledge in the form of constructs, techniques and methods – the know-how for creating artifacts to satisfy a given set of functional requirements [15].

Our research choice is the qualitative multi-method [16], i.e. we use a combination of several techniques in our research, namely: study of real world application scenarios, questionnaire, interviews and the procedure for ontology design and evaluation as proposed by Grüninger and Fox [17] (Fig. 1).

Since real world scenarios and ontology development play a pivotal role in our work we will briefly explain them in the following.

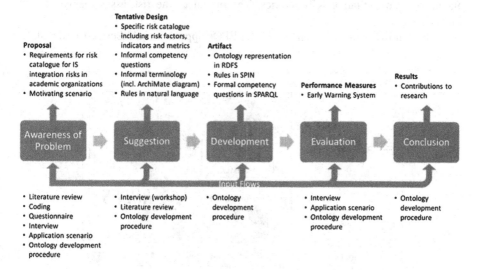

Fig. 1. This figure shows our research design. In the middle the DSR phases are depicted. Below each phase the applied techniques are listed; above each phase the achieved results are shown.

The scenario considered for our approach is the merger of various universities of applied sciences in Switzerland resulting in the now-called "Fachhochschule Nordwestschweiz" (FHNW; University of Applied Sciences and Arts Northwestern Switzerland). Multiple semi-formal interviews were conducted with the Head of ICT at FHNW, who had decision making competence during the merger. From the application scenario we derived a risk catalogue with corresponding KRI. We used the same scenario for evaluation complemented by a second one taken from the business sector.

Since ontology engineering is a challenging task, we followed Bertolazzi et al. [18], who argue that ontology engineering should not start from scratch but propose to use an already existing model, they call the Core Enterprise Ontology (CEO). A CEO should contain the most general business concepts, common to the majority of enterprises, independently of the specific activity field. For our work we use the ArchiMEO ontology [19, 20] as the CEO. What was not covered by ArchiMEO, namely concepts and relations related to the risk domain, were derived from the competency questions as introduced by Grüninger and Fox [17].

In addition McGrath's [4] framework for cognitive risk identification and measurement (CRIM) provided a solid fundament in the problem awareness and suggestions phase.

4 The EPOs Approach

Our main contribution, the EPOs ontology is in the center of our approach. As shown in Fig. 2, the EPOs ontology consists of the ArchiMEO ontology extended by the newly developed Risk Ontology (RO). Input for EPOs is a risk catalogue with 20 risk factors. Each risk factor is determined by one to six key risk indicators (KRI) with qualitative or quantitative KRI metrics. The output of the risk assessment is a risk exposure report.

In the following the components of the EPOs approach are described in detail.

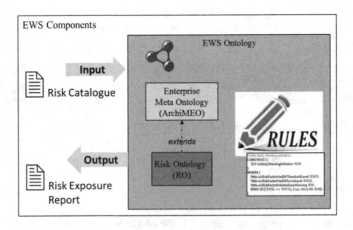

Fig. 2. This figure depicts the EPOs components.

4.1 Risk Catalogue

In our approach we focus on *one* risk, that is the risk of failure of the IS integration in the post-merger integration phase, in short: IS PMI. Founding on McGrath's [4] recommendations a risk catalogue with risk factors that impact this risk was elaborated based on the FHNW application scenario. The risk catalogue contains a set of risks factors relevant for IS PMI. For each risk factor one or more indicators are defined along with the metrics to measure them. If a certain threshold is reached, an early warning message is issued indicating that IS PMI might be on risk.

Risk Factors. Each risk factor is described by a metadata set providing details on the risk factors, such as description, domain information etc. and also information about significance, priority and mitigating efforts.

Risk factors are grouped into risk domains according to Alaranta [21] and Heller [22]: process domain risks (regarding the transformation of two separate IS configurations into one); content domain risks (regarding the new integrated IS configuration); and context domain risks (regarding the broader organizational, cultural, and geographical landscape in which the integration unfolds) (Table 1).

Table 1. This table shows the metadata set. The metadata elements are explained in the second row, while an example is given in the third row. The whole catalogue can be found at http://www.thoenssen.ch/Research.

Risk catalogue for the integration of information systems in academic M&A		
Risk factor ID	A unique identifier.	RF01
Short name	The name of the risk factor.	Process drift
Description	A detailed description of the risk factor.	The reality differs from the IS integration project plan.
Domain	The IS domain of the risk: Process, Content or Context.	Process
Probability	The probability of the risk: Probable, Likely, Less Likely or Unlikely.	Likely
Impact	The impact of the risk: Critical, Severe, Serious, Substantial, Moderate or Minor.	Critical
Significance	The multiplication of the probability and impact: High, Medium, Moderate or Low.	High
Mitigation	The mitigation level of the risk: Overly Mitigated, Well Mitigated or Negligently Mitigated.	Well mitigated
Mitigation E.	The mitigation effort of the risk: Significant, Moderate or Low.	Moderate
Priority	The multiplication of the significance and the mitigation effort: High, Medium or Low Priority.	High
Source	The source of the risk factor: Interview and/or questionnaire and/or internal document and/or literature.	Questionnaire

All risk factors are deliberately chosen to have no dependencies within the risk catalogue (however, the key risk indicators are related to each other). The purpose behind this approach is to clearly separate each risk factor from one another to be able to pinpoint – without any interferences – where the problems are. Hence, it is possible to freely extend the EWS with additional risk factors or modify it by deleting (or simply ignoring) unsuitable ones, without the need to change the internal structure of the calculations.

The calculations of the significance of a risk factor, as well as the mitigation priority, are based on McGrath's [4] risk probability, impact and mitigation aspects. However, he does not propose any specific formula to calculate the significance or priority. We adapted the calculation method to specify the exact range for each factor's risk significance and risk mitigation effort level based on the risk ranking method presented by Partnerships British Columbia [23].

Key Risk Indicators (KRI). In the risk catalogue to each risk factor one or more KRI including KRI metrics are assigned – These are the basis for all calculations regarding the initiation of early warnings. KRI in EPOs come in many variations: risk factors can be indicated only by quantitative KRI, only by qualitative KRI, or a mixture of both. This allows for a greater variety of monitoring and is in accordance with Grosse-Ruyken and Wagner [24] who state that a good balance between quantitative and qualitative measures, amongst others, is needed for an efficient and effective EWS monitoring. Each KRI in return consists of usually four KRI metrics with levels of 0, 1, 2 or 3. These KRI metrics are individual to each KRI and, depending on the nature of the KRI, can be anything from percentages, temporal values, monetary values or qualitative values. In some cases KRI can also have Boolean attributes which limits the levels of a KRI metric to either false or true (Table 2).

In the current prototype of EPOs information is not extracted automatically from sources, as done for example in the approach of Emmenegger et al. [25] and Laurenzi [26]. Data input concerning the KRI is entered manually by either a person in charge for risk management or by a group of persons, as done in the evaluation of our approach – an effort of two man hours is needed. The risk catalogue consists of 20 risk factors including a total of 52 key risk indicators with 188 KRI metrics.

4.2 Calculation of Early Warnings

As mentioned, each KRI consists of a KRI metric with a KRI metric score of 0, 1, 2 or 3. For each KRI a threshold was defined by when an early warning is triggered. The threshold is defined by the risk manager, or another responsible person, and represents the highest score he/she is willing to tolerate, i.e. the early warning threshold indicates the risk tolerance. In case a risk factor is determined by more than one KRI, each KRI has a percentage ratio of at least 0.1 (10 %) depending on its importance. Adding up the weightings of all KRI that determine a risk factor must result in 1 (100 %). As depicted in Fig. 3, the risk factor RF01 is determined by three KRI. KRI11: Process delay is considered more important with a weighting of 0.6 than KRI12 and KRI13 with weightings of 0.2 each.

Table 2. In this table 'EWSL' stands for early warning score level and is between 0 and 3; it is not indicated, since it is a dynamic value that is calculated by the early warning system. 'T' stands for the early warning threshold and is 1, 2 or 3 (i.e. low risk tolerance, medium risk tolerance and high risk tolerance). 'S' is the KRI score and also a dynamic value; it can be between 1 and 3. 'W' is the weighting of the KRI and is between 0.1 and 1. In the KRI metrics column it is specified which metric is applicable for the KRI. Conveniently to each risk factor and KRI an 'ID' is assigned. Refer to Sect. 4.2 for details on the calculation of 'EWSL' and 'S'.

EWSL	T	Risk Factor ID	S	W	KRI ID	KRI Metrics
	1	RF01: Process drift	.6		KRI11: Process delay	3: > 4 weeks
						2: > 2 - <=4 weeks
						1: > 1 - <=2 weeks
						0: <=1 week
			.2		KRI12: Increased costs	3: > 20 %
						2: > 15 % - <=20 %
						1: > 10 % - <=15 %
						0: <=10 %
			.2		KRI13: Functionalities that cannot be implemented	3: > 25 %
						2: > 20 % - <=25 %
						1: > 15 % - <=20 %
						0: <=15 %

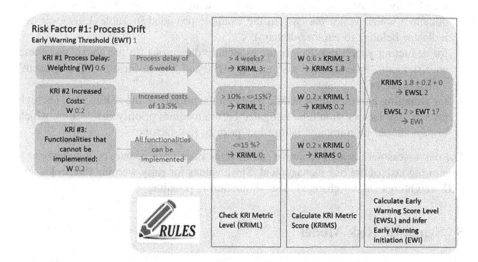

Fig. 3. The visualization an early warning initiation calculation.

In EPOs thresholds for early warnings as well as the percentage ratio for more than one KRI per risk factor were derived from findings gained from the study of the FHNW case. However, these values can be considered as default to be changed according to an end user's risk scoring and risk affinity before risk assessment is done.

Calculation of the Early Warning Score Level. The early warning score level is calculated by adding the KRI metric score formula $EWSL = KML * KW$ (where EWSL = early warning score level, KML = KRI metric level, and KW = KRI weighting) for each KRI of a risk factor. The resulting formula is: $EWSL = \sum_{i=1}^{n} KML_i \times KW_i$. The early warning score level can be between 0.0 and 3.0, i.e. the lowest possible level and the highest possible level.

Figure 3 details the calculation procedure of an early warning initiation. All rounded rectangles (incl. Risk Factor) represent concepts implemented in EPOs. The large arrows represent the input that is needed for the calculations. SPIN rules are depicted beneath the concepts. Exemplary values are indicated in light grey color.

4.3 Risk Exposure Report

The output of EPOs - in case at least one early warning is triggered - is one or more early warning messages. Each early warning message is structured as follows: Firstly, the risk factor denomination including its significance and priority are indicated. Secondly, it is indicated to which risk domain the risk factor belongs to. Thirdly, the early warning score (level) is shown and compared with the early warning threshold and it is stated that the early warning score is higher than the early warning threshold.

In the risk exposure report all imminent risks indicators are listed which allows

- for identifying main problem areas (in case many risk factors belong to the same domain - which is not the case in the example provided in Table 3, as each risk indicator belongs to another domain);
- for deducting which risk factors are especially dangerous (in the above case these are risk factor 'Process drift' with a score of 3.0 and 'No business and IS alignment' with a score of 2.7); and
- for arranging a mitigation schedule based on the priority of the risk factors, i.e. high priority risk factors should be mitigated before medium priority risk factors, which in turn should be mitigated before low priority risk factors.

The risk assessment described above can be done once or of course, continuously as a monitoring task of the risk management. Since values for early warning thresholds and KRI weightings can be changed, also simulation is possible.

5   Semantic Risk Model

In order to support the risk assessment and monitoring in a machine executable way that is also cognitively adequate for humans, the risk model is implemented using semantic technologies. Benefits of such an approach are widely accepted (cf. amongst others [20, 27, 28]). That is, concepts and relations are represented in a risk ontology depicted in Fig. 4.

Table 3. An example of such a risk exposure report is given in this table. In the early warning message all relevant information is summarized. The other rows display each element of the early warning message to provide a better overview.

Risk exposure report with initiated early warnings			
Early warning message	An early warning for risk factor 'Process drift' (High significance, High priority) is initiated. The risk factor belongs to the 'Process' domain. The early warning score is 3.0 and equal or higher than the early warning threshold of 1.	An early warning for risk factor 'No business and IS alignment' (low significance, low priority) is initiated. The risk factor belongs to the 'Content' domain. The early warning score is 2.7 and equal or higher than the early warning threshold of 2.	An early warning for risk factor 'Responsibilities bias' (moderate significance, medium priority) is initiated. The risk factor belongs to the 'Context' domain. The early warning score is 1.0 and equal or higher than the early warning threshold of 1.
Risk factor ID	RF01	RF12	RF16
Short name	Process drift	No business and IS alignment	Responsibilities bias
Domain	Process	Content	Context
Probability	Likely	Less Likely	Likely
Significance	High	Low	Moderate
Priority	High	Low	Medium
Early warning score	3.0	2.7	1.0
Early warning threshold	1	2	1

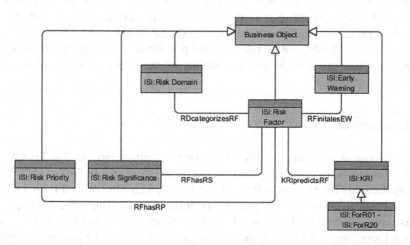

Fig. 4. This figure shows the concepts and sub-concepts for the risk ontology including relationships.

5.1 The Risk Ontology

Figure 4 shows the concepts and sub-concepts of the EPOs risk ontology. All concepts are specializations of the ArchiMEO 'Business Object' concept and share the namespace ISI (IS Integration). Each 'Risk Domain' categorizes one or many 'Risk Factors'. Each 'Risk Factor' initiates no or one 'Early Warning'. Each 'Risk Factor' is predicted by one or many 'KRI' (Key Risk Indicator); 'ForR01' to 'ForR20' are the specializations for 'KRI' – 20 sub-concepts are implemented. Each 'Risk Factor' has one 'Risk Significance'. Each 'Risk Factor' has one 'Risk Priority'.

5.2 Technical Implementation

EPOs is implemented as an extension of the ArchiMEO enterprise ontology [19]. ArchiMEO is a (re-)usable enterprise ontology based on the ArchiMate 2.1 standard [29]. ArchiMEO is provided by the University of Applied Sciences and Arts Northwestern Switzerland (FHNW) and is publicly available, licensed under a Creative Commons Attribution-ShareAlike 3.0 Unported License (URL: https://github.com/ikm-group/ArchiMEO). It is represented as an RDF(S)/OWL ontology, which allows for operative use in business applications [19]. EPOs is implemented using RDF Schema 3.0 and SPIN rules (inferencing rules) for calculating the various scores and initiating early warnings. For ontology and rule management we use TopBraid Composer™ Standard Edition from TopQuadrant [30].

Inferred properties are an integral part of EPOs and crucial not only for determining early warnings; inferencing is used as an intermediate to determine various calculations. SPIN rules) are executed by running the inference engine which results in creating new triples with deducted information from the original asserted data sets: the inferred properties of which the EWSL and the KRI scores are the most important.

All SPIN rules are attached to the concept classes and are thus part of the risk ontology. They are implemented as subclasses of spin:ConstructTemplates. This approach ensures firstly, that all rules are grouped together for easy identification and secondly, in most cases provides great flexibility when working with arguments instead of specific instances.

6 Evaluation

EPOs was evaluated within the context of two real-world scenarios. The FHNW scenario evaluated the EPOs approach in retro-perspective since the IS PMI of FHNW was already finished. The evaluation showed that the requirements (derived from the same scenario) were addressed correctly and the prototype was implemented accordingly. As it was a retro-perspective sight it could clearly be stated whether EPOs is able to weight KRIs appropriately and issue relevant warnings or not. It could be proved that a) with the help of EPOs imminent risk indicators can be identified and classified and b) that risk indicators that were implicit by the time of the merger, could have been made explicit with EPOs and hence, c) the risk of failure in IS PMI can be appropriately estimated.

Within the second scenario EPOs was evaluated in an on-going international joint venture between two companies of the private sector – a bank and a Swiss services/ software solutions provider. The goal was to evaluate the progress of the joint venture so far and to get an impartial basis for further decisions with the help of EPOs. Five interviews with employees were conducted in which the risk catalogues was perused and all risk indicators assessed. The answers of all interviewees were checked for contradictions and aggregated into one mean solution. Thresholds and weights remained unchanged. This evaluation proved that, although EPOs was intended to be used in academic M&A, it can also be used to assess Joint Ventures in the private sector. Even more, the evaluation showed that EPOs can also be applied in ongoing integration projects. Evaluating a case based on the assessment of KRI by a number of independent people has also a possible side effect: it allows to identify differences in ratings and hence, to become aware of potential hotspots and miscommunications. In addition changes to EPOs were made based on selected feedback.

The evaluation in both real world scenarios proved that EPOs' functionality of calculating KRI scores made the most severe problems explicit and hence, helps to reduce the risk of failure in post-merger IS integration.

7 Conclusion and Further Work

With our work we could close a gap in research concerning the risk of failure of the integration of information systems in the post-merger phase of M&A in academic organizations. Up to our knowledge no approach existed before that allows for identification, classification and assessment of IS PMI related factors that could indicate such a risk. The research result confirms that it is possible to identify and assess factors that influence the risk of information systems integration in M&A with the help of an early warning system.

As in our approach risk factors are considered independently from each other – what makes extensions easy and allows for changing thresholds and weights as desired – it would be worth to also research dependencies. As shown by Emmenegger et al. [25] risk factors could influence other risk factors positively or negatively and thus, strengthen or weaken warnings.

We could prove that our approach can also be used in other domains than academic M&A. Adding domain-specific risk catalogues, with respect to business sectors or subject of risk other than information system integration, could further support risk management in M&A.

References

1. IMMA.: Statistics (2014). http://www.imaa-institute.org/statistics-mergers-acquisitions.html. Accessed on 18 November 2014
2. Bruner, R.F.: Deals from Hell: M&A Lessons that Rise above the Ashes, p. 432. Wiley, Hoboken (2005)

3. Carleton, J.R., Lineberry, C.S.: Achieving Post-Merger Success: A Stakeholder's Guide to Cultural Due Diligence, Assessment and Integration, p. 240. Pfeiffer, San Francisco (2004)

4. McGrath, M.: Practical M&A Execution and Integration: A Step by Step Guide to Successful Strategy, Risk and Integration Management, p. 326. Wiley, Chichester (2011)

5. Alaranta, M., Henningsson, S.: An approach to analyzing and planning post-merger IS integration insights from two field studies. Inf. Syst. Front **2008**(10), 307–319 (2008)

6. Kovela, S., Skok, W.: Mergers and acquisitions in banking: understanding the IT integration perspective. Int. J. Bus. Manag. 7(18), 69–82 (2012)

7. Rentrop, C.E.: Informationsmanagement in der Post-Merger Integration. Erich Schmidt Verlag, Berlin (2005)

8. Wan, Y.: Managing Post-Merger Integration: A Case Study of a Merger in Chinese Higher Education. University of Michigan, Michigan (2008)

9. Mehta, M.C.: IT Integration Decisions During Mergers and Acquisitions. University of Houston, Houston (2005)

10. Lundqvist, S.: Searching for Keys to Successful Post-Merger Integration: A Longitudinal Case-Study Following a Public Sector Merger. Abo Akademi University Press, Abo (2012)

11. Lam, J.: Enterprise Risk Management: From Incentives to Controls. Wiley, Hoboken (2003)

12. The Institute of Operational Risk, Institute of Operational Risk Operational Risk Sound Practice Guidance Key Risk Indicators (2010)

13. Gilead, B.: Early Warning: Using Competitive Intelligence to Anticipate Market Shifts, Control Risk, and Create Powerful Strategies. American Management Association, New York (2004)

14. Hevner, A., Chatterjee, S.: Design Research in Information Systems, vol. 22, pp. 9–23. Springer, US (2010)

15. Vaishnavi, V., Kuechler, B.: Design Science Research in Information Systems (2004). Jacob-Burckhardt-Str. 1a. Accessed on 10 April 2014

16. Myers, M.D.: Qualitative Research In Business & Management, Reprint, p. 284. SAGE, London (2009)

17. Grüninger, M. Fox, M.S.: Methodology for the design and evaluation of ontologies. Toronto (1995)

18. Bertolazzi, P., Krusich, C., Missikoff, M.: An approach to the definition of a core enterprise ontology: CEO. In: International Workshop on Open Enterprise Solutions: Systems, Experiences, and Organizations, pp. 104–115 (2001)

19. Hinkelmann, K., Martin, A., Nikles, S., Thönssen, B.: ArchiMEO: representing archimate as an enterprise ontology. Olten (2014)

20. Hinkelmann, K., Maise, M., Thönssen, B.: Connecting enterprise architecture and information objects using an enterprise ontology (2013)

21. Alaranta, M., Mathiassen, L.: Managing Risks: Integration of Information Systems, IT Pro, pp. 30–40, January/February 2014

22. Heller, R.W.: Managing merger risk during the post-selection phase. Georgia State University (2013)

23. Partnerships British Columbia, Methodology for Quantitative Procurement Options Analysis (2011)

24. Grosse-Ruyken, P.T., Wagner, S.M.: APPRIS Project Report. Zürich (2011)

25. Emmenegger, S., Laurenzi, E., Thönssen, B.: Improving supply-chain-management based on semantically enriched risk descriptions. In: Proceedings of the International Conference on Knowledge Management and Information Sharing 2012. Emmerson (2011)

26. Laurenzi, E.: An Ontology and Inference Engine for the Assessment of Procurement Risk Management. University of Camerino, FHNW (2011)

27. Hinkelmann, K., Merelli, E., Thönssen, B.: The role of content and context in enterprise repositories (2010)
28. Uschold, M., Grüninger, M.: Ontologies: Principles, Methods and Applications. Edinburgh/ Toronto (1996)
29. The Open Group, ArchiMate® 2.1 Specification (2013). http://pubs.opengroup.org/ architecture/archimate2-doc/. Accessed on 04 July 2014
30. TopQuadrant Inc., TopBraid Composer™ Standard Edition. http://www.topquadrant.com/ tools/modeling-topbraid-composer-standard-edition/. Accessed on 30 September 2014

Open Data for BIS

The Adoption of Open Data and Open API Telecommunication Functions by Software Developers

Sebastian Grabowski[1(✉)], Maciej Grzenda[1,2], and Jarosław Legierski[1]

[1] R&D Center/Orange Labs, Orange Polska S.A,
ul. Obrzeżna 7, 02-691 Warsaw, Poland
{sebastian.grabowski,maciej.grzenda,
jaroslaw.legierski}@orange.com
[2] Faculty of Mathematics and Information Science,
Warsaw University of Technology, ul. Koszykowa 75, 00-662 Warsaw, Poland

Abstract. The primary objective of the work is the preliminary investigation of the adoption of Open Data and Open API telecommunication functions by software developers. The analysis is based on the statistical data collected during developer contests. Based on Open API Hackathon and Business Intelligence API Hackathon contests, the interest of software developers in telecommunication operator functions and Open Data exposed in Open APIs form is assessed. Conclusions on the categories of open city data attracting the attention of software developers are formulated.

Keywords: Open API · Open Data · API exposition · Contest · Hackathon

1 Introduction

In recent years, the idea of Open Data has gained particular attention. Open Data seen as a way of opening access to Public Sector Information (PSI), by publishing the data sets in machine readable format, has become a major concept for government administration. Many institutions and companies expose their data in an open form. Examples include portals such as: American data.gov registry [1], British data.gov.uk created in order to comply with the EU INSPIRE directive [3, 4], or the publicdata.eu portal of European Commission [2]. In business sector, Open Data portals include IATI Registry [5], Uber [6] or KPMG service [7, 8].

Key advantages of Open Data movement are enhanced public trust and confidence, which builds upon transparency [13]. However, Open Data realized at a local level such as a city level, provides also a platform for the promotion of proactive civic engagement [13]. It is important to note that the public advantages of Open Data are strongly related among other factors to local context. Among contextual information, having an impact on the success of Open Data policies, government structure, but also local culture are mentioned [16]. Hence, the observations made in one country may not directly transfer to another country with its own legal system and culture. This confirms the role of the studies on Open Data adoption at a country level.

© Springer International Publishing Switzerland 2015
W. Abramowicz (Ed.): BIS 2015, LNBIP 208, pp. 337–347, 2015.
DOI: 10.1007/978-3-319-19027-3_27

Importantly, recent studies emphasise the role of Open Data also as an enabler of innovation [14]. Among other purposes in this area, Open Data can be used for application development [14, 15]. In fact, some studies show even significant importance of Open Data for startup IT entrepreneurs. E. Lakomaa and J. Kallberg report that 43 % of surveyed Swedish startup IT entrepreneurs find Open Data essential for their business plans [14]. Except for the direct use of Open Data in innovative applications, its use to build evidence for the viability of the projects, provide information about potential market or enhance existing services e.g. by supplying reliable listing of correct delivery addresses are also observed [14].

At the same time, a growing number of Open Data sets is constantly published. However, this raises new questions about the way individual data sets are used and whether the data offerings match user needs. What is particularly interesting, is how to allocate the budget for the publication and updates of Open Data sets. A major input for the deciding upon the availability and frequency of the updates of individual data sets is the monitoring of the use of the Open Data [12]. From the innovation point of view, the use of the data for software application development is of particular interest.

The primary objective of this work is to investigate which Open Data sets are particularly interesting for software developers. This is one of key questions for institutions and companies making their data public. Another aspect is a technical one i.e. the category of API services to be used to publish the data. To provide empirical basis for such an analysis, the statistical data from developer contests (hackathons) describing the access to individual data sets is used. The analysis takes into account also the use of Open API exposed by a telecommunication operator during the contests. Hence, the empirical evidence arising from the events using both Open Data and commercial services, provides basis for this study. At the same time, these events directly corresponded to the idea of stimulating the use of Open Data, laid out among others by A. Zuiderwijk and M. Janssen [16].

As stated above, for software developers using Open Data, the way the data is published is of particular interest. For fine grained data access, the most desirable form of data exposure is the form of open programming interfaces – Open APIs. Open API interfaces are known in ICT and telecommunication sector for many years. Open API implementations have evolved over the last few years from emphasis on the SOA/SOAP (Service-Oriented Architecture/Simple Object Access Protocol), which is a standard in the corporate sector, towards the growing popularity of architectural model ROA/REST (Resource-Oriented Architecture/Representational State Transfer), used by many of the most popular Internet applications. The use of each of these two categories of Open APIs is also discussed below.

2 Open Data Exposed in Developers Contests

In the last two years, Research and Development Center of Orange Polska has organised contests dedicated for developers, namely Open API Hackathon organised in 2013, and Business Intelligence Hackathon API (BIHAPI) in 2014. Despite the fact that both contests included "hackathon" word in their official names, the duration of each contest (several weeks) was longer than the duration of a traditional hackathon.

This allowed for building more complicated applications than during one day hackathons.

2.1 Open API Hackathon

The first edition of a contest was conducted by Orange in cooperation with Oracle Communications, IQPartners and City of Warsaw in 2013. The main aim of this project was to promote the use in of both exposition APIs and data from different sources in individual applications developed within the contest. This included APIs from telecommunication area (Telco APIs), city area (Open Data APIs) and other open source projects available in Internet [9] (Fig. 1).

Fig. 1. Open API Hackathon web page [6.03.2015]

Open API Hackathon was divided into two phases:

- Phase I - collection of abstracts. The abstracts were proposing application ideas using Open APIs exposed in the contest.
- Phase II - application development. At the beginning of this phase, all ideas submitted in the form of the abstracts were evaluated by jury of the contest. Next, accepted ideas were qualified to development and were implemented by their authors.

A listing of Open APIs exposed for the contest is contained in Table 1.

During contest 18 APIs were exposed in the form of REST Web Services. As an exposition platform Oracle Communication Service Gatekeeper (OCSG) was used.

Table 1. Open API Hackathon APIs exposition

Lp	API	Description	API owner
1	Send SMS	Sending SMS messages	Orange
2	Receive SMS	Receiving SMS messages	Orange
3	Send USSD	Sending USSD messages	Orange
4	Receive USSD	Receiving USSD messages	Orange
5	Send MMS	Sending MMS messages	Orange
6	Get location	Returns value of subscriber location.	Orange
7	Get location random	Returns random value of subscriber location (for abroad users).	Orange
8	Payment	Allows to charge an amount to an end user account	Orange
9	Account management	Allows to query and modify the balance of a user account in Prepaid billing system.	Orange
10	Get time	Returns accurate time synchronized with atomic clock.	Orange
11	Terminal Status	Allows to check mobile terminal status (busy, free, logged out etc.).	Orange
12	Click 2 Call (Audio Call)	Allows to establish a voice call between mobile and fix subscribers.	Orange
13	Click 2 Announcement	Allows to establish a voice call form fix or mobile subscriber and play announcement	Orange
14	Click 2 conference	Allows to establish a conference call between three mobile and fix subscribers.	Orange
15	Number portability	Returns the information to which operator belongs a requested MSISDN number.	Orange
16	BusTimeTables	Allows to obtain timetables and bus lines in the city of Warsaw for chosen stops.	City of Warsaw
17	Bicycle station (Veturilo)	Allows to obtain information (including location) about Veturilo bicycle stations in the city of Warsaw for chosen area.	City of Warsaw
18	Park and ride locations	Allows to obtain information (including location) about P&R parking stations in the city of Warsaw for chosen area.	City of Warsaw

In the first phase of the contest, 73 proposals of new applications and services were submitted – 23 of them were individual ideas and 50 were group ideas. Two persons signed with more than one idea (3 and 5, respectively). Multiple ideas were also submitted by 11 teams (up to 6 ideas per a team). In the first stage, there were over 100 persons participating. For the second stage 56 ideas were qualified and 34 ideas finished that phase (11 individual and 23 group ideas). Over 30 persons were involved in the software development. The first place in the contest was taken by the application *Green Your Office with Orange,* which by using a number of different communication functions, allows its users to manage smart home environments.

The second prize was awarded to Veturilo Finder – the application, which displays on a map the location of all Veturilo city bike stations in a requested region. Among all the participants most popular API was Send SMS – it was used in 27 projects.

Other popular APIs were Location, Send USSD and Reception API. Detailed data on the API use are presented in Fig. 2.

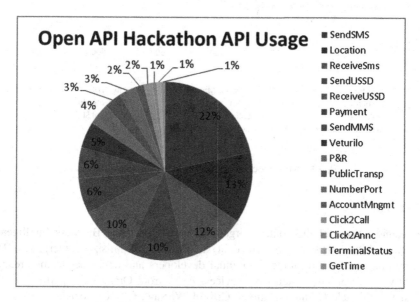

Fig. 2. API usage during Open API Hackathon

The most popular APIs was Send SMS, which was used for the initialization of the services and for notification containing particular service data. For similar purpose USSD was used. Those APIs were common and combined with many others. The second most frequently used API was Account Management, which gave its users the opportunity to check and update the balance of a user account in Prepaid system. The latter API was often used with Payment and charging operations. The third most popular APIs are City APIs such as Veturilo public bicycle and Public Transport. Those were mainly used with Location API for location based services. There were also several projects making use of Call Control functions, which focused on the handling of telephone connections, e.g. call establishing conference and playing announcements functionalities.

During 11 days of development phase, there were nearly 4000 API invocations which gives an average of over 110 invocations per a developer.

It is important to note that when the metrics showing the popularity of APIs is changed from the number of applications using the service to the number of calls to this service, the ranking of popular APIs may change. Still, what is shown in Fig. 3, SMS service and Location service, allowing to check the location of a mobile terminal based on the owner consent, were the dominating services also in the call count criterion. Nevertheless, the selection of an appropriate metric to measure the popularity of an API in view of innovative application potential requires further research.

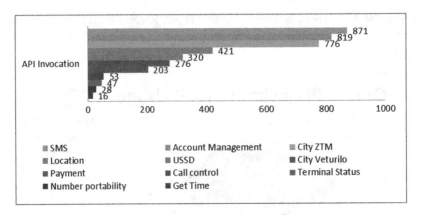

Fig. 3. API invocations during Open Api Hackathon

2.2   Bihapi

The second edition of the contest, organised under the name Business Intelligence Hackathon API (BIHAPI) was aimed at small and medium-sized enterprises. The primary objective was to attract individual developers and developer teams creating innovative applications, systems or services. Apart from Orange, the partners in the contest were Oracle Communications, City of Warsaw, City of Poznan and City of Gdańsk. The patron of the competition was the Ministry of Administration and Digitization of Poland. In BIHAPI out of 81 exposed APIs: 36 were exposed by Orange, 30 by the City of Warsaw, 11 from Poznań and 4 from Gdańsk. Hence, in this case, most of the APIs selected for the contest were Open Data APIs.

A complete list of all exposed APIs is shown in Table 2. It can be observed that the number and the content of individual data sets exposed by different cities largely varies, which makes the applications using Open Data largely linked to one city only. This illustrates the need for inter-city cooperation on defining the set of leading Open Data sets that ideally would be exposed by different cities in the same format. The business potential of individual applications will be increased once many cities decide to open the same data sets in the same formats. To attain such a standardisation, ground truth on the priority of the exposure of individual data sets is needed. The statistics on the use of individual data sets during the BIHAPI hackathon can largely help identify such key data sets with inter-city application potential in Poland (Fig. 4).

In BIHAPI, 101 ideas were submitted and 67 of them were accepted for the development phase. The winners of the contest were as follows:

- *CoSieStao* - mobile application for the residents of the city, which in a clear and simple way informs its users about extraordinary events.
- Via *CRM* - public transport passenger relationship management system.
- *Poczytaj mi* - distribution of illustrated fairy tales for children in conjunction with a new, innovative technology, being real-time communication – WebRTC.
- *Stacz kolejkowy* - mobile application for a queuing system used in city administration offices.

Table 2. Business Intelligence Hackathon API exposition

Warsaw	Poznan	Orange
City offices (WFS)	Streets (WFS)	Send_SMS
Cash machines (WFS)	Addresses (WFS)	Receive_SMS
Dormitory (WFS)	Cemeteries (WFS)	Send_MMS
Pharmacies (WFS)	Tombs (WFS)	Number_Portability
Hotels (WFS)	Ticket machines (WFS)	SIM_Status_HLR
Police stations (WFS)	Parking meters (WFS)	Prepaid_Recharge_Debit
Metro entrances (WFS)	Bus stops (WFS)	Get_Prepaid_Balance
Real estates for sale (WFS)	City transport lines (WFS)	Click2Call
Real estates for rent (WFS)	WiFi spots (WFS)	Click2Conference
Sport fields (WFS)	Bike paths (WFS)	Click2Announcement
Swimming pools (WFS)		Get_Time
Park and ride (WFS)	Cemetery Map (WFS)	Send_USSD
Squares (WFS)	News (WS)	Receive_USSD
Address points (WFS)	City events (RSS)	SIM_GeoLocalization (random)
Veturilo stations (WFS)		SIM_GeoLocalization
Streets (WFS)	Gdańsk	Call_Redirection
Theatres (WFS)	Standard Map (WMS)	WebRTC_Web_call
Bicycle roads (WFS)	Orthophotomap (WMS)	WebRTC_SIP_call
Passports queue (WS)	Hypsometric map (WMS)	SIM_Signal_Level
Standard maps (WMS)	POI map (WMS)	SIM_Cell_ID_ info
Cultural POI maps (WMS)		SIM_Access_Type
Commercial POI maps (WMS)		SIM_IMEI_number
OrtofotoMap (WMS)		SIM_Roaming_Country
Historical maps (WMS)		SIM_Activity_ Control
Sport POI maps (WMS)		SIM_Reset_Location
1939 maps (WMS)		SIM_Activity_Info
Transport POI maps (WMS)		Generate invoice
Education POI maps (WMS)		Get_My_Calendar_Event
ZTM (City Transport Office) (WMS)		Store_My_Calendar_Info
19115 intervention (WS)		Check_Presence
		Get_Address_Contact
		Store_Address_Contact

- *IfCity* - application used to explore the city and quickly locate interesting objects. The application is providing the advice on how to find the shortest route, choose the best property, hotel, or other places.

The types of data sources on the side of the cities are as follows:

- WMS - Web Map Service (WMS) - created by Open Geospatial Consortium (OGC) standard for bitmap-based exposition of maps. WMS service returns map sections in a bitmap form.

Fig. 4. Business Intelligence Hackathon API web page [18.03.2015]

- WFS - The OGC Web Feature Service Interface Standard – returns data in XML form i.e. as vector data.
- WS - Web Services in REST or RESTlike form.

Only two APIs – 19115 non-emergency issue reporting and events download, and News from Poznań were exposed in SOA/SOAP on City site. All of the APIs presented above were translated to RESTlike architecture style using contest platform (Oracle Communication Service Gatekeeper). Hence, from a developer standpoint, all the services took the form of RESTlike services, which simplified the development of applications consuming these services.

Some of the new APIs exposed in this contest by Orange were as follows:

- Call_Redirection - API allows to redirect the call (signed virtual MSISDN number) to the TTS (Text to speech) announcement or number.
- WebRTC_Web_call - API allows to establish audio/video call between two web browsers or between a web browser and a Session Initiation Protocol (SIP) client
- SIM_Signal_Level - API allows to get information about the 2G/3G signal strengths of visible Base Transceiver Stations from a SIM card.
- SIM_Cell_ID_ info - API allows to check the Local Area Code and Cell ID where the mobile SIM card is logged.
- SIM_Access_Type - API allows to check the type of network, which the SIM card is logged in. The information which is received is either 2G or 3G

- SIM_IMEI_number - API allows to get the International Mobile Station Equipment Identity (IMEI) number of the device in which the SIM card is placed in, together with the information about this device
- SIM_Roaming_Country - API allows to get the information about the country of the mobile network where the SIM card is logged in.
- SIM_Activity_ Control - API allows to check real mobile terminal availability information with using the paging procedure
- SIM_Reset_Location - API allows to delete the number from VLR and SGSN.
- SIM_Activity_Info - API allows to receive immediately selected information about the SIM operations (switch on/off, Local Area Code changes, MO/MT calls/sms etc.).

Table 3 presents the usage of APIs during BIHAPI development phase. The invocation to raster maps APIs (WMS type) were not logged and are not included in the statistics. Similarly to the previous contest, it can be observed that the largest number of calls was made to telecommunication services. This is in spite of much longer list of Open Data sets, made available especially by the City of Warsaw. This confirms the validity of the approach combining Open Data with Open APIs to promote the use of Open Data. Moreover, steep learning curve for some of the services has to be taken into account. For the users with no telecommunication background the use of some of the services, related to the configuration of audio and video calls or the status of a mobile terminal in a mobile network can be difficult not because of API exposition standards (SOAP vs. RESTful), but rather due to underlying telecommunication concepts and standards. Hence, inter-disciplinary teams are needed to fully exploit the potential of various Open Data standards and APIs. Such teams could include also domain experts to fully exploit domain knowledge related to domains such as public transport or real estate market. This is in line with the suggestions arising from other Open Data projects, including the conclusions summarised by D. Cowan and F. McGarry in [12].

Table 3. Usage of APIs in development phase

Total number of requests:	13 991
Orange API	8 011
Warsaw API *	5 044
Poznań API	936
Gdansk API	Not registered (WMS only)

Since the largest number of Open Data sets was made available for the project by the City of Warsaw, a more detailed analysis of their use is justified. For the Open Data published by the City of Warsaw, the total number of API calls was 5 044 requests. Key Open Data sets that attracted the largest number of calls are presented in Table 4.

It can be observed that the largest number of calls was made for city transport and real estates for rent. In fact, the data sets attracting the largest attention, are the data sets that reflect dynamically changing situation that is known to the city. This is definitely true for time tables, real estates offered by the city, queues in city offices or emergencies

Table 4. APIs usage statistics for the Open Data published by the City of Warsaw.

API	Number of calls	Number of calls [%]
City transport timetable	1380	27,36 %
Real estates for rent	597	11,84 %
Veturilo stations	459	9,10 %
Queue	340	6,74 %
Public transport bus stops	294	5,83 %
19115 - notifications	286	5,67 %
Public transport lines	285	5,65 %
Address points	261	5,17 %
Swimming pools	203	4,02 %
Real estates for sale	186	3,69 %
Metro entrances	127	2,52 %
Other	626	12,39 %

and disruptions to city services reported in 19115 database. Such data is hardly available from other data sources. In the case of vector data sources showing the location of various objects such as cycle stations or swimming pools, the Open Data sets published by the city are expected to be more reliable, up-to-date and integrated than unstructured content that could be collected and integrated from web resources. Hence, a preliminary observation can be made that not only the data available only for city halls, but also Open Data replacing publicly available data, present in noisy internet resources is of particular interest for software developers.

3 Summary

The work summarises the statistical data collected from developer contests combining the use of Open Data and Open API, providing an access to telecommunication functions. The data analysed in this study provide ground truth on the interest in individual data sets in varied software developer community, including both student and commercial development teams. The primary observations are that telecom services attract significant interest and when possible the use of Open Data is combined with such services. Moreover, key categories of Open Data, made public by the city offices were identified, based on the actual number of calls to individual data sources.

In the future further analysis of the statistical data collected during the contests is planned, including the cluster analysis of the data to search for typical patterns of use of multiple services. Equally importantly, the search for the metrics reflecting the interest in individual Open Data sets, and taking into account both the number of calls, and the number of applications using individual data sets is planned. This is to further refine guidelines for the organisations implementing Open Data exposition. An important aspect of these guidelines will be how to estimate the innovation potential caused by the opening of individual data sets.

References

1. Portal. http://data.gov/. 13 June 2014
2. Portal. http://publicdata.eu. 13 June 2014
3. Portal. http://inspire.ec.europa.eu/. 30 May 2014
4. Portal. UK Gov https://data.gov.uk. 21 March 2014
5. Portal. http://www.iatiregistry.org/. 21 March 2014
6. Portal. https://developer.uber.com/. 21 March 2014
7. Portal. http://www.opendata500.com/us/list/. 21 March 2014
8. Portal. http://www.kpmg.com. 21 March 2014
9. Portal. OpenAPIHackathon http://80.48.104.236/contest/. 06 March 2014]
10. Portal. BIHAPI http://bihapi.pl/. 21 March 2014
11. Portal. OpenMiddleware Community http://openmiddleware.pl. 21March 2014
12. Cowan, D., Alencar, P., McGarry, F.: Perspectives on open data: issues and opportunties. In: 2014 IEEE International Conference on Software Science, Technology and Engineering (SWSTE), pp. 24–33 (2014)
13. Kassen, M.: A promising phenomenon of open data: a case study of the chicago open data project. Gov. Inf. Q. **30**(4), 508–513 (2013)
14. Lakomaa, E., Kallberg, J.: Open data as a foundation for innovation: the enabling effect of free public sector information for entrepreneurs. IEEE Access **1**, 558–563 (2013)
15. Lindman, J., Kinnari, T., Rossi, M.: Industrial open data: case studies of early open data entrepreneurs. In: 2014 47th Hawaii International Conference on System Sciences (HICSS), pp. 739–748 (2014)
16. Zuiderwijk, A., Janssen, M.: Open data policies, their implementation and impact: a framework for comparison. Gov. Inf. Q. **31**(1), 17–29 (2014)

The Use of Open API in Travel Planning Application – Tap&Plan Use Case

Michał Karmelita[✉], Marcin Karmelita, Piotr Kałużny,
and Piotr Jankowiak

Uniwersytet Ekonomiczny w Poznaniu,
al. Niepodległości 10, 61-875 Poznań, Polska
michal.karmelita@kie.ue.poznan.pl

Abstract. The paper presents a new solution for the problem of getting around the foreign city using the mobile application as an assistant. The app uses data provided through Orange APIs, as well as other data sources available in the Internet to reach the goal.

Keywords: Open data · Orange API · Travel planning

1 Introduction

Online calendars and scheduling tools are widely used for day-to-day meeting planning. However, a gap in their functionality has been identified, i.e., travel planning between meetings. This is especially important when user has to travel between different event places or when he is on a business trip.

The main objective of the project is to provide users with a simple and intuitive way of planning a trip according to the tight schedule of the day, without excessive user's engagement. Existing solutions such as jakdojade.pl[1] or Google Transit do not provide complex customer service. They focus only on a single and simple queries without any possibilities for querying automation. Particularly in case of events sequence. One solution to address abovementioned problem is to use Open Data API in order to plan travel schedule and, in some cases, order a specified means of transportation (i.e., book tickets or order a taxi). In this paper, a possible solution to the problem is describe: mobile application Tap&Plan.

2 Used data sources

The application was developed for BIHAPI[2] competition, whose main idea was to engage Orange REST API's as well as other open data sources for improving the quality of life in Polish cities. Presented use case is based on four Orange APIs:

---

[1] http://jakdojade.pl/ - mobile service for travel planning in Polish cities. Service uses public transport schedules as well as local trains' schedules.
[2] http://www.bihapi.pl/.

© Springer International Publishing Switzerland 2015
W. Abramowicz (Ed.): BIS 2015, LNBIP 208, pp. 348–350, 2015.
DOI: 10.1007/978-3-319-19027-3_28

- "send SMS" and "receive SMS" - allow to send and receive text messages,
- "get my calendar event" and "store my calendar info" – provide functions to keep and edit user's calendar.

In addition, Google APIs that have been used: Google Transit[3] for single route determination and Google Calendar[4] to allow synchronization with existing calendars.

3 Proposed solution

Tap&Plan application enables a complex scheduling of user's journey. Instead of manually planning the trip, which is time and effort consuming, user can create a trip plan between his scheduled appointments. The advantage over existing solution is that Tap&Plan provides a graphical and textual information about meetings and allows to manage and plan the journey between meetings, without using different applications for specified means of transportation. The application generates the queries, which correspond to the user's schedule. That schedule could be synchronized with user's calendar through Google Calendar API services. The route determination for travelling by public transport, is based on the Google Transit API, while ordering taxi service uses the Orange send SMS API and receive SMS API. A determined route is presented using a GPS-like user interface. Tap&Plan is especially useful when user has a number of meetings in the foreign city. It also selects the appropriate mean of transport - the fastest, the cheapest, or the preferred one. After that, depending on the transport

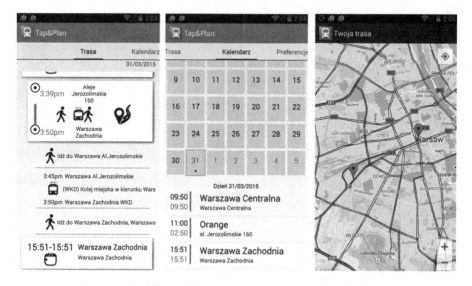

Fig. 1. The Tap&Plan user interface.

---

[3] https://developers.google.com/transit/.

[4] https://developers.google.com/google-apps/calendar/.

selected, a proper tool pops up and automatically orders a taxi or checks public transport timetable. If there will be a longer time interval between meetings, the application recommends a restaurant in the nearest area. The application's user interface is presented on Fig. 1. The right screen presents a sample view of single scheduled journey and the directions screen – in this case using public transport. The second screen presents the build-in calendar andscheduled appointment. The last one shows a route presentation on the map.

The future work on the presented solution will focus on algorithms improvement in order to increase efficiency and effectiveness of route computation. Another important research direction is to expand the functionality so the application supports delays of trains and public transport. Authors plan to systematically increase the number of supported cities.

Author Index

Printed in the United States
by Baker & Taylor

Printed in the United States
By Bookmasters